MEMBRANE TRANSPORT

New Comprehensive Biochemistry

Volume 2

General Editors

A. NEUBERGER
London

L.L.M. van DEENEN
Utrecht

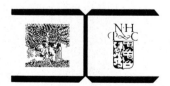

ELSEVIER/NORTH-HOLLAND BIOMEDICAL PRESS
AMSTERDAM·NEW YORK·OXFORD

Membrane transport

Editors

S.L. BONTING and J.J.H.H.M. de PONT

Nijmegen

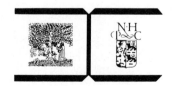

1981

ELSEVIER/NORTH-HOLLAND BIOMEDICAL PRESS
AMSTERDAM·NEW YORK·OXFORD

© Elsevier/North-Holland Biomedical Press, 1981
All rights reserved. No part of this publication may be reproduced, stored in a retrieval system, or transmitted, in any form or by any means, electronic, mechanical, photocopying, recording or otherwise without the prior permission of the copyright owner.

ISBN for the series: 0444 80303 3
ISBN for the volume: 0444 80307 6

Published by:
Elsevier/North-Holland Biomedical Press
1, Molenwerf, P.O. Box 1527
1000 BM Amsterdam, The Netherlands

Sole distributors for the U.S.A. and Canada:
Elsevier/North-Holland Inc.
52 Vanderbilt Avenue
New York, NY 10017

Library of Congress Cataloging in Publication Data

Printed in The Netherlands

Preface

Membrane transport is of crucial importance for all living cells and organisms. Nutrients must be taken up, waste products removed by passage through the cell membrane. The water content of the cell must be regulated. Ion gradients across the cell membrane are required to maintain membrane potentials, which play a crucial role in excitation processes, and to drive other transport processes. Transport processes play a role, not only across plasma membranes, but also across cell organelles like mitochondria. Membrane transport occurs by various mechanisms, passive and active transport, mediated and non-mediated transport. This book attempts to give a comprehensive, integrated and up to date account of all these aspects of this field.

After Chapter 1 on non-mediated transport of lipophilic compounds, Chapters 2 and 3 are devoted to the passive transport of water and other small polar molecules and to that of ions. Chapter 4 discusses the insertion of ionophores in lipid bilayers as model systems for carriers and channels in biological membranes. Chapter 5 treats the general principles of mediated transport. Chapters 6, 7 and 8 are devoted to the ATPases, which are involved in the primary active transport of Na^+, Ca^{2+} and H^+, respectively. After Chapters 9 and 10 on specific transport systems in mitochondria and bacteria, the book concludes with Chapters 11 and 12 on secondary active transport, the coupling of the transport of metabolites and water to that of ions.

The area of membrane transport has always been an interdisciplinary field. Physiologists, biochemists, biophysicists, cell biologists and pharmacologists have all made their contributions to the development of our knowledge in this field, often in collaborative studies. The appearance of this book in the series New Comprehensive Biochemistry is justified perhaps more by the future contributions to be expected from fundamental biochemistry than by the contributions made by biochemistry so far. Our biochemical understanding of the molecular structure and dynamics of the various transport systems is still in a primitive state compared to that for biomolecules like nucleic acids and water-soluble proteins. The editors hope that the publication of this volume may arouse the interest of many biochemists, especially the younger ones, for this field of biochemistry and thus contribute to its development.

S.L. Bonting
J.J.H.H.M. de Pont

Nijmegen, February 1981

Contents

Preface	v
Contents	vii

Chapter 1. Permeability for lipophilic molecules, by W.D. Stein — 1

1. Criteria for recognising simple diffusion	1
2. Gross effects of lipid solubility and molecular size	2
3. The human erythrocyte data	7
4. Systematic treatment of solvent properties and mass selectivity	11
5. Comparison of different model solvent systems	13
6. Permeation of large lipophilic molecules— steroid transport	16
7. What region of the cell membrane provides the major permeability barrier against lipophilic solutes?	18
8. Studies on artificial membranes	22
9. Partition coefficients of cell and artificial membranes	23
10. Affectors of membrane permeability	24
(a) Alcohols, anaesthetics and other fat-soluble additives	25
(b) Temperature	25
(c) Cholesterol content	25
(d) Fatty acid composition	26
11. Overall survey and conclusions	26
References	27

Chapter 2. Permeability for water and other polar molecules, by R.I. Sha'afi — 29

1. Introduction	29
2. Methods for measuring water and small polar nonelectrolyte movements across a barrier	30
(a) Radioactive tracer movements across a barrier which can be mounted between two solutions	30
(b) Net water flow under the influence of pressure	31
(c) Radioactive tracer movement in cell suspension studied with rapid flow technique	33
(d) Nuclear magnetic resonance technique	33
(e) Osmotic volume changes studied with stop-flow technique	34
(f) Unstirred layer effect	37
3. Water movements across membranes and tissues	38
(a) Relationship of water diffusion to osmotic flow	38
(b) Solvent drag, reflection coefficient and the "pore" concept	40
(c) Effect of temperature on permeability of membrane to water	43
(d) Effect of sulfhydryl-reactive reagents on water transport	45
(e) Effect of antidiuretic hormone on water transport	46
(f) Membrane cholesterol and the permeability to water	47
(g) Is there rectification of water flow?	48
(h) Miscellaneous factors	48
(i) Possible structural basis for the apparent presence of hydrophilic pathway for water transport	49
4. Permeability of membranes to small polar nonelectrolytes	51
(a) Is the mechanism by which small hydrophilic solutes permeate cell membranes similar to that used by large lipophilic molecules?	51

(b) What agents affect the movements of these small nonelectrolytes?	54
(c) Is there a carrier-mediated mechanism for urea transport?	55
(d) Are all the pathways used by water available for urea and other small polar nonelectrolytes?	57
5. Summary and conclusions	58
References	59

Chapter 3. Ion permeability, by E. Rojas — 61

1. Introduction	61
2. Theoretical basis for the concept of permeability	62
(a) The gradient of electrochemical potential as a force	62
(i) Chemical and electrochemical potential	62
(ii) The flux equations	63
(b) The Nernst–Planck equation	67
(c) Integration of the Nernst–Planck equation	68
(i) The constant-field hypothesis	68
(ii) Definition of permeability	69
(iii) Net ion fluxes	70
(iv) Unidirectional fluxes and the flux ration	70
(v) The Goldman equation	70
(vi) Net ionic fluxes and resting membrane conductance	71
3. The measurement of ionic permeabilities	72
(a) Cation permeabilities in resting electrically excitable cells	72
(i) Permeabilities from tracer fluxes	72
(ii) Permeabilities from the Goldman equation	75
(b) Permeability changes in stimulated electrically excitable cells	76
(i) Sodium inflow in axons under voltage clamp	76
(ii) Timing the flux during a rectangular voltage clamp pulse	77
(iii) Timing the sodium and potassium permeability changes during an action potential	77
(iv) Timing the sodium flux during an action potential	79
(c) Selectivity ratios for monovalent cations	79
(i) Permeability ratios from the reversal potential	79
(ii) Permeability ratios from tracer fluxes	84
4. Mechanisms	86
(a) The concept of ionic channel	86
(i) Hodgkin–Huxley channels	86
(ii) Molecular transitions associated with the activation of channels	90
(b) Mechanisms for permselectivity	94
(i) Hille's selectivity filter	94
(c) Gating mechanisms	95
(i) The two-state transition model	96
(ii) The aggregation-field effect model of Rojas	100
(iii) Channel counting and single-channel conductance	102
References	104

Chapter 4. Channels and carriers in lipid bilayers, by J.E. Hall — 107

1. Perspectives	107
(a) Terminology	107
(b) How to tell a channel from a carrier	107
(c) Ion selectivity and its consequences	111

2. Carriers and matters unique to them 112
   (a) The carrier model 112
   (b) Special limiting cases 114
3. Channels and matters unique to them 115
   (a) Gramicidin—the most studied channel 115
   (b) Voltage-dependent channels 118
       (i) Channels which turn on with voltage 118
       (ii) Channels which turn off with voltage 119
4. New directions 120
References 120

Chapter 5. Concepts of mediated transport, by W.D. Stein 123

1. Introduction 123
2. The kinetic analysis of facilitated diffusion 123
   (a) Description of the experimental procedures 124
       (i) The zero *trans* procedure 124
       (ii) The equilibrium exchange procedure 125
       (iii) The infinite *trans* procedure 126
       (iv) The infinite *cis* procedure 127
   (b) Some general considerations 127
3. First model for facilitated diffusion—the simple pore 129
   (a) Kinetic analysis 129
       (i) Zero *trans* procedure on the simple pore 132
       (ii) Equilibrium exchange on the simple pore 132
       (iii) The infinite *trans* procedure on the simple pore 132
       (iv) The infinite *cis* procedure on the simple pore 133
   (b) Some further tests for the simple pore 133
4. Second model for facilitated diffusion: the complex pore 135
5. Third model for facilitated diffusion: the simple carrier 136
   (a) Introduction 136
   (b) The zero *trans* and equilibrium exchange procedures on the simple carrier 138
   (c) The infinite *trans* procedure on the simple carrier 138
   (d) The infinite *cis* procedure on the simple carrier 139
   (e) The simple pore and simple carrier compared 140
   (f) Some further tests for the simple carrier 142
6. Fourth model for facilitated diffusion: the conventional carrier 142
7. A molecular interpretation of the transport parameters 143
   (a) R_{ij}—The resistance parameters 144
   (b) K—The intrinsic dissociation constant 144
   (c) The asymmetry parameter—R_{21}/R_{12} 145
8. Exchange diffusion and countertransport 146
   (a) Exchange diffusion 146
   (b) Countertransport 147
9. The kinetics of competition 151
10. Secondary active transport 152
11. Primary active transport 154
12. Design principles for active transport systems 155
13. Conclusion 156
References 157

Chapter 6. Sodium-potassium-activated adenosinetriphosphate, by F.MA.H. Schuurmans Stekhoven and S.L. Bonting   *159*

1. Introduction   159
   (a) Cation transport in cells   159
   (b) Relation to energy metabolism   159
   (c) Nature of the cation transport system   160
2. Reaction mechanism   161
   (a) Substrate binding   161
   (b) Phosphorylation of the enzyme   162
   (c) Transformation of the phosphoenzyme   163
   (d) Hydrolysis of the phosphoenzyme   164
   (e) Return to the native enzyme form   165
   (f) K^+-stimulated phosphatase activity   167
3. Structural aspects   168
   (a) Subunit structure and composition   168
   (b) Conformational states   170
4. Phospholipid involvement   171
   (a) Phospholipid headgroups   171
   (b) Fatty acid groups   173
   (c) Role of phospholipids   173
5. Transport mechanism   174
   (a) Normal and reversed Na^+-K^+ exchange transport   174
   (b) Na^+-Na^+ exchange transport   176
   (c) Uncoupled Na^+ efflux   177
   (d) K^+-K^+ exchange transport   177
   (e) Phosphate reaction as non-transporting system   178
6. Concluding remarks   179
References   179

Chapter 7. Calcium-activated ATPase of the sarcoplasmic reticulum membranes, by W. Hasselbach   *183*

1. Introduction   183
2. The sarcoplasmic reticulum membranes, a structural component of the muscle cell—organization, isolation and identification   184
3. Phenomenology of calcium movement   186
   (a) Energy-dependent calcium accumulation   186
   (b) Coupling between calcium accumulation and ATP splitting   187
   (c) Passive calcium efflux   189
   (d) Calcium efflux coupled to ATP synthesis   190
4. Reaction sequence: substrate binding   191
   (a) Calcium binding   191
   (b) ATP binding   193
   (c) ADP binding   195
   (d) Magnesium binding   195
   (e) Phosphate binding   197
5. Reaction sequence: phosphoryl transfer reaction   197
   (a) Phosphorylation of the transport protein in the forward and the reverse mode of the pump   197
   (b) $ATP-P_i$ exchange   198
   (c) Phosphate exchange between ATP and ADP   199
   (d) ADP-insensitive and ADP-sensitive phosphoprotein   200
   (e) Phosphoryl transfer and calcium movement   203
References   205

Chapter 8. Anion-sensitive ATPase and ($K^+ + H^+$)-ATPase, by
J.J.H.H.M. de Pont and S.L. Bonting   209

1. Introduction   209
2. Anion-sensitive ATPase   209
    (a) Definition and assay of enzyme activity   209
    (b) Effects of substrate, cations and pH   210
    (c) Anion dependence   212
    (d) Other properties of the enzyme   215
    (e) Localization   215
    (f) Brush border membranes   219
    (g) Erythrocytes   220
    (h) Transport function   221
3. ($K^+ + H^+$)-ATPase   222
    (a) Introduction   222
    (b) Purification   222
    (c) Structural and chemical properties   223
    (d) General enzymatic properties   224
    (e) Partial reactions   224
    (f) Activators and inhibitors   226
    (g) Phospholipid dependence   228
    (h) Vesicular transport   229
    (i) Role in gastric acid secretion   232
References   232

Chapter 9. Mitochondrial ion transport, by A.J. Meijer and K. van Dam   235

1. Mitochondrial metabolite transport   235
    (a) Introduction   235
    (b) Survey of the mitochondrial metabolite translocators   235
    (c) The use of mitochondrial transport inhibitors in metabolic studies   237
    (d) Kinetic properties of the individual translocators   238
    (e) Studies on the distribution of metabolites across the mitochondrial membrane in the intact cell   238
    (f) Recent developments in mitochondrial metabolite transport   241
       (i) Transport of adenine nucleotides   242
       (ii) Transport of pyruvate   244
       (iii) Transport of acylcarnitine   246
       (iv) The glutamate-aspartate translocator   246
       (v) Hormones and mitochondrial metabolite transport   248
       (vi) Isolation of translocators   249
2. Mitochondrial cation transport   249
    (a) H^+ 249
    (b) Ca^{2+} 251
    (c) Monovalent cations   252
References   252

Chapter 10. Transport across bacterial membranes, by W.N. Konings,
K.J. Hellingwerf and G.T. Robillard   257

1. Introduction   257
2. The chemiosmotic concept   259

3. Energy-transducing systems	260
(a) Cytochrome-linked electron transfer systems	260
(b) The ATPase complex	263
(c) Bacteriorhodopsin	265
4. Solute transport	267
5. Carriers for facilitated secondary transport	269
6. The phosphoenolpyruvate-dependent sugar phosphotransferase system	272
(a) Group translocation	272
(b) Purification and general properties	274
(c) Cellular localization of the PTS components	274
(d) The complexity of the PTS	276
(e) The specificity for phosphoenolpyruvate	276
7. Interaction between energy-transducing processes	277
8. Methods for the determination of transmembrane gradients	278
9. Model systems for transport studies	279
References	282

Chapter 11. Coupled transport of metabolites, by P. Geck and E. Heinz   285

1. Introduction	285
(a) Source of energy	285
(b) Principles of coupling	286
(c) Material for transport studies	288
(d) Electrochemical potential difference	289
2. Types of energization	289
(a) Primary active transport	289
(i) Phosphotransferase systems	290
(b) Secondary active transport	291
(i) Carrier model of cotransport	291
(ii) Kinetics of influx	293
(iii) Effect of membrane potential	294
(iv) Predictions of kinetic feature from a model	295
(v) Effects of electrical potential	295
(vi) Effect of cotransport on the membrane potential	295
(vii) Pseudocompetition	297
3. Special systems	298
(a) Nonpolarized cells	298
(b) Epithelia	301
(c) Cell-free system (vesicles, liposomes)	305
4. Conclusion	307
References	307

Chapter 12. The coupled transport of water, by A.M. Weinstein, J.L. Stephenson and K.R. Spring   311

1. Introduction	311
2. Phenomenological models of transport	315
(a) General formulation	315
(b) Thermodynamic formulation	319
(c) The effect of unstirred layers	323
3. Solute-solvent coupling in the lateral intercellular space	331
(a) Elementary compartment model	331

(b) Standing gradient interspace models	337
(c) Comprehensive interspace models	343
4. Conclusion	348
References	349
Subject Index	353

CHAPTER 1

Permeability for lipophilic molecules

W.D. STEIN

Department of Biological Chemistry, Institute of Life Sciences,
Hebrew University, Jerusalem, Israel

1. Criteria for recognising simple diffusion

The outer membrane of the cell is an organelle, richly endowed with receptors which recognise and react to signalling molecules from the environment, endowed with enzymes for the degradation and synthesis of nutrients within the cell and external to it, and bearing transport systems which control the entrance and egress of specific metabolites. Subsequent chapters of this series will deal with all aspects of this dynamic commerce of the cell membrane as mediated by these membrane proteins. The present chapter confines itself, however, to those properties of the cell membrane which arise from the lipid backbone or substructure into which these more dynamic proteins are embedded. For a nutrient or any foreign molecule which finds no specific membrane component with which to interact, the lipid bilayer of the membrane provides the barrier which determines whether the molecule in question can cross the membrane. How then does this cell membrane matrix discriminate between possible permeants? This question is the theme of the present chapter.

It is clear that the set of molecules that we will deal with, i.e. those for whom no specific transport system exists, will be defined by elimination. If there is no evidence for the existence of a specific mode of transport for the test molecule we will assume that this molecule crosses the cell membrane by simple diffusion, if it crosses the membrane at all. Such an assignment is, of course, temporary. As soon as evidence arises for the intervention of a specific system for the transport of our test molecule, this molecule will then be eliminated from the list of those entering by simple diffusion. (Some molecules will of course cross the membrane both by simple diffusion and by a specific system, in parallel.)

The surest evidence for the elimination of a substance from the list of those entering cells only by simple diffusion is the existence of a specific inhibitor of its transport. Thus, 10^{-6} M copper ions will slow down the entry of glycerol into the human red cell 100-fold, while having no effect on a host of other permeating substances [1]. Glucose entry into these cells is, likewise, specifically inhibited by 10^{-8} M of the drug cytochalasin B [2]. Inhibitors may be reagents reacting with membrane receptors or they may be merely substrates of the specific transport systems. Thus glucose will inhibit the entry of other sugar molecules, which will thus be eliminated from the simple diffusion list. Glucose also inhibits its own entry in that glucose uptake by human erythrocytes is not an increasing linear function of glucose concentration in the medium but rather reaches a saturation level as the glucose concentration is raised. (One expects the rate of entry by simple diffusion to be directly proportional to the concentration in the external medium.) Finally, the involvement of a specific transport system must be invoked if molecules of very similar structure, for example optical isomers (d- and l-glucose), enter the cell at different rates. The list of molecules not eliminated by any of these tests forms the set of molecules whose permeability is to be understood in terms of the non-specific properties of the membrane. To acquire such understanding we must set up models to account for the permeability properties of the membrane acting as a discriminator between possible permeants. Subsequent sections of this chapter are an attempt to provide such models.

2. Gross effects of lipid solubility and molecular size

The first factor which was realised to be a major determinant of permeability by simple diffusion was the lipid solubility of the permeating species. Fig. 1 shows classical data of Collander [3] in which the permeability of 54 different substances entering the plant cell *Nitella* is plotted against the solubility of that substance in olive oil. The approximate size of each permeant is indicated by the different symbols. On both axes of the figure the data are plotted on a logarithmic scale. Lipid solubility is expressed as the partition coefficient, that is, the ratio of the concentration of solute present in the organic phase to the concentration in the aqueous phase, at equilibrium distribution of the test substance. Extensive listings of such partition coefficients are available for a variety of organic solvents [4] and in a later section we will devote some effort to a comparison of the various solvents. The permeabilities in this and subsequent figures are expressed as the amount of solute crossing unit area of cell membrane in unit time under the influence of unit concentration difference. Permeabilities can be expressed in the same dimensions as velocity, that is in cm/s, according well with their intuitive meaning of a rate of movement of permeant.

Clearly, from Fig. 1, the solubility of a solute in an organic solvent correlates very well with the permeability of the *Nitella* membrane for that solute. But it is also clear that the correlation is only partial. Thus, of two solutes with the same partition coefficient the one with smaller molecular weight would seem to permeate faster. Solute size as well as lipid solubility are both important determinants of permeation rate. The particular solvent chosen, olive oil, seems however to be a very good model for the ability of the membrane barrier to discriminate between the various permeants, since the overall increase in permeability as the structure of the permeant is varied correlates closely with the increase in partition coefficient. Were the two parameters to be strictly linked all the data would fall on the line of unit slope in the figure, the line of identity. Later we shall see cases where the data do not support such a close similarity between certain membranes and model solvents.

Why should the permeabilities correlate so well with the lipid solubilities? Fig. 1

Permeability for lipophilic molecules

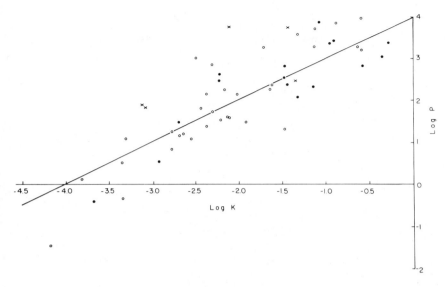

Fig. 1. Permeability of *Nitella* cell membranes to non-electrolytes in relation to the olive oil solubility of these solutes. On the ordinate the logarithm of the permeability (in units of 10^{-7} cm/sec); on the abscissa, the logarithm of the olive oil/water partition coefficient of the permeant. Data measured at 20°C. Crosses: molecular weight up to 50; open circles: molecular weights from 60 to 119; filled circles: molecular weights above 120. Data taken from Collander [3]. The straight line is of unit slope.

suggests strongly that the cell membrane acts as a barrier to solute permeation by presenting to these solutes a region with lipid solvent characteristics. There is little doubt that this region is composed of the lipid molecules, the phospholipids and cholesterol which make up the bulk of the membrane lipid of most cells. Modern views of membrane structure [5,6] see this lipid as organised into a bilayer some 40 to 50 Å thick, while embedded into this bilayer are the receptors, enzymes and transport systems comprising the protein constituents of the membrane.

How would a 40 Å thick layer of lipid molecules comprise a barrier to solute diffusion? Consider now Fig. 2, which represents a lipid phase separating two aqueous phases and acting as a permeability barrier between those two phases. The net flux of molecules from side 1 to side 2 of this barrier is v and, from the definition of permeability coefficient given above, is given by

$$v = PA(C_1^{aq} - C_2^{aq}) \tag{1}$$

where P is the trans-membrane permeability coefficient, C_1^{aq} and C_2^{aq} are the concentrations of permeant at sides 1 and 2 of the barrier and A is the cross-sectional area of the organic phase. If, now, the solute is in equilibrium across the

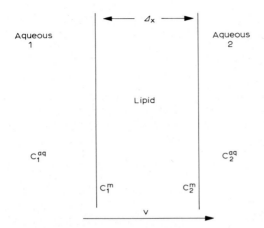

Fig. 2. Scheme for simple diffusion across a lipid layer. C_1^{aq} and C_1^m, C_2^m and C_2^{aq}, are the concentrations of permeant in the bulk aqueous phase on side 1; just inside the membrane at face 1; just within the membrane at face 2; and in the bulk aqueous phase on side 2, respectively. Δx is the thickness of the lipid layer; v the net flux of permeant.

solvent/aqueous phase barrier at both faces then we can write

$$K = \frac{C_1^m}{C_1^{aq}} = \frac{C_2^m}{C_2^{aq}} \qquad (2)$$

where K is the partition coefficient for the solute between the organic and aqueous phases, according to our definition. Substituting from (2) into (1), we obtain

$$v = \frac{PA}{K}(C_1^m - C_2^m) \qquad (3)$$

Once solute does not acccumulate within the membrane, i.e. at the steady state, the net flux of solute *across* the membrane (v, in Eqn. 3), must be equal to the net flux *within* the membrane given by

$$v = D_{mem} A (C_1^m - C_2^m) / \Delta x \qquad (4)$$

where Δx is the thickness of the barrier to diffusion. Eqn. 4 expresses Fick's Law for diffusion within a continuous phase, with D_{mem} as the diffusion coefficient for the solute in question, within that phase.

If we solve for P between Eqns. 3 and 4, we obtain

$$P = \frac{K D_{mem}}{\Delta x} \qquad (5)$$

an equation relating the measurable permeability coefficient across a membrane to the effective diffusion coefficient within the membrane, to the membrane/aqueous phase partition coefficient for the solute in question, and to the membrane thickness.

Eqn. 5 provides a very clear theoretical basis for the data of Fig. 1 (and similar data on other systems, as we shall see). The measured permeability coefficients for a set of solutes should parallel the measured partition coefficients, if the model solvent corresponds exactly in its solvent properties to the permeability barrier of the cell membrane. In addition, the molecular size of the solute is very likely to be an important factor as it will affect the diffusion coefficients D_{mem} within the membrane barrier phase. Data such as those of Fig. 1 will convince us that we have in our chosen solvent a good model for the solvent properties of the membrane's permeability barrier. We can now calculate values of $P\Delta x/K$ for the various solutes, and obtain estimated values of the intramembrane diffusion coefficient, and are in a position to study what variables influence this parameter. Fig. 3 is such a study in which data from Fig. 1 are plotted as the calculated values of $D_{mem/\Delta x}$ (calculated as P/K) against the molecular weight of the permeating solute. The log/log plot of the data has a slope of -1.22, which means that one can express the dependence of diffusion coefficient on molecular weight (M) in the form: $D_{mem} = D_0 M^{-1.22}$, where D_0 is the calculated diffusion coefficient for a solute of unit molecular weight.

This steep molecular weight dependence of the calculated intramembrane diffusion coefficient is somewhat unexpected. For diffusion in water, the experimental values of the diffusion coefficient bear an $M^{-1/2}$ or $M^{-1/3}$ dependence, a form which has a good theoretical basis. The accepted theory asserts that it is friction between the molecule and the medium that holds back a diffusing molecule. This friction will be proportional to the surface area of the diffusing particle and, hence, to the two-thirds power of the molecular weight. (The diffusion coefficient itself will be proportional to the square root of the friction [7].) Such a frictional model is applicable when the medium in which diffusion is occurring is continuous, that is, when the size of the diffusing molecule is large compared with the molecules comprising the medium in which diffusion is occurring. Apparently, from Fig. 3 which shows the high value found for the mass dependence of intramembrane diffusion coefficients, such is not the case for the cell membrane. A high dependence of the diffusion coefficient on the molecular weight of the diffusion is found for molecules diffusing within soft polymer matrices [8,9]. In the soft polymer field, the steep molecular weight dependence of the diffusion coefficient is attributed to the structured nature of the medium in which diffusion is occurring. The larger the diffusing molecule, the greater must be the energy invested to separate from one another the chain molecules of the medium, to provide a space for the diffusing molecule.

The data of Fig. 1 (replotted as Fig. 3) are consistent, therefore, with a very simple model: the membrane acts as a barrier by virtue of its possessing a region which discriminates between solutes as if it were an organic solvent as far as its partitioning properties are concerned, and as if it were a structured, soft polymer as far as its size-sieving properties are concerned. Both of these properties might be

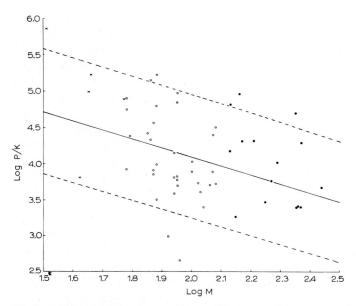

Fig. 3. Calculated relative intramembrane diffusion coefficients across the *Nitella* cell membrane as a function of molecular weight of the permeant. Ordinate: logarithm of the ratio of the permeability coefficient (in 10^{-7} cm/s) to the olive oil/water partition coefficient for the permeants of Fig. 1. Abscissa: logarithm of their molecular weights. The solid straight line is the linear regression of log(P/K) on log M with slope of -1.22. The dashed lines are at a distance of one standard deviation away from the regression line.

expected to be provided by the lipid region of the cell membrane, which we can therefore identify as the permeability barrier.

There is an important point that remains to be clarified. Eqn. 5, which seems to fit the data very well, tells us that the higher the partition coefficient between membrane and water, the faster will the relevant permeant cross the membrane. Some might find this result disturbing. It might be thought that a high partitioning ability of the membrane for the solute would ensure that the solute remains dissolved within the membrane, trapped there, and hence is unable to leave the membrane. But this mistaken view is based on a false conception of the nature of diffusion. Diffusion is a result of random non-directed movements of the diffusing molecule and appears to us to have direction merely because there are a greater number of (net) molecular displacements in a direction with the concentration gradient than against the gradient. In Fig. 2, a high partitioning into the membrane phase ensures a larger number of molecular impacts inside the membrane and a large number of net molecular displacements in the direction of the concentration gradient. A high partitioning increases the effective concentration gradient of the permeant and in this way increases the number of molecules moving in the direction of the gradient, and hence crossing the membrane. A small number of molecules will, indeed, be trapped within the membrane as a result of partitioning. This will *delay* the appearance of permeant at the opposite face of the membrane. But a steady state is reached when the intramembrane pool is filled with dissolved permeant and the rate of trans-membrane permeation is then constant and given by the product of membrane partition coefficient and intramembrane diffusion coefficient as in Eqn. 5.

We can quantify this argument in an illuminating way so as to give some numerical insight into the phenomenon. (i) The time lag for permeation across a membrane until the steady state is reached can be obtained [9] using Crank's formula [10] in the form

$$D_{mem} = l^2/6\lambda \tag{6}$$

where D_{mem} is as in Eqn. 5, l is the thickness of the membrane into which the permeant is partitioning and λ is the lag time. For the data of Figs. 1 and 3 the values of D_{mem} are some 1 to $10 \cdot 10^{-9}$ cm^2/s. With a membrane thickness of 40 Å, the lag time turns out to be some 3 to 30 µs, a value which would not be noticed in most current methods of measuring cell permeability. (ii) For a typical cell volume of some 10 pl, the relative volume of cell membrane to whole cell volume is some 10^{-3}. For many of the permeants depicted in Fig. 3, the partition coefficients range from 1 to 10^{-4}. Thus the amount of permeant trapped within the membrane by partitioning will be some 10^{-3} to 10^{-7} of that present inside the cell at equilibrium. This is also too small to be measured, except by very sophisticated methods. Thus both the time taken before the steady-state rate of permeation is reached and the amount of trapped permeant are too small to be noticeable in any permeation study as conventionally carried out.

3. The human erythrocyte data

The treatment of Collander's data on the permeation into giant algal cells has occasioned little controversy. But a similar treatment of the now extensive data on permeation into human red blood cells raises a number of problems on which researchers in the field have as yet not reached general agreement.

Fig. 4 depicts the data of Sha'afi et al. [11], of Savitz and Solomon [12] and of Klocke et al. [13] on the penetration of 23 different compounds into human red cells plotted as a log/log plot (cf. Fig. 1) against the relevant ether/water partition coefficient. The strong dependence of permeability on partition coefficient that was noted by Collander for plant cells is somewhat less obvious for the human erythrocyte. Values of the partition coefficient vary over a range of 10^3 for molecules which have the same permeation rate. Molecules with the same partition coefficient have permeation rates which vary over a range of 10^2. But a line of unit slope can be drawn through many of the points on the log/log plot showing the strong dependence of permeation coefficients on partition coefficients.

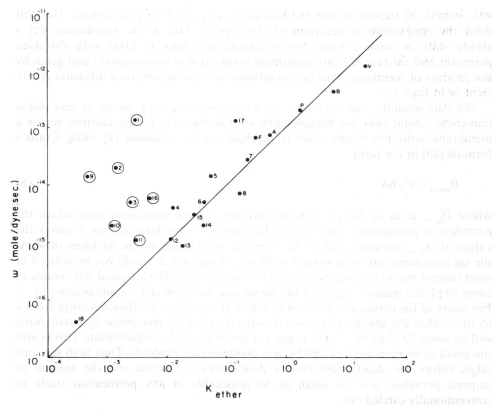

Fig. 4. Permeability of human red blood cell membranes to non-electrolytes in relation to ether/water partition coefficients. Ordinate: permeability on a logarithmic scale, in units of mol/dyne/s. Abscissa: partition coefficients for the ether/water system. Data measured at room temperature (20 to 21°C). Permeants are numbered as follows: 1, water; 2, formamide; 3, acetamide; 4, propionamide; 5, butyramide; 6, *iso*butyramide; 7, valeramide; 8, *iso*valeramide; 9, urea; 10, methylurea; 11, (1,3)-dimethylurea; 12, (1,3)-propandiol; 13, (1,4)-butandiol; 14, (1,3)-butandiol; 15, (2,3)-butandiol; 16, ethylene glycol; 17, methanol; 18, malonamide; A, acetic acid; B, butyric acid; F, formic acid; P, propionic acid; V, valeric acid. Data from Savitz and Solomon [12] for permeants 2, 9, 16, and 17; from Kocke et al. [13] for A through V; from Sha'afi et al. [11] for all others.

There are a number of reasons why the human red cell data are less tractable than those for plant cells. In the first place, the cells in question come from a species which is almost certainly more highly evolved than the primitive algae studied by Collander; the membranes of the human red cell contain numerous specialised transport systems. The data on specialised systems should be set apart from the remaining data if the interaction between solute structure and simple diffusion properties of the membrane is to be investigated. Excellent evidence exists that the uptake of glycerol into human red cells occurs by a specialised system [1,14] which can handle other simple glycols such as ethylene glycol and also the ester monacetin. The penetration of urea into human red blood cells has often been thought of as occurring by a specialised system, and the evidence for this is now strong [15]. Whether this system can also transport other ureas is less clear, however. Also, the penetration of water into this cell is firmly believed to occur by a pathway parallel to the lipid region, perhaps through aqueous pores (see Chapter 2). It has been reported that the penetration of the smaller amides formamide and acetamide, is somewhat inhibitable by specific inhibitors [16,17]. If one now excludes all these compounds (marked with a circle in Fig. 4) from consideration, the correlation of partition coefficient with permeation rate improves. This has occurred because one has removed a number of substances with low partition coefficients but with high (allegedly protein-mediated) permeation rates.

There is, however, another quite different way of interpreting these data. One can assume, for the moment, that ether is not a good model for the solvent properties of the human erythrocyte membrane. Let us try a more polar solvent such as *n*-octanol. Then one plots, by analogy with Fig. 3, the permeability coefficients for the permeants crossing the human red cell membrane divided by the relevant octanol/water partition coefficients, against the molecular weight of the permeant. We do this in Fig. 5. The data show an excellent fit to a straight line even when those permeants penetrating apparently by a specific system are included. In particular the amide series (permeants Nos. 2, 3, 4, 5, 6, 7, 8, and 18) seem to fit the straight line plot. If the very accurate data of Sha'afi et al. [11] on the red-cell permeability of the amides are plotted (as in Fig. 6), as permeability coefficient against the number of carbon atoms in the permeant, it is clear that for the human red cell and for the dog red cell (and for an artificial lipid bilayer system) there is a definite minimum permeability value in the region of the third carbon atom. On the other hand, the partition coefficients (Fig. 6) show no such minimum. The data might, at first sight, suggest that the small amides, penetrating relatively faster than the larger ones, might make use of some other parallel pathway for entering the red cell. (Such a pathway would allow the penetration of only small permeants and be perhaps some type of pore through hydrophilic regions of the membrane. This is, indeed, the contention of Sha'afi et al. [11].) But we saw in Fig. 5 that the dependence on molecular size of the (calculated) intramembrane diffusion coefficient for all of these amides is much the same as that for the other molecular species in Fig. 5. The same straight line fits the range of molecular weights from 18 through 102. The controversy as to whether the smaller amides move across the red cell membrane by pores or by dissolution in and diffusion across an organic, structured matrix thus resolves itself into the question as to whether or not we know enough about the problem of diffusion across structured phases as to demand a straight line plot for data such as those of Fig. 5.

Data on an ascending homologous series of permeants provide an instructive choice of model solutes since with such a series, it is merely the length of the molecule that increases with molecular weight, the cross-sectional area of the molecule remaining unaltered. It has been clearly established for the diffusion of

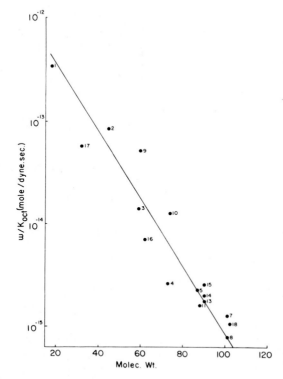

Fig. 5. Calculated relative intramembrane diffusion coefficients across the human red blood cell membrane as a function of molecular weight. Abscissa: molecular weight on a linear scale. Ordinate: permeability coefficients from Fig. 4 divided by relevant octanol/water partition coefficients (listed in [4]).

such homologous series of compounds within polymer phases that it is the smallest cross-sectional area of the diffusing molecule that determines the rate of diffusion. One might therefore expect that for intramembrane diffusion one would find a less steep dependence of rate on weight after the first few members of an ascending homologous series. Indeed, the fatty acid series studied by Klocke et al. [13] (Fig. 4) shows practically no dependence of the calculated intramembrane diffusion coefficient on molecular weight.

The data on the permeability of red blood cells can thus largely be accounted for on the model in which penetration occurs by dissolution into an organic phase with solution properties similar to n-octanol followed by diffusion across the organic phase which is structured and provides a steep molecular weight dependence of the diffusion rate constant. Parallel pathways through the membrane exist, however, for a variety of metabolites, including sugars, ions, amino acids, nucleosides and also small polyols. Water moves across specialised regions of the membrane (probably

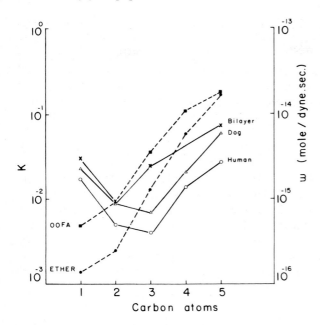

Fig. 6. Partition coefficients and permeability coefficients of amides as a function of carbon chain length. Dashed lines: partition coefficients for (olive oil + fatty acid)/water system (OOFA) or ether. Solid lines: permeabilities across red cell membranes of dog or man, as indicated; data from [11] or across artificial lipid bilayer membranes, data from Poznansky et al. [25].

protein) but the evidence that other small hydrophilic molecules cross using such parallel paths is still ambiguous.

4. Systematic treatment of solvent properties and mass selectivity

The reader will already have felt the need for a more systematic approach to the problem of sorting out the solvent and mass selectivity properties of the cell membrane. Such an approach [8,9] starts off with Eqn. 5, which can be rewritten in the form

$$P = P_0 K M^{-s_m} \tag{7}$$

where P_0 is a constant, being the term D_0 of the previous section divided by the membrane thickness while s_m is a parameter describing the mass selectivity of the permeability barrier. Taking logarithms of both sides of Eqn. 7 we obtain

$$\log P = \log P_0 + \log K - s_m \log M \tag{8}$$

The equation in this form cannot be used directly since we do not have experimental values for the partition coefficient of the permeability barrier within the membrane. What we do have for each permeant are values of K_{est}, partition coefficients for various model organic solvents which we want to test as descriptors of the cell membrane barrier, and hence as possible estimates of the K values. In order to test the reasonableness of a particular choice of organic solvent we transform Eqn. 8 into the generalised form

$$\log P = \log P_0 + s_k \log K_{est} - s_m \log M \qquad (9)$$

in terms of experimentally measurable parameters P, K_{est} and M. For a correct choice of organic solvent, the term s_k, the validity index, must be unity.

We can obtain sets of values of P, M and K_{est} for a particular cell membrane, using various organic solvents and the range of permeable solutes. We now insert these values into Eqn. 9 and solve by the method of least squares (multivariate regression), obtaining the parameters P_0, s_k and s_m. The parameter s_k tells us whether we have chosen the correct organic solvent as a model for the solvent

TABLE 1

Mass selectivity of intramembrane diffusion coefficients and solvent validity indices

When s_k is recorded, data were fitted to Eqn. 9. When just s_m is recorded, this is the slope of the regression line of (log of) intramembrane diffusion coefficient on log molecular weight.

Cell type	Model solvent	Validity index (s_k)	Mass selectivity index (s_m)
Biological membranes:			
Chara [8]	Olive oil	1.1 ±0.1	2.9±0.6
Nitella [8]	Olive oil	1.4 ±0.1	3.7±0.5
Phascolosoma [8]	Olive oil	1.0 ±0.4	5.1±1.7
Arbacia [8]	Olive oil	1.1 ±0.2	4.2±1.7
Beef red cell [8]	Olive oil	1.4 ±0.3	6.0±1.6
Human red cell [a]	Olive oil/oleic acid	0.99±0.1	5.0±0.4
Dog red cell [a]	Olive oil/oleic acid		4.5±0.8
Rabbit gall bladder [9]	Olive oil/oleic acid		2.8±0.5
Nil 8 hamster fibroblast [19]	n-Octanol		3.9
HTC rat hepatoma [19]	n-Octanol		3.7
Artificial membranes:			
Egg lecithin bilayers [26]	n-Decane	0.16	0.25
	Olive oil	0.56	2.2
	Ether	0.79	2.8
	Octanol	0.98	4.2
Liposomes from lecithin, cholesterol + phosphatidic acid [24]	Olive oil		3.6±0.4
Diffusion coefficients in water			0.5 to 0.7

[a] W.R. Lieb, personal communication.

properties of the permeability barrier. The parameter s_m tells us the mass dependence of the intramembrane diffusion coefficients, assuming the particular organic solvent in question, while P_0 is a measure of the overall "tightness" of the cell membrane. Applying this analysis to the permeation data for various cell systems we obtain the values listed in Table 1. Also included in the table are data obtained by various investigators from permeability measurements on artificial membranes.

Inspection of Table 1 shows that for many such membranes it is possible to find a solvent with a "validity index" close to unity. With such solvent, the mass selectivity parameter is generally a fairly large number, above three, suggesting that cell membranes in general discriminate sharply between solutes on the basis of their size and may well possess a fairly structured permeability barrier that behaves more like a soft polymer than like a simple fluid.

Quantitatively, what the results of Table 1 mean is this: cell membranes discriminate between various solutes both according to the solubility in organic solvents of the solutes... and according to solute size. Ether or olive oil are not bad models for the partitioning properties of cell membrane permeability barriers. Ether/water partition coefficients vary, for the most commonly tested permeants, by about 10^4 (say, from malonamide, $K_{ether} = 3 \cdot 10^{-4}$, to butanol, $K_{ether} = 7.7$). Molecular weights vary from methanol (32 M_r) to triacetin (218 M_r), a 7-fold range. With a mass selectivity parameter of 5, the variation in P from this source would be $7^5 = 1.68 \cdot 10^4$, comparable to that arising from the partitioning patterns. Thus the mass selectivity and the partitioning selectivity contribute to a comparable degree to the overall solute discrimination of cell membranes. Of course, with molecules as hydrophilic, for their size, as the sugars, or as large, for their hydrophobicity, as the steroids, partitioning or mass, respectively, may dominate the membrane selectivity behaviour.

5. Comparison of different model solvent systems

It may perhaps not be obvious why one should be so keen to find an accurate model solvent for a particular cell membrane. It should be pointed out first that different organic solvents do certainly have markedly differing solvent characteristics. (A prevailing conception that such solvents do not differ much in their solvent powers comes from an erroneous reading of the papers of Hansch and his colleagues [4], who have studied extensively the dependence of solubility characteristics on both solvent and solute structure. What they have found is a linear relationship when the logarithms of the partition coefficients for a set of solutes, in a particular solvent, are plotted against such parameters in a second solvent. This is very different from a linear relationship between the partition coefficients themselves.) Some data obtained by Collander [18], reproduced as Fig. 7, make this point clear. The partition coefficients between *iso*butanol and water, for 138 different compounds, plotted against the ether/water partition coefficients on a log/log scale, fall on quite a good straight line, showing clearly that the same factors determine the two partition

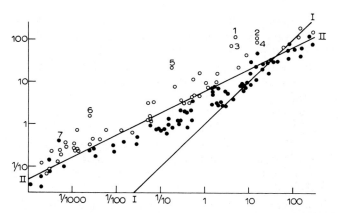

Fig. 7. Relation between ether/water partition coefficient (abscissa) and *iso*butanol/water partition coefficient (ordinate) for a large number of non-electrolytes. Figure taken with kind permission from Collander [18].

coefficients. Nevertheless, the selectivity of the two organic solvents is very different, since the SLOPE of the best fit line in Fig. 7 is far from unity, being indeed about 0.5. This means that if the partition coefficients of two substances are in the ratio of, say, 1 : 10000 in ether, they will be in the ratio of 1 : 100 in *iso*butanol. Ether is far more discriminating. Molecules having partition coefficients in the range 10 to 100 for ether (ethyl acetate, benzoic acid, caproic acid) will have very similar values in *iso*butanol, but a compound such as glycerol ($K_{\text{ether}} = 7 \cdot 10^{-4}$, $K_{iso\text{butanol}} = 0.1$) will be very much discriminated against by ether.

A plot of some partition coefficients for steroids of different structure, where the decane partition coefficients are plotted against those for octanol, is given in Fig. 8. Decane discriminates far more effectively between these steroids than does n-octanol, a 10-fold range of K values in octanol being transformed into a 1000-fold range in decane.

We need to know how effectively cell membranes discriminate between various solutes — is ether, or is a more polar solvent such as *iso*butanol, or perhaps the less polar decane, a better model? A second reason for determining the solvent characteristics of the permeability barrier is that this approach will help us to identify the region of the cell membrane which indeed forms the major permeability barrier. Is it the cell membrane interior region, or is it a region close to the membrane surface? We will return to this point in following sections of this chapter.

Meanwhile, how can we decide which solvent is the best model for any particular cell membrane? One simple approach is based on a consideration of Fig. 7. The data here fit fairly well the relationship

$$\log (K_{iso\text{bu}}) = 0.8 + 0.5 \log (K_{\text{ether}}) \tag{10}$$

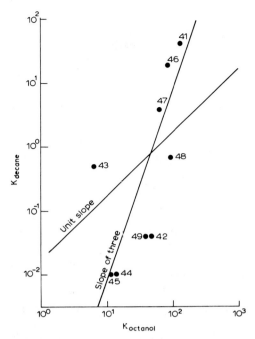

Fig. 8. Relation between decane/water partition coefficient (abscissa) and octanol/water partition coefficient (ordinate) for some steroids. Data from Giorgi and Stein [19]. The straight lines have slopes as indicated. Steroids are numbered as follows: 41, progesterone; 42, corticosterone; 43, cortisone; 44, cortisol; 45, dexamethasone; 46, (5,α)-dihydrotestosterone; 47, testosterone; 48, (17,β)-estradiol; 49, oestriol.

or

$$K_{isobu} = 6.3 \cdot K_{ether}^{0.5} \tag{11}$$

The slope of the log/log plot in Eqn. 10 or the index to which the referent partition coefficient must be raised in Eqn. 11 is a measure of the relative solute discriminatory powers of the two organic solvents. If different organic solvents are so compared against a single reference solvent, slopes such as in Fig. 7 can be used to grade the different solvents in ascending order of discriminatory power from zero (for water) through unity for a solvent which is as discriminating as the reference solvent, through greater than unity for more discriminating solvents.

The basis for the differing discriminatory powers of the various organic solvents lies in their polarity. A very non-polar solvent such as hexadecane is a very effective discriminator between solutes; olive oil and ether form a closely matched intermediate grouping; n-octanol is less discriminating, while isobutanol is the least so of the various commonly used test solvents. Discrimination between various solutes almost always arises from the differing number of hydrophilic groupings possessed by the different solutes (hydroxyl groups, amine and amide groups, for instance). These hydrophilic groupings form hydrogen bonds with the water molecules. To transfer such a grouping from water to an organic phase involves breaking such bonds and hence requires energy. Partition coefficients are low if a good deal of energy is required for such a transfer. If however the organic solvent itself is fairly polar, the energy of transfer of a hydrophilic grouping will be low since the solute can make alternate bonds with the solvent molecule and the relevant partition coefficient will be low.

6. Permeation of large lipophilic molecules—steroid transport

Some data recently accumulated by Giorgi [19] provide a very clear example of how one can use permeability measurements to characterise the permeability barrier of cell membranes. Giorgi took advantage of the fact that the steroids represent a family of substances which differ only moderately in molecular weight although considerably in partitioning behaviour. This property enables one directly to judge the relative importance of size and partitioning selectivity as factors of the membrane permeability barrier. Steroid permeabilities were measured across the membranes of two cell lines in cell culture for nine different steroids, and partition coefficients into four different organic solvents were determined for these steroids. Some of the data are plotted in Fig. 9 as log/log plots of partition coefficients against permeability coefficients. The data in Fig. 9 fall fairly well on straight lines, indicating that the same factors that affect partitioning behaviour affect also permeation. However, the slopes of the straight lines differ considerably. As we have seen, a slope of unity in such a plot as that of Fig. 9 indicates that the model organic solvent and the membrane permeability barrier discriminate equally effectively between the various test solutes. This seems to be the case for n-octanol as the model solvent (upper line in Fig. 9) but is not the case for n-decane (lower line). For the solvents ether, olive oil, and n-decane (in that order) the membrane and model solvent increasingly differ in their discriminatory power. In simple terms, the range of partition coefficients for these nine steroids is some 4000-fold for n-decane, some 700-fold for ether, 80-fold for olive oil and only 20-fold for n-octanol and for the cell membranes. Thus it seems that the barrier in the cell membrane which discriminates among the various steroids is better modelled by a fairly polar solvent such as n-octanol than by a very non-polar solvent such as n-decane.

To test whether the barrier for steroids was the same as the barrier for the smaller solutes, Giorgi also measured the permeabilities and partition coefficients for some smaller molecules. Some of these data are reproduced in Fig. 10 as a log/log plot of permeability coefficients against n-octanol partition coefficients. Clearly the data for the small solutes and for the steroids fall on a fairly good straight line, suggesting that the same rate law holds for the permeation of the steroids as for that of the smaller solutes. All seem to encounter a rate-limiting permeability barrier which has

Permeability for lipophilic molecules

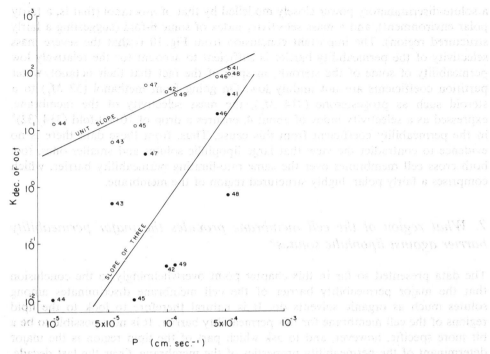

Fig. 9. Relation between partition coefficients in various solvent systems (abscissa) and permeability coefficients (ordinate). The permeants are numbered as in Fig. 8, and the data refer to the study of Giorgi and Stein [19] on a rat hepatoma cell line (HTC cells) grown in culture. The lower line and filled circles are for decane. Straight lines are drawn through the data, with slopes as indicated.

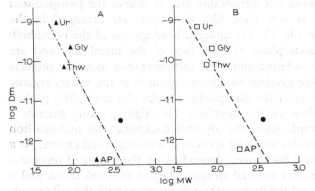

Fig. 10. Calculated intramembrane diffusion coefficients across membranes of (A) Nil 8 hamster fibroblasts and (B) the HTC rat hepatoma cell line, as a function of molecular weight of the permeant. Ordinate: logarithm of membrane thickness, taken as $50 \cdot 10^{-8}$ cm, multiplied by the permeability coefficient (in cm/sec) and divided by the octanol/water partition coefficient. Abscissa: logarithm of molecular weight. Data taken from Giorgi and Stein [19]. Permeants indicated as follows: large circle, mean value for all the steroids of Fig. 8; Ur, urea; Thw, thiourea; Gly, glycerol; Ap, antipyrine. Regression lines through the points have slopes of -3.9 for the Nil 8 cells and -3.7 for the HTC cells.

a solute-discriminatory power closely modelled by that of *n*-octanol (that is, a fairly polar environment), and a mass selectivity index of some 6-fold (suggesting a fairly structured region). The important conclusion from Fig. 10 is that the severe mass selectivity of the permeability barrier is sufficient to account for the relatively low permeability of some of the steroids, in spite of the fact that their octanol/water partition coefficients are not unduly low. On going from methanol (32 M_r) to a steroid such as progesterone (314 M_r), the mass selectivity of the membrane, expressed as a selectivity index of about 4, ensures a drop of 10 000-fold $(314/32)^4$ in the permeability coefficient from this cause. Thus, from these data there is no evidence to contradict the view that large lipophilic solutes and smaller ones, too, both cross cell membranes over the same rate-limiting permeability barrier, which comprises a fairly polar, highly structured region of the membrane.

7. What region of the cell membrane provides the major permeability barrier against lipophilic solutes?

The data presented so far in this chapter point overwhelmingly to the conclusion that the major permeability barrier of the cell membrane discriminates among solutes much as organic solvents do. It is natural therefore to look to the lipid regions of the cell membrane for this permeability barrier. It is now possible to be a bit more specific, however, and to ask which part of the lipid region is the major determinant of the permeability properties of the membrane. Over the last decade, physicochemical studies of natural and artificial membranes have provided a substantial body of information enabling this question to be clearly formulated and investigated.

The major components of most cell membranes are, of course, the phospholipid and cholesterol molecules. It is now clear [20] that these are arranged in the membrane much as depicted in Fig. 11. The charged head groups of the phospholipids are exposed to the aqueous phase at each face of the membrane and are relatively mobile in that phase, rotating about the carbon-carbon axis that projects from the glycerol backbone. The glycerol backbone region is at the water/organic phase interface, and these sectors of the phospholipid molecules are tightly packed against one another in the plane of the interface. This tight packing extends a number of carbon atoms inwards from the glycerol backbone, the hydrocarbon chains here together with the cholesterol ring forming a well-organised matrix. At a certain distance from the glycerol backbone, (depending on the degree of unsaturation of the hydrocarbon chain) this ordered structure begins to break down and a gradient of increasing flexibility of the hydrocarbon chain arises with the tail regions of the phospholipid hydrocarbons and the cholesterol side chain being the most highly flexible portion of the whole structure. Thus the membrane can be thought of, roughly, as consisting of five regions or sheets arranged in a series fashion (Fig. 11). On proceeding from the external face of the membrane inwards, one encounters successively (i) the outer region of the charged head groups, highly polar and highly

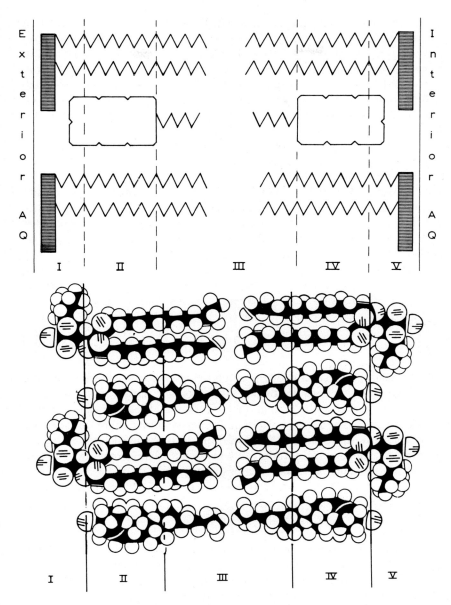

Fig. 11. (a) Schematic diagram for the arrangement of cholesterol and phospholipid in lipid membranes. The shaded rectangles represent the charged head groups of the phospholipids; the jagged lines represent hydrocarbon chains of the phospholipids and cholesterol; the large unshaded rectangle represents the steroid ring of cholesterol. Regions I and V are the outer region, highly charged and flexible; regions II and IV are less polar and very tightly packed; region III is extremely non-polar but very loosely packed. (b) More realistic model, adapted, with permission from N.P. Franks and W.R. Lieb, J. Mol. Biol. (1980) 141, 43–61.

flexible, (ii) the cholesterol ring, the glycerol backbone and first few hydrocarbon groupings, a less polar but increasingly non-polar section, highly inflexible and tightly packed, (iii) the tail groups of the hydrocarbon chain ends of the two faces of the membrane, an extremely non-polar but very flexible, loosely packed region, followed by (iv) and (v) which are very similar to (ii) and (i), respectively, although almost certainly somewhat different in packing and polarity, inasmuch as the cell membrane is in general asymmetrical in its phospholipid composition at the two membrane faces [21].

Which of these successive regions contributes most to the barrier properties of the cell membrane? Is the same region indeed the barrier for every type of solute molecule?

A molecule crossing the cell membrane from one side to the other will encounter this series of membrane regions and within each region it will experience a resistance to movement. The overall resistance to movement (the reciprocal of which will be the membrane permeability) will be given by the sum of these resistances. It may well be that one of these resistances is so strong that it quite overshadows the remaining resistances, which we can therefore ignore. Or it may be, at the other extreme, that all the five resistances are of comparable magnitude so that none can be neglected. Within each region j the resistance R_j will be given by a term such as

$$R_j = \frac{\Delta x_j}{K_j \cdot D_j} \tag{12}$$

where K_j is the partition coefficient for the solute in question distributing between the j-th region and water, Δx_j is the thickness of this j-th region, while D_j is the diffusion coefficient of the solute within the j-th region. Clearly each region is considered to have such an average regional property, although in reality such properties may certainly vary within the region. Eqn. 12 is the local form of Eqn. 5, for the region in question, permeabilities and resistances being reciprocal to one another. The overall resistance R of the membrane is given by

$$R\left(=\frac{1}{P}\right) = \sum_{j=1}^{5} R_j \tag{13}$$

Of the five regions considered in our "equivalent membrane", none is markedly thicker than any other, so the Δx_j terms need not concern us much. This leaves the K_j and D_j terms for each region as the major determinants of the differential resistance. These terms enter the equation as products showing that the overall resistance of a highly flexible region (high D_j) may be counteracted by a low partition (small K_j) to give a resistance comparable to that of a low D_j-high K_j region, as we shall see below.

Now, of the five regions in our equivalent membrane of Fig. 11, regions (i) and (v) may be expected, for uncharged solutes, to have K values close to unity since they are highly polar and water-like in their solvent characteristics. These regions are also highly flexible so that the D values will be high. The product $K \cdot D$ will be high and little resistance will be experienced by a non-electrolyte solute transversing these regions. Regions (ii) and (iv) are tightly packed, highly structured. Here D values will be low and severely mass-dependent, these regions behaving like soft polymeric sheets. Partition coefficients (K values) will not necessarily be very low, however, since the region is fairly polar at the glycerol backbone, presumably becoming increasingly non-polar as the hydrocarbon chain is descended. If the K values are comparable to those of fairly polar solvents (for example, n-octanol), the range of partition coefficient will also not be large. Thus *mass and not partitioning behaviour will be the major determinant of the differential selectivity of these regions.* Region (iii) is only loosely packed. D values will be high and not highly mass-selective. The region is very non-polar and partition coefficients of polar solutes will be very low. The range of K values will be wide. *In this region partitioning and not mass selectivity will be the major determinant of the resistance.*

It is clear, therefore, that if one considers a wide range of solutes on the model here proposed, different regions of the membrane may well become the rate-limiting barrier for the trans-membrane movement, according to the properties of the solute. Small, very polar solutes will be held back mostly by their low partitioning into the middle region (iii), their small size enabling them to slip between the cholesterol ring, the rigid chains of the hydrocarbons and glycerol backbone of regions (ii) and (iv). Large molecules will, however, find regions (ii) and (iv) to present overwhelming barriers to transit and the polarity of such large solutes will be of secondary concern. Finally, when different cell membranes are considered, composition differences with regard to the amount of cholesterol and the nature of the phospholipid molecules will affect the partitioning characteristics of the various regions and, much more so, the differences in packing of the regions. A particular solute might, therefore, find a different region of the membrane to be the rate-limiting barrier to diffusion in different cells and also under different physiological conditions in the same cell. These aspects will be briefly considered in a later section of this chapter.

Clearly, the data on steroid movements discussed in the preceding section are consistent with the view that it is the tightly packed regions (ii) and (iv) that provide the major barrier to movement for these large, not very polar molecules. Also for the range of cell types, data for which are collected in Table 1, the steep mass-selectivity dependence and the accuracy with which ether or olive oil models the partitioning behaviour of those membranes, suggest that, for the particular range of solutes chosen for study, regions (ii) and (iv), rather than the deep interior of the membrane, form the major permeability barrier. However, if, of the solutes chosen for Table 1, one had considered only the small, highly polar molecules, it may well be that the properties of the interior region would have proved to dominate. Such an analysis seems not to have been published, however, and on present analyses we must nominate the tightly packed proximal regions of the hydrocarbon chains and cholesterol ring together with the glycerol backbones of the phospholipids as the major barrier to the trans-membrane movement of solutes. The deep interior of the cell membrane comprising the distal regions of the hydrocarbon chains and the cholesterol side chain has not yet been noticed in current permeability studies of the cell membranes.

8. Studies on artificial membranes

Artificial lipid bilayer membranes can be made [22,23] either by coating an orifice separating two compartments with a thin layer of dissolved lipid (which afterwards drains to form a bilayered structure—the so-called "black film") or by merely shaking a suspension of phospholipid in water until an emulsion of submicroscopic particles is obtained—the so-called "liposome". Treatment of such an emulsion by sonication can convert it from a collection of concentric multilayers to single-walled bilayers. Bilayers may also be blown at the end of a capillary tube. Such bilayer preparations have been very heavily studied as models for cell membranes. They have the advantage that their composition can be controlled and the effect of various phospholipid components and of cholesterol on membrane properties can be examined. Such preparations focus attention on the lipid components of the membrane for investigation, without the complication of protein carriers or pore-forming molecules. Finally, the solutions at the two membrane interfaces can readily be manipulated. Many, but not all, of the studies on artificial membranes support the view developed in the previous sections of this chapter that membranes behave in terms of their permeability properties as fairly structured and by no means extremely non-polar sheets of barrier molecules.

Cohen and Bangham [24] determined the permeability of a substantial number of the classically studied permeants, across liposomes prepared in the form of multilayered vesicles. Their results were plotted in the form of a plot of the relative intramembrane diffusion coefficient (that is, the permeability divided by the olive oil/water partition coefficient), against the molecular weight of the permeant. The data fell on a fairly good straight line suggesting that the solvent olive oil is a good model for the discriminatory properties of this model membrane. The mass selectivity parameter, s_m of Eqn. 8, obtained from the data of this figure showed that diffusion rates varied with the -3.6 power of the mass, indicating a relatively structured region as being the major permeability barrier. Poznansky et al. [25] measured very carefully the permeabilities of four amides and urea across bilayers blown at the end of a capillary tube and into liposomes. Some of their data are included in Fig. 6. Their data showed a very clear dependence of intramembrane diffusion coefficient (calculated using ether/water partition coefficients) on the size of the solute. Poznansky et al. felt that their data indicated an essential difference as between these artificial membranes and red-cell membranes in their permeability behaviour towards the smallest amides, in that formamide and acetamide might enter red cells through a pathway parallel to the lipid. It seems hard to argue from Fig. 6, however, that there is an essential difference in this case between the behaviour of the model and of the cell membrane. Wolosin and Ginsburg [26] measured permeability coefficients for almost twenty fatty acids across black films (but the adequacy of the corrections they used for the effect of unstirred layers could be questioned). A comparison of their data with partition coefficients of these fatty acids into *n*-decane, olive oil, ether and *n*-octanol, analysed by the systematic treatment of Lieb and Stein (Section 4) indicated (see Table 1) that octanol was an adequate model for the solute discriminatory powers of these artificial membranes. The mass dependence of the intramembrane diffusion coefficients using *n*-octanol as the model solvent was steep, a selectivity index of about four being obtained. A subsequent analysis of these data [27], and some twenty other permeability measurements of different solutes across black films of phospholipid also gave a reasonably good straight line when plotted on a log/log scale as intramembrane diffusion coefficients (calculated using *n*-octanol as model solvent) against molecular volume. The slope of this plot, giving the mass selectivity index, is about four, once again. (The permeabilities of the faster penetrating fatty acids (Nos. 4–10 in their figure) may have been underestimated, if here the unstirred layer correction was indeed inadequate, but their omission from the plot makes little difference to the assessment of the adequacy of the straight line.)

Finally, Orbach and Finkelstein [28] have measured permeabilities across phospholipid black films (for a small but well-chosen range of solutes). Their data clearly showed that for this set of solutes, *n*-decane is an adequate model for the solute-discriminatory power of these membranes. The calculated mass selectivity parameter was small and statistically not very significant. Orbach and Finkelstein show also that the membrane resistances (calculated on the assumption that the membrane's partition behaviour is well modelled by decane while intramembrane diffusion coefficients have values as in free water) are sufficient to account for much of the membrane resistance for these solutes. These authors are convinced, therefore, that the membrane interior (region (iii) in Fig. 11) presents the major barrier to permeation of solutes. This indeed seems to be true for the membranes they studied and for the range of solutes whose permeabilities they measured. We have seen, however, that this conclusion is not valid for numerous cell membranes and some artificial membranes, including black lipid films rather similar to those studied by Orbach and Finkelstein. An extended study of a range of other solutes, covering a variety of sizes and polarities might well reveal the roles of either the inner or the outer membrane regions (depending on circumstances) as determinants of the barrier properties of the membrane.

9. Partition coefficients of cell and artificial membranes

It has been shown above how an analysis of permeability data can help one choose a solvent which will be a good model for the solute-discriminatory power of a cell membrane. It is conceivable, however, that the cell membrane in question and the chosen model solvent display the same range of partitioning capacity for a variety of solvents yet the absolute value of these partition coefficients, over the whole range of solutes, might be very different. A tightly packed membrane might present an overall resistance to the entry of all solutes yet discriminate between these solutes as would a simple liquid of equivalent polarity. What is needed to test this point are direct measurements of partitioning into the membrane and into model solvents. Seeman [29] found that for many solutes, octanol/water partition coefficients were about five times the corresponding values for the erythrocyte membrane/water system. In an extensive study, Katz and Diamond [30] measured the partitioning of a large number of solutes into the membranes of liposome preparations. Values found were very close to those for the solvent *n*-octanol, which modelled accurately the partitioning behaviour of these artificial membranes, both relatively as between the different solutes and also absolutely. In a later study, using spin-labeled derivatives of nitroxides, Dix and co-workers [31] found changes of (only) up to 2-fold in partition coefficients when artificial membranes were frozen by reducing the temperature.

Which regions of the membrane are probed by partitioning studies such as those of Katz and Diamond [30]? We saw previously that permeation studies probe that region of the membrane which is the major barrier to permeation and hence has a minimum value for the product of K and D, the local partition coefficient and regional diffusion coefficient, respectively. Partition studies, on the contrary, are weighted according to which region of the membrane has the greatest capacity for solute uptake, that is, which region has the greatest values for K. If the three regions of Fig. 11, namely (i) + (v), (ii) + (iv) and (iii), have approximately the same volume, an overall membrane partition coefficient for any solute will be determined by which region has the greatest value for the partition coefficient, since most of the solute will be found in this region. On the whole, the more polar regions will be expected to have the highest partition coefficients for the general run of solutes, hence the more polar regions will be weighted, probed and revealed in a partition study. Therefore, it is not surprising that Katz and Diamond found partitioning behaviour of their membrane preparations to follow the behaviour of a fairly polar model solvent such as *n*-octanol.

What *is* surprising is that permeability data should be so closely modelled by the more polar model solvents. This paradox indicates that the diffusion resistance contributes overwhelmingly to the overall resistance of the major permeability barrier. For the many solutes and membrane classes described in the preceding sections, the tight packing of the proximal regions of the hydrocarbon chains ensures that this region is the major permeability barrier in spite of the relative ease with which these solutes partition into this region. (For small and highly polar solutes, this conclusion need not be valid, however.) The close match between absolute partition coefficients of membranes and good model solvents suggests that we will not be far wrong if we take the absolute values of partition coefficients for chosen model solvents as sound predictors for the partitioning behaviour of the permeability barrier of the membrane.

10. Affectors of membrane permeability

A number of additives and treatments have been shown to affect membrane permeability, both in cell membranes and in artificial membranes. Much of the information can be readily understood in terms of the concept of "the tightly packed, not very non-polar" membrane barrier that we have been considering.

(a) Alcohols, anaesthetics and other fat-soluble additives

These compounds increase the permeability of solutes that permeate by simple diffusion, in proportion to the concentration and the partition coefficient of the individual additive [14]. (These same compounds *decrease* the rate of trans-membrane movement of many solutes that cross cell membranes by carrier systems or by other mediated pathways.) This effect is precisely paralleled by the behaviour of the "plasticisers" used in polymer membrane work, which act by dissolving in the polymer membrane, distorting its rigid structure and hence allowing the freer movement of diffusing molecules. Exactly this mechanism has been postulated to account for the effect of the lipid-soluble additives in cell membrane studies [9]. These agents will dissolve in the lipid phase (according to their concentration and partition coefficient) and break up the rigid, ordered structure which forms the major permeability barrier. Distorting the tightly packed hydrocarbon chains increases the free volume available for solute diffusion between the chains. No such effect would be expected (or is found) for diffusion in unstructured media such as free water or simple hydrocarbons. Any such effect found in biological membranes is good evidence for the hypothesis that the major permeability barrier for trans-membrane transport is a tightly packed region of the membrane, displaying low diffusion coefficients in the absence of additives.

(b) Temperature

Temperature coefficients of membrane diffusion can be very large [14]. Activation energies of 10 to 40 kcals/mol, corresponding to Q_{10} values of 1.7 to 9.0 are by no means uncommon. High temperature coefficients for partition coefficients are found, however, only for very non-polar solvents. Partition coefficients measured on membrane system have low temperature coefficients. Diffusion in relatively unstructured systems such as water has a low temperature coefficient, activation energies of 4 to 6 kcals/mol being the rule. But high temperature coefficients are generally found for diffusion in structured, polymer systems. These high activation energies arise from the necessity for breaking numerous bonds between the molecules comprising the matrix in which diffusion is occurring. Thus the high temperature coefficients of trans-membrane movement of many solutes suggests once again that the rate limiting barrier for permeation is a structured region.

(c) Cholesterol content

The cholesterol content of a membrane will affect its permeability [14]. The cholesterol content of cell membranes can be varied by incubating the cells in media containing liposomes either enriched with cholesterol (which then adds to the cholesterol content of the cell membrane), or without cholesterol (when cholesterol is removed from the cell membrane). Permeability measurements on such cells with an altered cholesterol content show clearly that permeability is increased when the cholesterol content of the membrane is reduced while permeability decreases when the cholesterol of the membrane increases. Similar results are obtained when artificial membrane systems are studied, and the effect of cholesterol can be followed in such systems by physical methods. Cholesterol addition is known to increase the tightness of packing of cell membranes.

The studies confirm that the close-packed region of the membrane forms the rate-limiting barrier for trans-membrane movement by simple diffusion. This situation may be physiologically relevant in that the leakiness of the cell membrane can clearly be affected by physiological control of the cholesterol content of the cell membrane. Bacterial and plant cell membranes which do not contain cholesterol and which are not particularly leaky must have other compositional peculiarities to provide well-sealed membranes.

(d) Fatty acid composition

The fatty acid content of a membrane will affect its permeability [14]. The fatty acid composition of the phospholipid chains in the cell membrane can be varied by modifying the diet of the organism from which the membranes are derived, or by the addition of foreign phospholipids to the isolated cells. Free fatty acids foreign to the cell can be added to a cell preparation and incorporated into the membranes. Finally, membranes from different organisms differ in their content of fatty acids. In comparative studies of permeability and lipid composition, the effect of this latter variable can be evaluated. The available data give a clear answer: increasing the degree of unsaturation of the fatty acid chains of the hydrocarbons and decreasing the length of the hydrocarbon chains leads to an increased permeability. Increasing the number of *cis*-double bonds in the fatty acid chains will decrease the tight-packing of the hydrocarbon chains and would be expected, therefore, to lower the permeability barrier. Decreasing the length of the hydrocarbon chain also decreases the tight-packing of the hydrocarbon chains and thus influences permeability very dramatically.

11. Overall survey and conclusions

Available data on the permeability of cell membranes are consistent with the view that the vast majority of substances find the major barrier to their trans-membrane movement to be the tightly packed proximal chains of the phospholipid hydrocarbons, together with the cholesterol moieties and the glycerol backbone. This region has solvent properties well modelled by the less non-polar organic solvents (ether and even *n*-octanol). The data reveal a steep mass selectivity (three to six power inverse dependence of intramembrane diffusion coefficient on molecular weight or volume). The region acts as a barrier by virtue of being tightly packed, and modifications of this tight-packing by additives and alteration of membrane composition can reduce its effectiveness as a barrier. Water and perhaps some of the smallest hydrophilic solutes will find the interior of the cell membrane, the highly non-polar but freely mobile region of the distal portions of the hydrocarbon chains, to be the major barrier. (For these solutes parallel paths for trans-membrane movement, not crossing through the lipid phase, might also be present.) If the permeability of a particular compound has not been measured for a particular cell, and some idea of the probable value of this parameter needs to be estimated (the compound may be a new drug or newly found metabolite), calculations may be attempted on the tightly packed hydrocarbon chain model. The partition coefficient of the compound in ether or octanol can be measured or even calculated [4] from its composition. The intramembrane diffusion coefficient can be calculated by assuming a mass selectivity factor of between three and six (if the number is not known for the cell in question) and by assuming that water will diffuse through the tight-packed region about one thousand times as slowly as it diffuses in free water. A membrane thickness of some 40 Å is not unreasonable. The agreement between calculated and measured permeabilities is really quite good. The model of the cell membrane as a tightly packed region of intermediate polarity does a good job of accounting for much of the permeability data for biological cell membranes.

Acknowledgements

I am very grateful to William Lieb, Hagai Ginsburg and Chana Stein for their helpful comments on an earlier version of this chapter, and to Michal Razin and Chana Stein, again, for drawing the original figures.

References

1. Jacobs, M.H., Glassman, H.N. and Parpart, A.K. (1935) J. Cell. Comp. Physiol. 7, 197–225.
2. Lin, S. and Spudich, J.A. (1974) J. Biol. Chem. 249, 5778–5783.
3. Collander, R. (1954) Physiol. Plant. 7, 420–445.
4. Leo, A., Hansch, C. and Elkins, D. (1971) Chem. Rev. 71, 525–616.
5. Stein, W.D. (1969) J. Gen. Physiol. 54, 81s–90s.
6. Singer, S.J. and Nicolson, G.L. (1972) Science 175, 720–731.
7. Stein, W.D. (1967) in The Movement of Molecules Across Cell Membranes, Academic Press, New York.
8. Lieb, W.R. and Stein, W.D. (1969) Nature 224, 240–243.
9. Lieb, W.R. and Stein, W.D. (1971) in F. Bronner and A. Kleinzeller (Eds.), Current Topics in Membranes and Transport, Vol. 2, Academic Press, New York, pp. 1–39.
10. Crank, J. (1957) The Mathematics of Diffusion, Oxford University Press, London.
11. Sha'afi, R.I., Gary-Bobo, C.M. and Solomon, A.K. (1971) J. Gen. Physiol. 58, 238–258.
12. Savitz, D. and Solomon, A.K. (1971) J. Gen. Physiol. 58, 259–266.
13. Klocke, R.A., Andersson, K.K., Rotman, H.H. and Forster, R.E. (1972) J. Physiol. (Lond.) 222, 1004–1013.

14 Deuticke, B. (1977) Rev. Physiol. Biochem. Pharmacol. 78, 1–97.
15 Kaplan, M.A., Hays, L. and Hays, R.M. (1974) Am. J. Physiol. 226, 1327–1332.
16 Naccache, P. and Sha'afi, R.I. (1973) J. Gen. Physiol. 62, 714–736.
17 Owen, J.D. and Solomon, A.K. (1972) Biochim. Biophys. Acta 290, 414–418.
18 Collander, R. (1950) Acta Chem. Scand. 4, 1085–1098.
19 Giorgi, E.P. and Stein, W.D. (1981) Endocrinology (in press).
20 Franks, N.P. and Lieb, W.R. (1979) J. Mol. Biol. 133, 469–500.
21 Rothman, J.E. and Lenard, J. (1977) Science 195, 743–753.
22 Fettiplace, R., Gordon, L.G.M., Hladky, S.B., Requena, J., Zingsheim, H.P. and Haydon, D.A. (1975) in E.D. Korn (Ed.), Methods in Membrane Biology, Vol.4, Plenum, New York, pp. 1–76.
23 Huang, C. and Thompson, T.E. (1974) in S. Fleischer and L. Packer (Eds.), Methods in Enzymology, Vol. 32, Academic Press, New York, pp. 485–489.
24 Cohen, B.E. and Bangham, A.D. (1972) Nature 236, 173–174.
25 Poznansky, M., Tong, S., White, P.C., Milgram, J.M. and Solomon, A.K. (1976) J. Gen. Physiol. 67, 45–66.
26 Wolosin, J.M. and Ginsburg, H. (1975) Biochim. Biophys. Acta 389, 20–33.
27 Wolosin, J.M., Ginsburg, H., Lieb, W.R. and Stein, W.D. (1978) J. Gen. Physiol. 71, 93–100.
28 Orbach, E. and Finkelstein, A. (1980) J. Gen. Physiol. 75, 427–436.
29 Seeman, P. (1972) Pharmacol. Rev. 24, 583–655.
30 Katz, Y. and Diamond, J.M. (1974) J. Membr. Biol. 17, 101–120.
31 Dix, J.A., Kivelson, D. and Diamond, J.M. (1978) J. Membr. Biol. 40, 315–342.

CHAPTER 2

Permeability for water and other polar molecules

R.I. SHA'AFI

Department of Physiology, University of Connecticut Health Center, Farmington, CT 06032, U.S.A.

1. Introduction

One of the main functions of the plasma membranes of living cells is to control the transport processes into and out of the cells of many substances and thus to regulate the composition of the intracellular fluid. The fluid usually contains solutes at concentrations which are quite different from their corresponding values in the bathing medium. This is achieved by the ability of the membrane to discriminate among various solutes so that some are allowed through, others are kept inside or outside the cell, and still others are carried actively. In addition, important processes such as oxidative metabolism, protein synthesis and several other synthetic processes are intimately connected with and dependent on membrane processes. In fact, continued existence of the cell is critically dependent on its having a functional plasma membrane.

Studies of permeability characteristics of the cell membrane have been of considerable interest to cell physiologists, since these characteristics help to define functional and structural properties of the plasma membrane and help elucidate the factors that determine the rate of movement of different substances into and out of various tissues in the body. Much of our present understanding of the cell membrane structure has been derived from the early work of Overton [1] on the movement of water and nonelectrolytes across cell membranes. Aside from being of considerable theoretical importance, the process of water transport across biological membranes and the effect of certain hormones on this process in some tissues is of practical importance.

Our present view of the cell membrane structure is that it is analogous to a two-dimensional oriented solution of globular lipoproteins dispersed in a discontinuous fluid bilayer of lipid solvent [2]. Phospholipids constitute the bulk of the lipids. Both components of the membrane (proteins and lipid) are free to some degree to have lateral mobility in the plane of the membrane and in some cases are asymmetrically distributed across the two halves of the bilayer [2]. Some proteins are peripherally attached to either of the two faces while other polypeptide chains may span the whole thickness of the membrane [2]. These latter proteins probably correspond to the intramembranous particles seen in freeze-etch images and freeze-fracture experiments [3]. Accordingly, these polypeptide chains may provide a hydrophilic environment for the transport of water and certain solutes. In fact, some of them have been implicated in the transport of water [4], anions [5,6] and most probably glucose.

2. Methods for measuring water and small polar nonelectrolyte movements across a barrier

The permeability coefficient of a given barrier, either a simple cell membrane or series of membranes, to water and nonelectrolytes can be measured either under steady-state conditions or under a net flow of the substance under study. Since the movements of water and small nonelectrolytes across biological barriers are rapid, the main experimental difficulty is in designing methods for monitoring the changes. In general, there are two types of experimental arrangements. First, the barrier is mounted in a chamber between two solutions to which the investigator has access (frog skin, gallbladder, small intestine, black lipid bilayer, etc.). Secondly, the system is composed of a suspension of cells (red cells, white cells, liposomes, etc.).

(a) Radioactive tracer movements across a barrier which can be mounted between two solutions

In general, the experimental procedure is to mount the membrane or the tissue which may be composed of many layers of cells between two compartments as shown in Fig. 1. The two solutions which are homogeneous and of identical composition, are stirred constantly. The labelled substance (THO, or radioactive nonelectrolyte) is

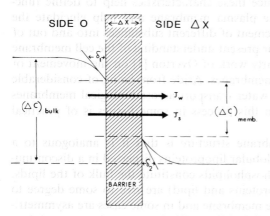

Fig. 1. Concentration profile in the solutions adjacent to the permeability barriers. $(\Delta C)_{bulk}$, $(\Delta C)_{membrane}$, are the concentration gradients between the two bulk phases and the two sides of the membranes, respectively. δ_1 and δ_2 are the thickness of the unstirred layers.

added to one compartment. At a preset time, samples are removed from the second compartment and counted. The total radioactivity in this second compartment is plotted against time. Initially there is a good linear relationship between the total radioactivity and time, and the slope of this linear relation is related to the flux. The system is in steady state since no net movement, except for the radioactive substance, occurs from one compartment to another.

In the case of the movement of labelled water (THO), the diffusional permeability coefficient, P_d, in cm/s can be calculated as follows:

$$J_{THO} = -D_{THO}(dC_{THO})/dx \tag{1}$$

Where
J_{THO} = the flux of radioactive water per unit area and time (mol/cm²/s)
D_{THO} = the apparent diffusion coefficient of THO (cm²/s)
C_{THO} = is the concentration of THO at any point X in the membrane (mol/cm³)
X = distance across the membrane (cm).

Under steady-state conditions, J_{THO} is constant and Eqn. 1 can be integrated to give

$$J_{THO} = -P_d[C_{THO}^{\Delta x} - C_{THO}^0] \tag{2}$$

where $P_d = \gamma D_{THO}/\Delta x$; C_{THO}^0, $C_{THO}^{\Delta x}$ are the concentrations of THO in bulk phases; Δx is the thickness of the pathway, and γ the partition coefficient. It is assumed that the concentration in the membrane is related to the concentration in the bulk phase by the partition coefficient γ and it is further assumed that the partition coefficient is identical for the two sides of the membrane. Under the usual experimental conditions $C_{THO}^{\Delta x} \ll C_{THO}^0$, and therefore P_d can be calculated from the experimentally determined J_{THO} and C_{THO}^0. The surface area available for diffusion and the volumes of the two compartments must be taken into consideration in calculating P_d.

Basically, identical theoretical analysis as well as experimental procedures are used in the case of studying the movement of small polar nonelectrolytes.

(b) Net water flow under the influence of pressure

Net water flow is usually measured by monitoring the time course of the change in the volume of a given compartment in response to a pressure gradient (usually the pressure gradient is generated by a difference in the concentration of impermeable solute). The basic equation which governs the volume flow of water (J_v expressed as volume per unit area per unit time) in response to the osmotic pressure generated by the solute (the barrier is assumed to be completely impermeable to the solute) is given by

$$J_v = -L_p(\Delta U_w/\tilde{V}_w) \tag{3}$$

For dilute and ideal solution $(U_w/\tilde{V}_w) = [\phi RT/\tilde{V}_w]\ln X_w = -\phi RTC_{s(m)}$, and Eqn. 3 can be rewritten as

$$J_v = -L_p RT\phi \Delta C_{s(m)} \ldots \quad (4)$$

Eqn. 4 can be rewritten to give:

$$J_w = P_f[\phi C_s^0 - \phi C_s^{\Delta x}] \ldots \quad (5)$$

Where
$P_f = \gamma L_p RT/\tilde{V}_w$
$J_v = J_w \tilde{V}_w$ = volume flow per unit area and time (cm^3/cm^2/s)
L_p = the hydraulic water conductivity (cm^3/dyne/s)
ΔU_w = the difference in the chemical potential for water
$\tilde{V}_w$ = the partial molar volume of water (mol/cm^3)
ϕ = osmotic coefficient
X_w = mole fraction of water
R = gas constant
T = absolute temperature
$\Delta C_{s(m)}$ = the concentration difference of the impermeable solute across the membrane (mol/cm^3)
$C_s^0, C_s^{\Delta x}$ = the concentrations of the solute in the bulk phases.

It is assumed that the concentration of the solute in the membrane is related to the concentration of the solute in the bulk phase by the partition coefficient γ and it is further assumed that γ is the same for both sides of the membrane. The filtration permeability coefficient, P_f, in cm/s can be calculated from the experimentally determined J_v and the difference in the concentration of the impermeable solute in the two compartments.

If the membrane system consists of more than one permeability barrier, then the overall permeability of the system, P_t, will be given by $P_t = \Sigma_{i=1}^{\eta} P_i$ for a parallel system, $1/P_t = \Sigma_{i=1}^{\eta} 1/P_i$ for a series system where P_i represents the permeability coefficient of the ith barrier. In the case of a single system where the flow of water across it is composed of diffusional and of viscous components, the value of the permeability coefficient $P_t \equiv P_d + P_v$ where P_v is the permeability coefficient of the portion of the system which contains the "pores". Using this analysis the value of P_f used in Eqn. 5 can be divided into a P_d component and a P_v component. Based on this analysis the inequality $P_f/P_d - 1 > 0$ must imply flow through "channels". The ratio P_f/P_d is commonly designated by $g(P_f/P_d \equiv g)$ where g is called the porosity factor (i.e. $g > 1$ implies the presence of pores). Moreover, if these "channels" are large enough so that Poiseuille's law of volume flow can be applied, then the radius of these "channels" can be calculated by the following relationship:

$$r = \sqrt{(8 D_w \eta_w \tilde{V}_w / RT)(g-1)}$$

Where D_w and η_w are the diffusion coefficient (cm²/s) and viscosity (poise) of water in these channels.

(c) Radioactive tracer movement in cell suspension studied with rapid flow technique

The theoretical analysis and experimental manipulation will be discussed using the red cells as an illustrative example. In this method, a suspension of red cells at a relatively high hematocrit is mixed in a rapid-flow mixing chamber with isotonic buffered solution containing the labelled substance, and the mixture is forced down a tube with ports that permit axial sampling of the suspension medium. These ports are covered with filter paper through which red cells cannot pass. Conversion of distance to time by means of velocity is the underlying principle. Because a certain minimum velocity must be exceeded in order to have efficient mixing between the red cells and the buffered solution, it becomes quite impractical to study the permeability coefficients of substances of low exchange rate (>1.0 sec) [7,8].

The basic equation which governs tracer diffusion between two well-stirred compartments is

$$d(v_p p)/dt = -k_1' p + k_1' q \tag{6}$$

Where
- v_p = the volume of the suspending medium (cm³)
- p, q = the specific activity in the suspending medium and the cell respectively (cpm/ml H_2O)
- k_1' = a proportionality constant related to the permeability coefficient
- t = time.

Under the usual experimental conditions when no net flow of either water or solute occurs (steady-state), the solution for Eqn. 6 is:

$$(p/P_\infty - 1) - (p_0/P_\infty - 1) \exp(-k_1' p_0 t / V_0' P_\infty) \tag{7}$$

in which subscripts 0 and ∞ refer to zero and infinite time. V_0' is the volume of cell water at time equal to zero.

(d) Nuclear magnetic resonance technique

This is a simple and rapid technique for measuring the time-course for water movement across mammalian erythrocytes [9]. The basis of this method is the fact that water protons can absorb energy from a radio-frequency magnetic field if the cells are placed in a static magnetic field. The decay time, which is commonly known as "spin-spin" relaxing time (T_2) of this coherent energy can be measured by standard NMR techniques. By adding a suitable concentration of a paramagnetic material such as manganese to the plasma, the spontaneous decay time (T_2) of the label for water in the plasma can be made very short (<2 ms). Since the decay time of the label for water inside is long (140 ms), the decay time within the cell will be dominated by the rate of water exit ($\simeq 8$ ms). The measurements are done under steady-state conditions and the only drawback of this technique is that only the rate of water exit can be measured. This is a much superior technique in measuring P_d than one outlined in Section 2.3, and can be used to measure water movement across membranes of other cells (for more details of this and related techniques, see [10]).

(e) Osmotic volume changes studied with stop-flow technique

The principle of this method is based on the use of light scattering or transmission as an index for measuring rapid changes in cell volume. Two solutions, one containing a dilute suspension of cells and the other containing the permeating molecule, are rapidly mixed. When steady flow is achieved, the flow is stopped abruptly and the fluid is isolated in an observation tube through which light passes. The time-course of cell volume changes can be measured indirectly from the changes in the intensity of either 90° scattered light or 180° transmitted light. In the application of this principle, various approaches have been developed for rapid mixing, and for recording [10–14]. It must be pointed out that when employing this technique the experimental conditions must be carefully chosen so that the light intensity is linearly related to cell volume. Usually this relationship holds true when the changes in cell volume are relatively small. Failure to take this into consideration could lead to erroneous conclusions.

When red cells are placed in a medium containing an iso-osmolal concentration of impermeant solute together with a suitable concentration of a permeant nonelectrolyte which enters the cell less rapidly than water, the cell initially shrinks and then returns to its initial volume after passing through a well-defined minimum. In the shrinking phase, water moves out of the cell owing to the excess of osmotically active material externally, while the nonelectrolyte moves down its concentration gradient into the cell. At the minimum point, the inward volume flow of the solute is exactly balanced by the outward volume flow of water. Subsequently, the osmotic pressure gradient reverses its direction, in part because the impermeant solute is now more concentrated in the cell than in the medium. Water enters the cell along with the permeant solute, and the volume change ends when the permeating nonelectrolyte is equally distributed between the cells and the medium. The cell volume will respond to changes in medium osmolality (hyperosmolar) in one of four ways: (a) it will decrease to a new equilibrium value, (b) it will decrease and then increase to the initial equilibrium value, (c) no changes in volume take place, or (d) the cell swells and then shrinks. A mirror image is obtained when the solution is hypo-osmolar. Condition (a) is obtained when the cell membrane is completely impermeable to the solute, (b) when the solute is permeable but less so than water, (c) when both the solute and water are equally permeable, and condition (d) when the solute is more permeable than water (usually called anomalous osmosis). We will consider only the first two conditions since the last two are less informative.

In analysis of osmotic flow and net solute movement, the two basic equations which are invariably used are those given by Kedem and Katchalsky [15] and Katchalsky and Curran [16]. Kedem and Katchalsky give the following relations to express the relative fluxes:

$$J_v = -L_p \Delta\pi_i + L_{pd}\Delta\pi_s \tag{8}$$

$$J_d = -L_{dp}\Delta\pi_i + L_d\Delta\pi_s \tag{9}$$

Again the same convention as described earlier is used, and flow into the cell is considered to be in the positive direction. The osmotic pressure due to the permeant solute is denoted by $\Delta\pi_s$, which is defined as $\Delta\pi_s = RT(C_s^0 - C_s^{\Delta x})$; $\Delta\pi$ has units of dyne/cm². In these equations we speak of differences in concentrations in bulk phases since the partition coefficient which relates the concentration in the membrane phase to that of the bulk phase is incorporated in the permeability coefficients. The subscripts i and s refer to impermeant and permeant solute respectively. L_{pd} is the cross-coefficient for the volume flow arising from differences in the osmotic pressure of the permeant solute, $\Delta\pi_s$, when there is no difference of either *hydrostatic* or osmotic pressure produced by impermeant solutes ($\Delta\pi_i = 0$). L_{dp} is the relative diffusional solute mobility per unit *hydrostatic* (or impermeant solute) pressure difference when $\Delta\pi_s = 0$. Although L_p is always positive, L_{pd} and L_{dp} are both negative and have the same units as L_p. If the Onsager reciprocal relation holds, then $L_{pd} = L_{dp}$. J_d is the diffusional flow and is a measure of the relative velocity of solute to solvent flux. L_d is a phenomenological coefficient, related to the permeability coefficient, and describes solute movement down its own concentration gradient in the absence of hydrostatic or osmotic pressure differences due to impermeant solutes. Rather than to measure L_d directly, it is much more convenient experimentally to measure ω_s, the permeability coefficient. From Eqns. 8 and 9, one can derive the following equations [16]:

$$J_s = (1 - \sigma)\bar{C}_s J_v + \omega_s \Delta\pi_s. \tag{10}$$

ω_s is related to the phenomenological coefficients by

$$\omega_s = (L_d L_p - L_{pd}L_{dp})\bar{C}_s/L_p \tag{11}$$

and $\bar{C}_s$ is defined by

$$\bar{C}_s = (C_s^0 - C_s^{\Delta x})\text{Ln } C_s^0/C_s^{\Delta x}. \tag{12}$$

In the notation of Sha'afi et al. [14], $C_s^{\Delta x} = n_s/V'$ and $C_i^{\Delta x} = C_{i,0}^{\Delta x}V_0'/V'$ in which n_s is the amount of permeant solute in the cell water whose volume is V'. V is cell volume and b is the sum of the volume of fixed framework and solute dry weight so that $V' = V - b$. $C_{i,0}^{\Delta x}$ is the intracellular concentration of the impermeant solute at $t = 0$. Now, J_v can be expressed as $J_v = (1/A)(dV'/dt)$, and $J_s = (1/A)(dn_s/dt)$. A is the cell area which in the case of red cells is assumed to remain constant [14]. Rewriting Eqns. 8 and 10 one obtains

$$v(dv/d\tau) = -(1+a)v + (\beta + as) \tag{13}$$

$$(ds/d\tau) = (1-\sigma)(\bar{C}_s/C_s^0)(dv/d\tau) + r(v-s)/v \tag{14}$$

where

$$v = V'/V_0', \tau = AL_p RT C_i^0 t/V_0', a = \sigma C_s^0/C_i^0, \beta = C_{i,0}^{\Delta x}/C_i^0$$

$$\sigma \equiv -L_{pd}/L_p \equiv -L_{dp}/L_p, s = n_s/C_s^0 V_0', r = \omega_s/L_p C_i^0$$

When the membrane is completely impermeable to the solute (case a), Eqn. 14 reduces to zero. In Eqn. 13, s will be equal to 0 and $a = C_s^0/C_i^0$, since by definition the reflection coefficient, σ, is equal to unity when the membrane is completely impermeable to the substance. The solution of Eqn. 13 for the boundary conditions $\tau = 0$, $v = 1.0$ is given by Eqn. 15:

$$\tau = (1+a)^{-2} \mathrm{Ln} \frac{(1+a)-\beta}{(1+a)v-\beta} + (1-a)^{-1}(1-v) \tag{15}$$

The hydraulic conductivity, L_p, can be calculated by Eqn. 15 from the experimentally measured time course of cell volume change. In this solution, the assumption is made that the concentration of the impermeant solute in the bathing medium remains constant. This assumption is justified since in actual experimental conditions the volume occupied by the red cells is only 1% of the total volume of the suspension. Furthermore, L_p is assumed to remain constant during the time course of volume change, but it is not necessarily independent of medium osmolarity or direction of water movement [10,11,17]. On the basis of perturbation analysis an approximate solution for Eqn. 13 was worked out [10]. In this solution, which has the advantage of being easier to use, a normalized time constant is inversely proportional to the square of a normalized external pressure. The proportionality constant is directly related to L_p [10]. This approach is quite satisfactory when the total change in cell volume is not too large.

When the membrane is permeable to the solute under study, the reflection coefficient σ can be calculated by the zero time method developed by Goldstein and Solomon [18]. In this method, use is made of volume change at $\tau = 0$, when $v \to 1.0$ and $s \to 0$ so that

$$\lim_{\tau \to 0} (dv/d\tau) = \beta - (1+a).$$

Solving for s from Eqn. 13, differentiating with respect to τ to obtain $ds/d\tau$ and then equating $ds/d\tau$ with Eqn. 14, one obtains the following differential equation which describes the changes in cell volume when the membrane is permeable to the solute under study (for simplicity $\beta = 1.0$):

$$(v^2)(d^2v/d\tau^2) + (v)(dv/d\tau)^2 + (1 + a + r)(v)(dv/d\tau)$$
$$- (a)(v)(1 - \sigma)(\bar{C}_s/C_s^0)(dv/d\tau) + r(v - 1) = 0 \qquad (16)$$

An exact solution for this differential equation cannot be found. The presence of the time dependent $\bar{C}_s$, and of $(dv/d\tau)^2$, contributes to the difficulty of integrating this equation. Fortunately, the permeability coefficient, ω_s, of the solute under study can be obtained from this equation by the minimum volume method [13]. At the minimum $\tau = \tau_m$, $v = v_m$ and $(dv/d\tau)_m = 0$. Under these conditions, one obtains

$$r = (\omega_s/L_p C_i^0) = (v_m^2)(d^2v/d\tau^2)_m/(1 - v_m) \qquad (17)$$

Eqn. 17 has been used to calculate the permeability coefficients of mammalian red cell membranes to various solutes [19]. Using second-order perturbation analysis an approximate solution for the differential Eqn. 16 has been worked out [20].

The experimental technique outlined here can be used also to measure water and other nonelectrolyte movements across the membrane of other cells. The surface area A of mammalian red cells remains constant for small changes in cell volume. Because of this condition the mathematical analysis is considerably simplified. In case of other cells, the variation of the surface area during the time course of cell volume change must be taken into consideration in the theoretical analysis. Normally, the cell under study is assumed to be spherical and the surface area A is expressed in terms of the cell volume V [12,21,22].

(f) Unstirred layer effect

Noyes, and Whitney and Nernst were the first to introduce and develop the concept of unstirred layer (quoted in [23]). According to the theoretical views of Nernst, there is a layer of static fluid present at a solid-liquid interface. The concentration of a given substance in this layer is not equal to that in the bulk solution and diffusion is the primary mode of transport within the unstirred layer. The thickness of the unstirred layer, though undetermined, is critically dependent on the rate of stirring in the bulk phase.

One of the earliest theoretical objections to the calculation of the water permeability coefficients of biological membranes was advanced by Dainty [22] in his excellent review on water relations in plant cells. He argues—and correctly so—that all the equations used in calculating permeability coefficients are based on the implicit assumption that the aqueous solutions on both sides of the membrane are so well stirred that the concentrations at the membrane faces are the same as the bulk concentrations. It is well recognized that such a situation is impossible to achieve experimentally and, with the usual rates of stirring used in biological studies an appreciable unstirred layer can exist on either side of the membrane (Fig. 1). Theoretically, the presence of such a layer adjoining the membrane will lead to an underestimation of both ω_{THO} and L_p. Each permeability coefficient is affected to quite a different degree. As was pointed out by Dainty [22], ω_{THO} is affected much more than L_p.

The presence of an unstirred layer which may adhere to a given cell membrane can be treated operationally as a barrier with its own permeability property in series with the actual membrane. Its importance in membrane transport processes depends essentially on the permeability of the membrane itself relative to that of the unstirred layer to the particular molecule being transported. Consequently, only molecules which permeate membranes at high rates are affected, since diffusion in the unstirred layer is quite rapid. Water transfer across human red cell membrane and those of most other cells and tissues studied falls within this category. Dainty [22] has given the following equation by which the apparent diffusion permeability coefficient may be corrected for the effect of an unstirred layer of thickness, δ:

$$(\omega_{THO}RT)^{-1} = (\omega'_{THO}RT)^{-1} + \delta/D \tag{18}$$

in which ω_{THO} and ω'_{THO} are the measured and the true diffusional permeability coefficients of the membrane respectively. D is the diffusion coefficient in the unstirred layer in cm^2/s. The thickness δ is an operational one. In Dainty's notation, $\omega_{THO}RT = P_d$ and $\omega'_{THO}RT = P_d$ true. In some cases, the question of an unstirred layer is a critical one and the deviation of the porosity factor (g) from unity may be due totally to underestimation of ω_{THO}.

Because a correct molecular interpretation of the permeability coefficients of a membrane to a given substance rests on determining the contribution of the unstirred layer, it is usually necessary to measure the thickness of the unstirred layer that may be present under a given experiment condition. Such measurements have been made [14,23].

3. Water movements across membranes and tissues

The rates of water movement across biological membranes, tissues and artificial membranes cover a wide range [23]. The permeability coefficients of representative barriers for water transport are summarized in Table 1 (for more information see [23]).

(a) Relationship of water diffusion to osmotic flow

As stated earlier, the permeability of a given barrier whether it is a single membrane or a tissue to water can be measured either under steady-state conditions or as a bulk flow. The difference between these two types of measurements is not merely one of technique, for there is a fundamental molecular difference between the movement of water under these two conditions. The permeability coefficient measured under bulk flow is referred to as the hydraulic water-permeability coefficient (L_p), which has the units of volume per force per time ($cm^3/dyne/s$). Other terminologies are often used, such as filtration coefficient, $P_f = RTL_p/\tilde{V}_w$ in cm/s, and osmotic coefficient, $P_w = RTL_p$ in $cm^4/osmol/s$. $\tilde{V}_w$ is the partial molar volume for water. The rate of water movement under steady-state conditions is usually referred to as the diffusional permeability coefficient (ω_{THO}), which has the units of mol/dyne/s (the symbol $P_d = \omega_{THO} RT$ is often used to denote the same). As shown in Table 1, when both coefficients are expressed in the same units, it is generally found that $P_f \geqslant P_d$. This kind of comparison has since been made by numerous investigators using various kinds of cells, tissues and artificial membranes (for details, see [6,20,23]).

In many cases, the difference between P_f and P_d can be attributed to the presence of an unstirred layer. On the other hand, in the case of human red cell membrane the difference cannot be accounted for by the presence of an unstirred layer [14]. Additional support for this difference between P_f and P_d comes from the studies dealing with the effect of antidiuretic hormone (ADH) on water movement across epithelia. It was found that in the presence of ADH, the osmotic permeability coefficient of this tissue to water is 120 times greater than the diffusional coefficient (see Table 9.5 in [23]).

In classical terms, any value of $P_f/P_d = g$ greater than unity may be taken as evidence for bulk flow through channels, and the equality sign holds for truly

TABLE 1
Permeability coefficients of representative membranes and tissues to water under both osmotic and diffusional flows [a]

Cell	Under osmotic flow $P_f \cdot 10^4$ (cm/s)	Under diffusional flow $P_d \cdot 10^4$ (cm/s)	$g = P_f/P_d$
Human red cells	173.0	53.0	3.3
Toad urinary bladder	3.9	1.3	3.0
Zebra fish egg	0.47	0.36	1.3
Marine alga			
(*Valonia ventricosa*)	0.24	0.24	1.0
Crab muscle	97.2	1.20	81.0
Squid axon	10.9	1.4	7.8
Artificial lipid membrane	2.0	2.0	1.0
Artificial lipid membrane			
(treated with Amphotericin A)	18.0	6.0	3.0
Artificial lipid membrane			
(treated with nystatin)	39.7	12.0	3.3
Amoeba (*Chaos chaos*)	0.36	0.23	1.6
Human neutrophils	13.6	–	–
Leukemic cells	8.2	–	–
Chicken erythrocytes	8.3	–	–

[a] For original citation and more details see [4,10,23,59].

nonporous membranes. Furthermore, if Poiseuille's law dealing with bulk flow down right-cylindrical pores is assumed, then the value of g is related to the average pore radius (r) [24]. However, classical interpretation of water movement across biological membranes cannot be truly applied due to the small size of these proposed pores. In fact, Levitt [25] has calculated that the average number of water molecules per "pore" is about five.

The situation is quite different when water movement is treated in terms of irreversible thermodynamics. In terms of the frictional analysis of irreversible thermodynamics, g is given by the following relation [16]:

$$g = 1 + f^c ww / f^c wm, \qquad (19)$$

in which $f^c ww$ and $f^c wm$ are the coefficients of friction in the pore between water-water and water-membrane, respectively. Friction coefficients are positive and have units of dyne second per mol centimetre. A value for g greater than unity corresponds to transport in situations in which water-water friction in the membrane is more important than water-membrane friction.

Values of g greater than unity can be found in membranes in which the presence of porous channels is improbable. In these specialized membranes, there appear to be ways to transport water molecules in small clusters other than by bulk flow [20].

Even though the criticisms outlined in the preceding few paragraphs do not invalidate completely the criterion that $g > 1$ implies the existence of water-filled channels in the cell membrane, they do weaken it considerably, and its use in calculating average "pore" size becomes completely invalid. At present, the best one can say is that the intermediate values of g for red cell membranes strongly support, but do not necessarily prove, the pore model hypothesis. It is interesting to note that the values of the two permeabilities P_f and P_d are identical in the case of artificial lipid membrane. On the other hand, the value of the ratio P_f/P_d is significantly greater than one when these membranes are treated with certain antibiotics that apparently make these membranes porous.

(b) Solvent drag, reflection coefficient and the "pore" concept

Viscous flow of water through a porous membrane will affect the unidirectional fluxes to small polar solutes as well as tracer water passing through these pores. For example, the unidirectional fluxes of small polar solutes or tracer water in the same direction as bulk water flow will be enhanced while that in the opposite direction will be hindered. Solvent drag refers to the effect of bulk water movement on the unidirectional fluxes of either tracer, water or small polar solutes. The term $(1 - \sigma)\bar{C}_s J_v$ in Eqn. 10 represents the component of the overall unidirectional flux of the solute which is due to solvent drag (it is positive when J_s and J_v are in the same direction and negative otherwise).

In order to discuss the concept of solvent drag in quantitative terms let us consider a membrane barrier separating two dilute solutions of uncharged solute S.

Permeability for water; other molecules

As shown in Fig. 1, the first compartment will be denoted by O and the second compartment will be denoted by Δx. Let us assume that a small amount of the isotope of this substance is added to the first compartment (O) to give a final concentration of C_{s*}. The unidirectional flux of this substance from $o \to \Delta x$ in the absence of a volume flow of water is given by $J_s^{0 \to \Delta x} = \omega_s RT \Delta C_{s*}$. On the other hand, if there is a bulk water flow in the same direction $(o \to \Delta x)$ then the unidirectional flux of the substance will be given by

$$J_s^{0 \to \Delta x} = (1-\sigma)C_{s*}J_v + \omega_s RT\Delta C_{s*} \tag{20}$$

Since initially

$$C_{s*}^0 \gg C_{s*}^{\Delta x} \text{ then } \Delta C_{s*} \simeq C_{s*}^0,$$

$\bar{C}_{s*} = 1/2\, C_{s*}^0$ and Eqn. 20 can be rewritten as

$$J_s^{0 \to \Delta x} = 1/2(1-\sigma)C_{s*}^0 J_v + \omega_s RT C_{s*}^0 \tag{21}$$

Using the same analysis, if the isotope was added to the second compartment (Δx) instead of the first one (0) and bulk water flow was still going in the same direction as in the previous case $(0 \to \Delta x)$ then the unidirectional flux of the substance will be given by

$$J_s^{\Delta x \to 0} = -1/2(1-\sigma)C_{s*}^{\Delta x}J_v + \omega_s RT C_{s*}^{\Delta x} \tag{22}$$

If the concentration of the isotope added to the two sides are the same, i.e., $C_{s*}^0 = C_{s*}^{\Delta x}$ then the flux ratio is given by

$$\frac{J^{0 \to \Delta x}}{J^{\Delta x \to 0}} = \left[\frac{2\omega_s RT}{(1-\sigma)J_v} + 1\right] / \left[\frac{2\omega_s RT}{(1-\sigma)J_v} - 1\right] \tag{23}$$

Since for large x, $\ln(x+1)/(x-1) = 2/x$ Eqn. 23 can be rewritten as

$$\ln\left(\frac{J^{0 \to \Delta x}}{J^{\Delta x \to 0}}\right) = (1-\sigma)J_v/\omega_s RT \tag{24}$$

Eqn. 24 predicts that if bulk water flow exerts any influence on the movement of a substance (i.e. presence of solvent drag) then the natural logarithm of the flux ratio which can be experimentally measured should vary as a function of the rate of volume flow of water J_v. Moreover, this variation should be linear and the slope of the line determine the degree of coupling between bulk water flow and solute movement (i.e. the larger the slope the greater the coupling). Similar analysis can be used in the case of tracer water (THO) movement.

Since first introduced, the reflection coefficient (σ), of a membrane for a particular solute has played an important role in the field of permeability studies. It represents the discriminatory power of the membrane between water and solute. σ is defined as the ratio of the osmotic flow produced by a concentration gradient of test solute to the flow produced by the same gradient of an impermeable solute.

In red cells, this coefficient is usually measured by the zero time method [18] in which a suspension of cells in isotonic buffer is mixed rapidly with various test solutions of the probing molecule under study. The initial cell volume change ($\lim_{t \to 0} dV/dt$) is plotted against the various concentrations of the probing molecule. The concentration of the probing molecule that gives no volume flow can be found by interpolation. The reflection coefficient (σ) can be calculated from the relation $\sigma = -\Delta\pi_i/\Delta\pi_s$. This work has recently been criticized on the ground that the experimental procedure of these investigators can lead to large errors in the evaluation of σ and that σ values cannot be used to calculate "pore" size (for review, see [10]). The basic criticism centers around the difficulty in measuring ($\lim_{t \to 0} dV/dt$). This work indeed suffers from this experimental difficulty, but the error is not as severe as claimed. The conclusion that σ values cannot be used to calculate "pore" radius has been reached previously by others (for reviews, see [4,20]).

In 1963 Dainty and Ginzburg [26] showed that the reflection coefficient of a given solute is related to the various frictions in the following manner:

$$1 - \sigma - \omega_s \tilde{V}_s/L_p = \frac{K_s^c f_{sw}^c}{f_{sw}^c + f_{sm}^c} = \frac{A_{st}}{A_{wt}} \ldots \tag{25}$$

in which ω_s and $\tilde{V}_s$ are the diffusional permeability coefficient and partial molar volume of the solute; f_{sw}^c and f_{sm}^c are the coefficients of friction in the pore between solute-water and solute-membrane, respectively; K_s^c is the partition coefficient for the solute between the water in the pore and the external solutions; A_{st} and A_{wt} are the apparent areas for the filtration of solute and water, respectively. These authors also pointed out that if small hydrophilic solute and water molecules permeate cell membrane via commonly shared water-filled "pores", then there should be a frictional interaction between the permeating solute and the water, i.e. $f_{sw}^c > 0$. This gives rise to the phenomenon known as solvent drag effect. Consequently, the inequality given in Eqn. 26 must be satisfied:

$$\sigma < 1 - \omega_s \tilde{V}_s/L_p. \tag{26}$$

This relationship was found to hold true for five small hydrophilic solutes in human and dog red cells [20]. The presence of solvent drag has been reported in other tissues [27].

In general, the idea that a demonstration of solvent drag effect on solute movement implies transport across "aqueous pores" is thermodynamically sound. On the other hand, the extension of this idea to the studies on red cell, or even worse, to the studies on other tissues and the use of σ to calculate "pore" size should be re-evaluated and perhaps totally rejected for the following reasons: (a) Demonstration of the existence of solvent drag by means of a relation such as that given in Eqn. 26 is open to question. For one thing, this is an extremely indirect method; furthermore, the relationship is necessary, but not sufficient to justify the presence of water-filled channels. It is conceivable to have a case in which $f_{sw}^c > 0$ and hence $\sigma < 1 - \omega_s \tilde{V}_s / L_p$ without needing to postulate aqueous channels which pierce the membrane from one side to another. (b) The expression given in Eqn. 26 holds true only when the solute under study and water permeate only through a common "pore". If the solute can partially permeate by dissolution in the membrane and/or if water molecules permeate through pores not available to the solute, then the relationship shown in Eqn. 26 is considerably modified [26]. It is becoming increasingly evident as will be shown later, that only a very small population of the postulated "pores" through which water permeates is available for the transport of small hydrophilic nonelectrolytes. Although this modification does not invalidate the inequality given in Eqn. 26, it renders the use of σ for calculating "pore" size completely useless. (c) The inequality which is shown in Eqn. 26 only proves that water molecules can use the pathway available for small hydrophilic solutes, but not the converse.

(c) Effect of temperature on permeability of membrane to water

Relevant information concerning the physical state of membrane water and the nature of membrane–water interaction can be obtained from studying the energetics of water permeation across biological membranes. In the last four decades several articles appeared on the subject dealing with the influence of temperature on water transfer across cellular membranes and tissues (for reviews, see [4,6,10,23,28]). Based on these studies the values of the activation energies vary from 17 kcal/mol for sea urchin eggs to about 4 kcal/mol for most mammalian red cells (for more details, see [23], Tables 5, 6). High value for the activation energy reflects a relatively high water-membrane interaction.

Probably the most systematic and complete study on the influence of temperature on water transfer has been performed on mammalian red cells [10,20,28]. The dependence on temperature of both the tracer diffusional permeability coefficient (ω_{THO}) and the hydraulic conductivity (L_p) of water in human and dog red-cell membranes have been studied. The apparent activation energies calculated from these results for both processes are given in Table 2. The values for the apparent activation energies for water self-diffusion and for water transport in a lipid bilayer are also included in the table. For dog red cells, the value of 4.9 kcal/mol is not significantly different from that of 4.6–4.8 kcal/mol for the apparent activation energy of the water diffusion coefficient (D_w) in free solution. Furthermore, it can be shown that the product ($L_p - {}^\omega THO\tilde{V}_w)\eta_w$, where $\tilde{V}_w$ is the partial molar volume of water and η_w the viscosity of water remains virtually independent of temperature for dog, but not for the human red-cell membrane [20]. The similarity of the transmembrane diffusion with bulk water diffusion and the invariance of the

TABLE 2

Comparison of apparent activation energies for water fluxes in various systems [a]

System	Activation energies in kcal/mol	
	Under osmotic flow	Under diffusional flow
Human red cell membrane	3.3	6.0
Dog red cell membrane	3.7	4.9
Self diffusion	–	4.8
Viscous flow in water	4.2	–
Lecithin/cholesterol bilayer	14.6	–
Human red cell membrane (treated with PCMBS)	11.5	–
Nitella translucens	8.5	–
Chicken erythrocyte	11.4	–
Sea urchin egg	17.0	–
Barnacle muscle	7.5	–
Squid nerve	–	3.0
Ehrlich ascites tumor cell	9.6	–
Human neutrophil	18.4	–
Chronic leukemic lymphocytes	16.3	–

[a] For original citation see [4,10,23,59].

product $(L_p - {^\omega}THO\tilde{V}_w)\eta_w$ with temperature in dog red-cell membrane suggests that water in these membranes behaves operationally as in bulk solution. However, in the case of the human red-cell membrane, the results suggest that water–membrane interaction is relatively high. This indicates that in these cells the charged groups of membrane protein and phospholipid molecules impart unusual properties to water molecules.

The fact that the value of the apparent activation energy for water diffusion in lipid bilayer membranes is considerably higher than the corresponding values for red cell membranes strongly suggests that water molecules encounter a hydrophilic rather than hydrophobic environment while crossing the latter membranes. In fact, as will be discussed later, when water transport through "pores" is inhibited by certain mercury compounds, the apparent activation energy for osmotic water flow increases to 11.5 kcal/mol [4,10]. This value is in good agreement with the value of 12.4 kcal/mol calculated, assuming that water dissolves in the membrane as discrete molecules and moves across it by diffusion (for reviews, see [4,10]).

The temperature dependence of the diffusion coefficient (D_w) and the viscosity (η_w) of water are given by

$$D_w = \lambda_4^2(kT/h)(F^{\#}/F)\exp(-\Delta E/RT) \tag{27}$$

$$\eta_w = (\lambda_1/\lambda_2\lambda_3)(h\lambda_4^2)(F/F^{\#})\exp(\Delta E/RT) \tag{28}$$

in which h is Planck's constant; k, Boltzmann's constant; λ, is the distance between two equilibrium positions in the direction of motion; λ_1, the distance between two layers of molecules; λ_2 the distances between two adjacent molecules at right angles to direction of motion; λ_3, the distance between neighboring molecules in the same direction, F and $F^{\#}$ are partition functions of the molecule in the initial and activated states; R and T have their usual meanings and the activation energy $\Delta E = \Delta H^{\#} - T\Delta S^{\#}$ where $\Delta S^{\#}$ is the entropy and $\Delta H^{\#}$ is the enthalpy of activation. In theory it is possible to analyse the data concerning the temperature effect on water transport in red cells and other membranes in terms of the apparent enthalpy and entropy of activation. Such analysis in biological systems, however, involves parameters whose values cannot be determined with certainty. Accordingly, the values reported in the literature for the apparent change in entropy of activation for water movement suffers from a great deal of inherent inaccuracy and must be taken with reservation. These parameters can be eliminated if the temperature effect on diffusion is coupled with that on viscous flow ($D_w \lambda_w / T = \lambda_1 K / \lambda_2 . \lambda_3$).

$$\frac{-d\,Ln\,D_w}{d(1/T)} + \frac{d\,Ln\,\eta_w}{d(1/T)} = \frac{d\,Ln\,T}{d(1/T)} \tag{29}$$

$$R\left[\frac{-d\,Ln\,D_w}{d(1/T)} + \frac{d\,Ln\,1/\eta_w}{d(1/T)}\right] = RT \simeq 0.6 \text{ kcal/mol} \tag{30}$$

In the case of the dog red cell, the difference in apparent activation energies of osmotic and diffusion flow is 1.2 kcal/mol, which is in reasonable agreement in both magnitude and direction with the value 0.6 kcal/mol given in Eqn. 30.

(d) Effect of sulfhydryl-reactive reagents on water transport

The transport of water in certain mammalian red cells as well as other cells and tissues has been shown to be inhibited by certain sulfhydryl-reactive reagents [4,10,29]. A summary of these studies is given in Table 3. Half-maximal inhibition is produced at a concentration of 0.02 mM. With certain inhibitors this inhibition develops slowly and is fully reversible by treatment with excess amount of cysteine or glutathione. As stated earlier, the finding that $P_f/P_d > 1$ is one of the most compelling pieces of evidence for the existence of "pores". In the presence of PCMBS, $P_f = P_d$, suggesting that this compound acts by shutting off these "pores".

Recently the characteristics of the inhibitory site have been determined by comparing the inhibitory potency of a large number of sulfhydryl-reactive reagents (for reviews, see [4,10,39]). Based on these studies, the following observation can be made: (1) Mercury-containing compounds are more effective inhibitors than disulfide reagents. (2) The presence of a nitrogen atom on the ring, in addition to an NO_2 group on the side, increases the potency of the disulfide reagent as a water-transport inhibitor. (3) The effectiveness of the reagents as inhibitors of water movement

TABLE 3

Effect of *p*-chloromercuriphenyl sulfonate (PCMBS) on the permeability coefficients of membranes and tissues to water [a]

System	Permeability coefficients (cm/s)·10^4 ratio	
	Under osmotic flow, P_f	Under diffusional flow, P_d
Human red-cell membrane	173	53
+ 1 mM PCMBS	2.0	1.8
Human red-cell ghost	17.9	–
+ 1 mM PCMBS	≪17.9	–
Cat red-cell membrane	340	–
+ 1 mM PCMBS	170	–
Camel red cells	54.3	–
+ 1 mM PCMBS	48.9	–
Rabbit neutrophils	90	–
+ 1 mM PCMBS	60	–
Turtle intestine (proximal ileum)	9.2	–
+ 1 mM PCMBS (added to serosal) [b]	5.9	–

[a] For original citation see [43].
[b] It has no effect when added to mucosa.

parallels the electron-withdrawing capacity of the substituents. (4) Membrane sulfhydryl groups which are involved in the control of water transport are less reactive than those of a small SH-containing molecule, such as cysteine.

The permeability coefficients of most mammalian red cell membranes to water under control conditions are normally 10 times higher than the corresponding value for lecithin cholesterol bilayers [30]. In the presence of sulfhydryl-reactive reagents, the permeability coefficients of the two membranes are practically identical, which suggests that these reagents act by blocking the proposed "aqueous pores" leaving the lipid bilayer of the mammalian red-cell membranes as the only alternative route for water transport.

(e) Effect of antidiuretic hormone on water transport

Water movement in distal and collecting portions of mammalian kidney nephrons as well as in other similar systems (such as toad urinary bladder) can be significantly increased by a substance called antidiuretic hormone (ADH). Cyclic AMP (cAMP) mimics the action of ADH. There is strong evidence to suggest that the effect of ADH on membrane permeabilities to water is through specific membrane receptors and is mediated by a rise in the cell level of cAMP through the activation of the enzyme adenylate cyclase. The exact nature of the changes in the membrane that give rise to this increase in its permeability to water is not fully understood. One attractive hypothesis is that this rise in cAMP either alone or in conjunction with calcium ions and the calcium-dependent regulator protein (calmodulin) allows "pores" formation in the lipid membrane by membrane proteins. It is generally found that the permeability coefficient measured under bulk flow, P_f, is much more affected by ADH than that measured under diffusional flow, P_d. This means that the ratio $P_f/P_d \gg 1$. The recent finding that the commonly observed rapid reversibility of the ADH-dependent increase in permeability when ADH is removed can be significantly delayed by low temperature, inhibition of either glycolysis or oxidative phosphorylation and the addition of cytochalasin B is consistent with the protein "pores" theory [31]. It is important to point out that the effects of ADH and cAMP on water transport are not general. For example, water movement across human red-cell membrane is unaltered in the presence of pitressin or cAMP [10].

(f) Membrane cholesterol and the permeability to water

Membrane lipids, and particularly cholesterol, are instrumental not only in the control of diffusion across biological membranes but also in the determination of the activity of membrane-bound enzymes, their modulation by hormones and other agents, and the determination of membrane fluidity (for original references, see [4,6]). It is generally accepted that incorporation of cholesterol in a lipid bilayer membrane tends to decrease significantly the permeability of these membranes to water. Movement of water across these membranes occurs primarily by dissolution in the membrane matrix. The decrease in the rate of water transport as a result of cholesterol incorporation is due mainly to a decrease in membrane fluidity. As a general rule, it is found that the presence of cholesterol in membranes or the incorporation of cholesterol into dispersions composed of phosphatidylserine or ganglioside lead to a decrease in the fluidity of the hydrocarbon chains of lipid membranes which are in the liquid-crystalline state [4,20].

In contrast to lipid bilayer membranes, it has been found [4] that the permeability coefficient of the human red-cell membrane to water did not change when the free cholesterol content in the membrane was varied from 0.84 to 1.87 mg/ml cells. Furthermore, the permeability of the human red-cell membrane to sulfate and some nonelectrolytes remained constant when membrane cholesterol was partially removed (for review, see [6]). These results, however, should not be taken as evidence that water transport in human red cells is independent of membrane cholesterol, since this degree of variation may be insufficient to produce alteration. In fact, extensive depletion of membrane cholesterol induces a marked increase in nonelectrolyte permeability. The effect of membrane cholesterol on the transport of water is also found in other membrane systems. For example, the polyene antibiotic, Amphotericin B, which interacts specifically with sterol-containing membranes, increases the permeability of the mucosal but not the serosal membrane of toad bladder to water and other solutes [32]. It is possible that membrane cholesterol only effects the movements through the lipid bilayer pathway. This may explain the findings that the permeability coefficient of the human red cell membrane to water which most likely crosses these membranes through "aqueous pathways" did not change when the free cholesterol content in these membranes was varied. Since PCMBS seems to close the "aqueous pathway", it would be interesting to study the effect of varying the cholesterol of human red-cell membrane on the membrane permeability to water in the presence of PCMBS.

(g) Is there rectification of water flow?

Polarity is defined as the dependence of the true membrane permeability coefficient, in this case to water, on the direction of movement. Similar, but far from identical, is the apparent difference commonly found in the rate of water movement in response to an equal but opposite perturbation from iso-osmotic condition [10,11,17,22,33]. In fact, the two are quite different; directional dependent permeability to water is due to basic membrane structure whereas the second phenomenon may arise from experimental conditions. For example, it is generally observed that the rates of inward movement of water in some plants such as seedcoats and many fruits are much higher than the corresponding rates of outward movement of water under apparently the same driving force. A common experimental design in which apparent difference in the rates of water movement is observed is as follows: the osmotic pressure of the solution bathing one side of the membrane is either increased to cause water to move to this side, or decreased to the same extent. For a given driving force, it is generally found that the rate of water movement is greater in the second case.

The most thorough studies dealing with this question are those using human red-cell membranes as a model system [4,10]. It is commonly observed that the rate of red-cell swelling is faster than the rate of red-cell shrinkage. This apparent polarity of water movement is demonstrated only when net water flow is occurring and not under diffusional flow conditions [10]. Initially, it was suggested that these results were due to a dependence of the osmotic water permeability coefficient (L_p) on medium osmolality. Later, however, it was suggested that this behavior was due not to a change in medium osmolality, but rather to a rectification of water flow (inward rate being greater than outward rate). Recently, this problem has been reexamined, and it was concluded that L_p is essentially independent of both medium osmolarity and flow direction [34]. The most likely explanation for the discrepancy is that in the previous studies no account has been taken for the possible nonlinearity in the relationship between light intensity and cell volume.

(h) Miscellaneous factors

One of the first parameters which comes to mind among those factors which may affect membrane permeability properties in any system is pH. A study of pH effects may give some information about the role of ionizable protein groups in membrane structure on its permeability. It has been shown that the hydraulic water conductivity (L_p) is independent of pH in the range from 6 to 8 in human red cells, initial cell volume, the osmolality of the internal medium, membrane flexibility, the presence of compounds such as tetrodotoxin and valinomycin, which are known to produce selective actions on sodium and potassium transport across both biological and artificial membranes, external calcium and various other agents [4,10].

(i) Possible structural basis for the apparent presence of hydrophilic pathway for water transport

There is a large body of evidence, some of which has been discussed already, which supports the view that membranes in general and mammalian red-cell membranes in particular contain specialized "apparent polar pathways" for the transport of water. Fig. 2 depicts schematically two possible mechanisms for water transport. Movement of water by either of these two mechanisms will have similar characteristics as movement through "aqueous pores". Although the two models shown in Fig. 2 are entirely different from the structural point of view, they both give rise to many of the observed properties of water transport in the human red-cell membrane.

The hydrophilic pathway shown in Fig. 2A is highly schematic and should not be taken to mean that it is a fixed right cylindrical "pore" with static structure and dimensions. It could be assembled from membrane integral proteins which are probably aggregates of identical or nonidentical sub-units. It is generally agreed that there are two known major proteins which span the human red-cell membrane. They are known as band 3 (95 000– 100 000 M_r) and PSA-1 [35,36]. There are three lines of evidence which suggest that one or both of these two glycoproteins may be involved in the formation of hydrophilic pathways for water transport: (a) Marchesi et al. [35] have presented evidence which indicates that glycophorin is a component of the intramembranous particles seen in freeze-fracture experiments. It has been suggested that these membrane-intercalated particles seen in human red-cell membrane could

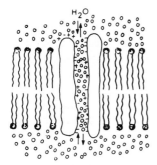
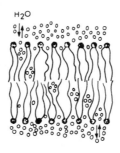

Fig. 2. Schematic representation of two possible models which may give rise to the apparent presence of hydrophilic pathways for water transport in human red-cell membrane. In Fig. 2A on the left, the pores are assembled from membrane integral proteins which are aggregates of identical or nonidentical sub-units. In Fig. 2B on the right, water molecules move across the membrane by jumping into the free volume (kinks) generated by the thermal fluctuations in membrane lipid. (From [4].)

provide a structural basis for the hydrophilic pathway [3,37]. (b) Incorporation of glycophorin prepared by trypsin hydrolysis of human erythrocytes in black lipid membranes increase significantly the permeability of these membranes to water [38]. This may be a nonspecific increase. (c) Using polyacrylamide gel electrophoresis, it has been shown that a band which contains band 3 and glycophorin can be selectively labelled by a water-transport inhibitor [39].

If the "aqueous pores" are made exclusively from glycophorin, then it would be difficult to understand why sulfhydryl-reactive reagents inhibit water transport. Glycophorin has been purified from human red-cell membrane and its amino acid composition is determined [35]. It has no cysteine and, therefore, it is difficult to see how sulfhydryl-reactive reagents will interact with this protein to inhibit water transport. It must be made quite clear that band 3 is quite dispersed and that it may comprise more than one polypeptide species. This band has been reported to contain a phosphorylated intermediate of the (Na^+-K^+)-ATPase [40]. It also binds 4,4'-diisothiocyano-2,2'-dithiostilbene-disulphonate (DIDS), a specific inhibitor of anion movement [5]. It may also be involved in the transport of glucose. The heterogeneity of protein band 3 is also indicated by the observation of Cabantchik and Rothstein [41] that after pronase treatment the staining of this band decreases and three individual bands are revealed.

Such a mechanism for water transport with a slight modification to account for hormone-sensitive water movement is probably common to other membrane systems. In the case of the collecting duct of mammalian kidney, for example, it is conceivable to imagine that there is an interruption somewhere along the pore. This interruption can be temporarily removed through a transient conformational change brought about by ADH-induced changes in the level of cAMP.

The model depicted in Fig. 2B is based on the idea that thermal fluctuations in membrane lipid can cause conformational changes in the hydrocarbon chains which lead to the generation of mobile structural defects known as "kinks" [42]. These kinks can be initiated only on either side of the membrane and then migrate to the other side, giving rise to mobile packets of free volume. Water molecules on either side of the membrane can jump into this free volume and "hitch a ride" to the other side. The kink is a thermodynamically stable structure and the free volume generated by it can be of different sizes [42]. Although it is not intuitively obvious how such a mechanism of water movement will give rise to the finding that $P_f/P_d > 1$, it cannot be ruled out on theoretical grounds ($f_{ww} > o$ in such a system). Accordingly, if water movement across the red-cell membrane is to be explained in terms of the "kinks" hypothesis, then one would have to postulate that the free volume generated by the "kinks" formation should be large enough so that $f_{ww} \gg o$. Not only that, but one must also postulate that in a black lipid bilayer, the free volume is very small so that no water filtration can take place since $P_f = P_d$ in these membranes. The inhibitory effect of sulfhydryl-reactive reagents on water movement in red-cell membrane is not entirely inconsistent with the "kink" hypothesis. It is not unreasonable to expect that interaction with membrane protein may produce a decrease either in the "kinks" concentration and/or a decrease in the free volume formed by the "kinks". Even though the "kink" hypothesis is an attractive one and certainly a very good mechanism for water diffusion in a lipid bilayer, it probably accounts for only 10% of water movement in the human red-cell membrane. This 10% is the component of water which crosses the cell membrane via the lipid region. For one thing, the permeability coefficient for water calculated on the basis of the "kink" hypothesis is one order of magnitude less than the experimentally determined value in the human red-cell membrane. Moreover, the "kink" hypothesis has to be drastically modified to account for the various properties of water transport in human red cells ($P_f/P_d = 3.3$, inhibition of water transport by various sulfhydryl-reactive reagents, $\Delta E = 4$ kcal/mol, etc.).

4. Permeability of membranes to small polar nonelectrolytes

In this section we will restrict our discussion to small polar (oil:water partition coefficient $K_{ether} \leq 0.003$) nonelectrolytes. However, before discussing this question in detail, it is worth noting that the value of the permeability coefficient of membranes for a large lipophilic molecule is determined by the molecule lipid solubility, its chemical nature, its molecule size and shape, and the number of hydrogen bonds (N_H) it is able to form with water. The permeability coefficient increases with increasing lipid solubility and decreasing N_H, whereas it decreases with increasing molecular size and degree of branching. For example, replacement of a hydroxyl group on the molecule by a carbonyl group or an amide group tends to decrease the permeability coefficient (for comprehensive reviews, see [6,10,20,23,43–45]). In the case of small hydrophilic molecules, we will restrict our discussion to the following points: (a) Is the mechanism by which these solutes permeate cell membranes similar to that for large lipophilic molecules? (b) Which inhibitors affect the movements of these small nonelectrolytes? (c) Is there a carrier-mediated mechanism for urea transport? (d) Do small molecules use the same pathways as water?

(a) Is the mechanism by which small hydrophilic solutes permeate cell membranes similar to that used by large lipophilic molecules?

Here the problem is to decide whether these solutes permeate by dissolution in the membrane fabric or by crossing the membrane through polar pathways "aqueous pores". It is worth pointing out from the start that there is no conclusive evidence which enables one to decide unequivocally between these two possibilities. As will become clearer by the end of this discussion, the data at hand in certain cases (such as mammalian red cells) are merely more consistent with the postulate that these molecules permeate partly through polar pathways and not entirely by dissolution in the membrane fabric. Probably the strongest evidence for this hypothesis is summarized in Table 4. The values of the diffusional permeability coefficient for egg lecithin spherical bilayers are included in the table for comparison. Based on their lipid solubility these molecules should be quite impermeable. In fact, in certain

TABLE 4

Permeability coefficients of representative membranes and tissues to certain small polar nonelectrolytes [a]

Barrier	Permeability coefficients in cm/s · 10^5				
	Formamide	Acetamide	Urea	Methyl urea	1,3-Dimethyl-urea
Human red cells	43.7	12.2	36.5	4.9	2.7
Rabbit gall bladder	–	7.0	8.9	–	–
Toad urinary bladder	–	0.16	0.14	–	–
Rabbit neutrophils	–	–	0.30	0.45	–
Characean plant	0.76	0.66	0.013	0.032	0.012
Gall bladder of					
Bufo bufo	46.6	3.19	5.14	–	–
Small intestine of					
Testudo hermanni	4.75	2.01	0.69	–	–
Large intestine of					
Testudo hermanni	1.80	0.87	0.263	–	–
Ventral skin of					
Rana esculenta	0.96	0.219	0.053	–	–
Urinary bladder of					
Bufo bufo	1.309	0.333	0.189	–	–
Spherical bilayer	7.8	2.4	0.49	–	–
Lecithin bilayer	16.0	14.5	0.365	–	–

[a] For original citation see [4,45,54].

membranes as well as in the lipid bilayer, where movement of nonelectrolytes is primarily by dissolution in the membrane fabric, urea permeates these membranes very slowly. The ratio P_d : urea/P_d : H_2O is 0.0031—more than an order of magnitude smaller than the 0.11 ratio for P_d : urea/P_d : H_2O in human red cells. On the other hand, when the lipid bilayer membrane is treated with the antibiotics nystatin or Amphotericin B, the ratio P_d : urea/P_d : H_2O increases to 0.11 [46]. It has been proposed that Amphotericin B and cholesterol are complexed in lipid bilayers to form an aqueous pore of 4 Å radius [47,48]. It is clear from Table 4 that these small solutes permeate certain membranes at very high rates and that there is no apparent correlation between lipid solubility and permeability. In this case, the primary parameter which is of overwhelming importance in determining the permeability coefficient of these membranes to these small hydrophilic solutes is steric hindrance. There are three parameters which are commonly used when considering steric hindrance factors [4,10,44]. The cylindrical radius of the permeating molecule, a measure of molecular size, has been shown to be paramount in the steric interactions governing the values of the reflection coefficient (σ) in mammalian red cells. When purely geometrical factors are dominant in the members of a given homologous series, the cylindrical radius was found to be the parameter of choice in studies of permeation of small nonelectrolytes. However, for small hydrophilic solutes that are members of different homologous series, a better index of geometrical factors is the molar volume, which is equal to the molecular weight divided by the density of the pure compound. The molecular weight may be construed as a measure of molecular size based on the spherical model. Division by the density modifies the strictly geometrical interpretation by introduction of the hydrogen bonding of the molecule with water because, as Pimental and McClennan [49] have pointed out, hydrogen bonding generally increases the density and lowers the molar volume. The correlation of hydrogen bonding with density is best illustrated in series such as butanediols, propanediols, pentanediols and others. The density of the members of each series decreases as the positions of OH groups approach each other and the ability to hydrogen bond with other molecules decreases. Thus, the molar volume as a mixed parameter—a geometrical construct modified by chemical properties, primarily hydrogen bonding—is a more preferable index of geometrical factors. In fact, it provides a very good fit for the permeability coefficients of human red-cell membranes to small hydrophilic solutes [4,20]. It is interesting to note that a similar relationship is found for lipid bilayer membranes treated with the antibiotics nystatin or Amphotericin B [28,46]. A third index which may be used in these considerations is solute molecular weight. However, molecular weight does not differentiate between the size of two isomers. The importance of steric factors is not restricted to solutes which permeate through polar routes; it also applies to lipophilic solutes which penetrate by dissolving in the membrane fabric.

A geometrical factor which may not be related to steric effects is the surface contact area between the permeating molecule and the membrane. this is probably more important for lipophilic solutes since lipid : lipid forces are predominantly short-range van der Waals forces. Based on these results, one cannot escape the conclusion that these small polar molecules will not be able to cross the membrane at such a high rate if the only mechanism is by dissolving in the membrane fabric. One is forced to conclude that they must permeate either by a specialized membrane transport system, such as a carrier-mediated mechanism, and/or through a specialized polar pathway. The temperature dependence of nonelectrolyte permeation across red-cell and certain other membranes is consistent with this conclusion. It is generally found that the values of apparent activation energies for the permeation of small hydrophilic solutes are significantly lower than the corresponding values for lipophilic solutes [50,51].

A second set of experimental evidence which supports the hypothesis that small hydrophilic solutes permeate by a mechanism different than that used by lipophilic ones is the finding that for a given homologous series of small solutes, the lipid solubility of the molecule (K_{ether}) and the number of hydrogen bonds it is able to form with water (N_H) seem to exercise a very small effect on the rate of penetration [4,20,44]. For example, the first member of the series is much more permeable than the remaining members even though it has the lowest lipid solubility and the highest value for N_H. This is also true, but to a lesser extent, for small lipophilic solutes (see [4], Table 6). The most important parameters that seem to determine the rates of movement of these polar nonelectrolytes is the molecular size. The value of the permeability coefficient decreases sharply with increasing size.

In recent studies, using antidiuretic hormone-treated toad urinary bladder, Levine and Worthington [52] have been able to demonstrate experimentally the presence of co-transport of labeled methylurea and, to a lesser degree, acetamide and urea with unlabeled methylurea. They concluded that the demonstrations of co-transport is consistent with the presence of ADH-sensitive amide-selective "channels" rather than a mobile carrier.

(b) What agents affect the movements of these small nonelectrolytes?

Movements of small hydrophilic solutes across human and probably other biological membranes can be affected by certain sulfhydryl-reactive reagents such as PCMBS and by phloretin. In the case of human red-cell membrane, a system which has been thoroughly studied in this respect, either one of these two compounds inhibits drastically the movements of these molecules. As is evident in Table 5, the inhibitory effect on the permeability decreases as the value K_{ether} increases. Whereas the compound PCMBS has no significant effect on the permeability coefficient of the human red-cell membrane to lipophilic nonelectrolytes, the compound phloretin enhances the rates of movement of these solutes. The compound phloretin seems to inhibit movement across some "polar pathways" and enhance movement through hydrophobic pathways (dissolution in the membrane). In fact, phloretin has been shown to increase significantly the permeability coefficient of pure lipid spherical bilayer (egg lecithin) to small polar solutes such as formamide [53]. Movement across this membrane occurs only by dissolution in the membrane fabric. It is possible that this is due to an increase in the mobility of membrane lipid. The exact nature of the phloretin effect is not fully understood. It is interesting to note that phloretin has been found to inhibit the movements of small polar nonelectrolytes only in red cells

TABLE 5

Effect of PCMBS and phloretin on the permeability coefficients of human red-cell membrane to various nonelectrolytes [a]

Molecule	K_{ether}	Relative permeability in the presence of	
		PCMBS	Phloretin
Urea	0.00047	0.10	0.34
Methylurea	0.0012	0.15	0.30
Formamide	0.0014	0.17	0.55
Acetamide	0.0025	0.28	0.80
1,3-Dimethylurea	0.003	1.00	1.00
Ethylurea	0.004	0.90	1.00
1,3-Propanediol	0.004	1.00	1.30
Ethylene glycol	0.005	1.00	1.80
Methanol	0.140	0.90	–
Ethanol	0.260	0.90	–
Water	0.003	0.15	1.30

[a] For original citation see [4].

obtained from certain classes (Mammalia, Reptilia and Amphibia) but not from lower classes (Agnatha, Aves., etc.) [55]. Other agents such as tannic acid are also known to decrease the permeability coefficients of certain mammalian red-cell membranes to various small nonelectrolytes [56].

In certain cases such as toad urinary bladder and mammalian kidney, the movement of small polar molecules, especially urea, are significantly increased by antidiuretic hormones [52]. Recently, it has been found that urea and water transport across the toad bladder can be separately activated by low concentration of vasopressin or 8 Br-cAMP [57]. Based on these studies it was concluded that membrane "channels" for water and small polar nonelectrolytes differ significantly in both their dimensions and densities. The solute "channels" are limited in number, have relatively large radii and carry only a small fraction of water flow. On the other hand, the water channels have small radii. These findings provide strong experimental support for the concept of "membrane pores" which we have been advocating (see Fig. 5 in [4]). In this respect it is not unreasonable to expect that PCMBS and phloretin would also inhibit the ADH-sensitive increase in the permeability of these systems to urea and other small polar nonelectrolytes.

(c) Is there a carrier-mediated mechanism for urea transport?

The movement of urea across various membranes including those of mammalian red cells has been studied and discussed extensively (for reviews, see [4,10]). The mechanism for urea transport in these cells has been subject to considerable debate [4,10]. This controversy stems from the fact that urea is a small polar molecule which is able to form five hydrogen bonds with water and yet in spite of these properties it is extremely permeable across some (mammalian red cells) but not all biological membranes. The simplest and most straightforward explanation for the observed high rate of urea movement is to postulate that those membranes behave operationally as a mosaic structure containing both lipid and polar regions. Accordingly, small hydrophilic solutes such as urea permeate through the polar route. An alternative explanation is to postulate that urea moves across these and some other biological membranes by means of a specialized carrier-mediated mechanism. The idea that urea may be transported across human red-cell membranes by means of a carrier-mediated mechanism was first advanced by Hunter et al. [56]. His conclusion was based on the finding that the movement of urea across these membranes can be inhibited by tannic acid. Recent evidence seems to support this conclusion. It has been found that the compound phloretin, a known inhibitor of facilitated transport systems, is a potent inhibitor of urea movement. Further evidence to support this view can be obtained from studies with sulfhydryl-reactive reagents. It was found that the permeability coefficient movement of urea across human red-cell membranes shows saturation kinetics and is significantly reduced in the presence of PCMBS [4,10]. On the surface, the evidence seems to be overwhelmingly in support of a carrier-mediated mechanism for urea transport. However, these findings are also consistent with the conclusion that urea permeates membranes through specialized "aqueous pores". Even if the conclusion is correct that the movement of urea shows saturation kinetics, it implies only that the number of these pathways through which urea molecules move is finite. If one accepts the view that the high permeability of mammalian red-cell membranes to urea is due to the presence of mobile carriers for its transport, then one is forced to postulate the presence of mobile carriers for the other highly permeable small hydrophilic solutes such as formamide. In addition, the properties of these carriers must be similar to those pathways found in antibiotic-treated lipid bilayer membranes since, as discussed earlier, there is similarity between urea transport in human red cell and lipid bilayer membrane which has been treated with antibiotics. Since there are consistent sets of arguments for the presence of "aqueous pathways" in mammalian red-cell membranes, it is not possible to decide which of these two mechanisms is responsible for the high rate of urea transport.

In order to obtain a better insight into this question, we have investigated the permeability characteristics of rabbit polymorphonuclear leukocyte membranes (PMNs) to urea, methylurea and thiourea [58]. These cells were chosen for two reasons. First, the value of the hydraulic permeability coefficient of the membranes to water is much lower than the corresponding value for mammalian red-cell membranes [13,59]. Secondly, the value of the apparent activation energy for water transport in the former cells is much higher than the corresponding value for the latter [59]. These two findings suggest that rabbit polymorphonuclear leukocyte membranes do not act as a molecular sieve (absence of equivalent pores). Accordingly, it is feasible by using these membranes to decide between the two possible mechanisms for urea transport. For example, if urea is transported across biological membranes by a carrier-mediated mechanism, then the rate of urea movement across PMN membranes would be very high. On the other hand, if urea permeates through "aqueous pores", then the permeability coefficient of these membranes to urea would be very small. Furthermore, the rates of permeation of urea, methylurea and thiourea across PMN membranes would be determined mainly by the lipid solubility of each solute. The results of these studies are summarized in Table 6. The values for K_{ether}, N_H and molar volume are included in the table. The permeability coefficients of human red-cell membranes to these solutes are also included in the table for comparison. It is very clear from the table that the two membranes behave quite differently with respect to the rates of permeation of these nonelectrolytes. This difference cannot be accounted for on the basis of differences in species. The pattern in the permeability coefficients of rabbit red-cell membranes to urea, methylurea and thiourea is similar to that found for human erythrocyte membranes. Three conclusions can be drawn from these studies. First, there is no need to postulate a specialized mechanism (carrier-mediated) for urea transport across rabbit PMN membranes. This conclusion can probably be extended to human red-cell and other biological membranes. It is very difficult to imagine why red-cell membranes would have a carrier-mediated mechanism for urea transport and PMN membranes would not. Secondly, there is no need to postulate that PMN membranes contain "aqueous pores" for the transport of small hydrophilic nonelectrolytes. In these cells one can

TABLE 6
Permeability coefficients of rabbit PMN and human red-blood-cell membranes to small nonelectrolytes [a]

Solute	Molar volume	K_{ether}	N_H	Permeability coefficient · 10^5 (cm/s)	
				Rabbit PMN	Human RBC
Urea	45	0.00047	5	0.30	31.2
Methylurea	61.5	0.0012	4	0.45	4.8
Thiourea	54.2	0.0063	5	0.87	0.07

[a] For original citation see [4].

postulate that the membrane is homogeneous and permeation occurs only by dissolution in the membrane fabric. Thirdly, the idea of "aqueous pores" is not a general property of biological membranes, but is restricted to certain types.

These findings are not restricted to the transport of urea across leukocyte membranes. It is well known that in some red cells (fish and birds) urea is less permeable than thiourea, while in others (reptiles and mammals) the converse is true. It has been suggested [20] that in the former cells, the membrane is homogeneous and permeation occurs only by dissolution in the membrane matrix, while in the latter cells the membrane contains both lipid and polar regions. Accordingly, thiourea (having a higher partition coefficient and lower N_H) will permeate faster than urea in the former cells, whereas in the latter cells urea is faster, owing to its smaller molecular size. The observation that phloretin enhances the movement of urea in red cells from fish and birds and inhibits the transport of urea in red cells from reptiles and mammals [55] is consistent with this view.

Finally, using ADH-treated toad urinary bladder, Levine and Worthington [52] have been able to demonstrate experimentally the presence in this system of co-transport of labeled methylurea and, to a lesser degree, acetamide and urea with unlabeled methylurea. They concluded that these results are consistent with the presence of ADH-sensitive amide selective channels rather than a mobile carrier.

(d) Are all the pathways used by water available for urea and other small polar nonelectrolytes?

If one accepts the conclusion that urea and other small hydrophilic solutes permeate cell membrane through specialized "polar pathways" and not by mobile carriers, then the next question to be answered is whether or not these molecules use most of the pathways available for water movement. Considering the data available, it is hard to escape the conclusion that only a small fraction of these postulated pathways is available for the movement of urea and other small solutes. This is based on the observation that it is possible to dissociate water and solute transport in human red-cell membranes as well as other systems. It is found (see Table 5) that while the compound phloretin significantly reduced the transport of urea and other small

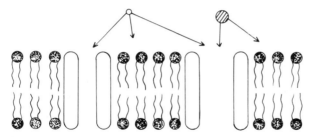

Fig. 3. Schematic representation of the various possible pathways available to water and small hydrophilic solutes for crossing cell membranes. The open circle represents water molecule and the hatched circle represents the solute molecule. (From [4].)

hydrophilic solutes, it enhances slightly the movement of water in human red-cell membranes [4,60].

Fig. 3 depicts schematically the author's view of how water and small polar nonelectrolytes move across membranes which behave operationally as both selective solvents and molecular sieves. The basic features of this model are as follows:

(1) Most of the water molecules permeate through "aqueous pores" with only a small component crossing through the membrane matrix.

(2) The radii of these channels are not uniform and only a small fraction of these pores is large enough to permit the transport of small hydrophilic solutes. Some of these "pores" may be large enough to accommodate glucose molecules. This will imply that glucose transport across human red-cell membrane is not by means of a mobile carrier mechanism – a not unlikely possibility.

(3) A fraction of the total movement of the solute takes place through the membrane matrix. This fraction increases with increasing lipid solubility.

(4) Certain sulfhydryl-reactive reagents totally abolish the movement through the "aqueous pathways" but do not affect the movement through the membrane matrix.

(5) The compound phloretin is able to react with membrane lipid and/or membrane proteins. This interaction affects those pathways which are large enough to permit solute penetration. This interaction can cause configurational changes in the proteins which form these pathways resulting in a significant decrease in the size of these "aqueous pores" and a change in the physical and chemical nature of the membrane matrix. The latter change can lead to an increase in the rate of solute movement though the lipid matrix.

5. Summary and conclusions

The results presented and discussed in this chapter suggest that: (a) Certain biological membranes act both as a selective solvent and as a molecular sieve. The latter property is due to the presence in the membrane structure of polar regions. In the case of mammalian red-cell membranes, these polar regions are probably assembled from membrane integral proteins. (b) In these membranes, small polar nonelectrolytes permeate through some of these specialized channels and not by mobile carriers. (c) Water molecules utilize most if not all these polar channels for permeation, whereas small polar molecules use only a very limited and specialized portion. (d) Under normal conditions, the concept of "aqueous polar regions" for the transport of water and polar nonelectrolytes is not general but limited to certain classes of membranes. (e) In the case of transport across hormonally sensitive membranes and tissues, it is possible that the ultimate molecular change in the membrane in the presence of the hormone is the generation of specialized "polar channels".

Acknowledgements

Unpublished studies cited here from the author's laboratory are supported in part by an NIH research grant. The author wishes to express his great appreciation to Dr. P. Naccache for his help, and to Ms. Joan Jannace for her infinite patience in typing the manuscript.

References

1 Overton, E. (1896) Vierteljahrschr. Naturforsch. Ges. Zurich 41, 388–406.
2 Singer, S.J. (1974) Annu. Rev. Biochem. 43, 805–832.
3 Da Silva, P.P. (1973) Proc. Natl. Acad. Sci. USA 70, 1339–1343.
4 Sha'afi, R.I. (1977) in J.C. Ellory and V.L. Lew (Eds.), Membrane Transport in Red Cells, Academic Press, London, New York, pp. 221–256.
5 Cabantchik, Z.I. and Rothstein, A.J. (1974) J. Membr. Biol. 15, 207–226.
6 Deuticke, B. (1977) Rev. Physiol. Biochem. Pharmacol. 78, 1–97.
7 Paganelli, C.V. and Solomon, A.K. (1957) J. Gen. Physiol. 41, 259–277.
8 Tosteson, D.C. (1959) Acta Physiol. Scand. 46, 19–41.
9 Conlon, T. and Outhred, R. (1972) Biochim. Biophys. Acta 288, 354–361.
10 Macey, R.I. (1979) in G. Giebisch, D.C. Tosteson and H.H. Ussing (Eds.), Membrane Transport in Biology, Vol. II, Springer-Verlag, Berlin, pp. 1–57.
11 Blum, R.M. and Forster, P.E. (1970) Biochim. Biophys. Acta 20, 410–423.
12 Hempling, H.G. (1967) J. Cell. Physiol. 70, 237–256.
13 Sha'afi, R.I., Rich, G.T., Mikulecky, D.C. and Solomon, A.K. (1970) J. Gen. Physiol. 55, 427–450.
14 Sha'afi, R.I., Rich, G.T., Sidel, V.W., Bossert, W. and Solomon, A.K. (1967) J. Gen. Physiol. 50, 1377–1399.
15 Kedem, O. and Katchalsky, A. (1958) Biochim. Biophys. Acta 27, 229–246.
16 Katchalsky, A. and Curran, P.F. (1965) Nonequilibrium Thermodynamics in Biophysics, 1st ed., Harvard University Press. Cambridge, MA.
17 Rich, G.T., Sha'afi, R.I., Romualdez, A. and Solomon, A.K. (1968) J. Gen. Physiol. 52, 941–954.
18 Goldstein, D.A. and Solomon, A.K. (1960) J. Gen. Physiol. 44, 1–17.
19 Sha'afi, R.I., Gary-Bobo, C.M. and Solomon, A.K. (1971) J. Gen. Physiol. 58, 238–258.
20 Sha'afi, R.I. and Gary-Bobo, C.M. (1973) Prog. Biophys. Mol. Biol. 26, 103–146.
21 Bangham, A.D., Hill, M.W. and Miller, N.G.A. (1974) in E.D. Korn (Ed.), Methods in Membrane Biology, Vol. 1, Plenum, New York, pp. 1–68.

22 Dainty, J. (1963) Adv. Botan. Res. 1, 279–326.
23 House, C.R. (1974) Water Transport in Cells and Tissues, Williams and Wilkins, London
24 Solomon, A.K. (1968) J. Gen. Physiol. 51, 335s–364s.
25 Levitt, D.G. (1974) Biochim. Biophys. Acta 373, 115–131.
26 Dainty, J. and Ginzburg, B.Z. (1963) J. Theor. Biol. 5, 256–265.
27 Diamond, J.M. and Wright, W.R. (1969) Ann. Rev. Physiol. 31, 581–646.
28 Solomon, A.K. (1973) in F. Kreuzer and J.F.G. Slegers (Eds.), Passive Permeability of Cell Membranes, Vol. 3, Plenum, New York, pp. 299–330.
29 Naccache, P. and Sha'afi, R.I. (1974) J. Cell. Physiol. 83, 449–456.
30 Haydon, D.A. (1969) in D.C. Tosteson (Ed.), The Molecular Bases of Membrane Function, Prentice Hall, Englewood Cliffs, NJ.
31 Masters, B.P. and Fanestil, D.D. (1979) J. Membr. Biol. 48, 237–247.
32 Lichtenstein, N.S. and Leaf, A. (1965) J. Clin. Invest. 44, 1328–1342.
33 Loesche, K., Benzel, C.J. and Csaky, T.Z. (1970) Am. J. Physiol. 218, 1723–1731.
34 Terwilliger, T.C. and Solomon, A.K. (1980) submitted for publication.
35 Marchesi, V.T. (1975) in G. Weissmann (Ed.), Cell Membrane, H.P. Publishing, New York, pp. 45–53.
36 Steele, T.L., Fairbanks, G. and Wallach, D.F.H. (1971) Biochemistry 10, 2617–2624.
37 Singer, S.J. (1975) in G. Weissmann (Ed.), Cell Membrane, H.P. Publishing, New York, pp. 35–46.
38 Lea, E.J.A., Rich, G.T. and Segrest, J.P. (1975) Biochim. Biophys. Acta 382, 41–50.
39 Sha'afi, R.I. and Feinstein, M. (1977) Adv. Exp. Med. Biol. 84, 67–83.
40 Knauf, P.A., Proverbio, F. and Hoffman, J.F. (1974) J. Gen. Physiol. 63, 305–323.
41 Cabantchik, Z.I. and Rothstein, A.J. (1974) J. Membr. Biol. 15, 227–248.
42 Traüble, H. (1972) in F. Kreuzer and J.F.G. Slegers (Eds.), Passive Permeability of Cell Membranes, Vol. 3, Plenum, New York, pp. 197–227.
43 Naccache, P. and Sha'afi, R.I. (1973) J. Gen. Physiol. 62, 714–436.
44 Wright, E.M. and Diamond, J.M. (1969) Proc. Roy. Soc. B. 172, 227–271.
45 Wright, E.M. (1976) in P.L. Altman and D.D. Katz (Eds.), Cell Biology, Vol. 1, Federation of American Societies for Experimental Biology, pp. 127–131.
46 Holz, R. and Finkelstein, A. (1970) J. Gen. Physiol. 56, 125–145.
47 Andreoli, T.E. (1974) Ann. N.Y. Acad. Sci. 235, 448–468.
48 DeKruyff, B. and Demel, R.A. (1974) Biochim. Biophys. Acta 338, 57–70.
49 Pimental, G.C. and McClellan (1960) The Hydrogen Bond, Freeman, San Francisco, p. 52.
50 Sha'afi, R.I. and Volpi, M. (unpublished data).
51 Galey, W.R., Owen, J.D. and Solomon, A.K. (1973) J. Gen. Physiol. 61, 727–747.
52 Levine, S.D. and Worthington, R.E. (1976) J. Membr. Biol. 26, 91–107.
53 Poznansky, M., Tong, S., White, P.C., Milgram, J.M. and Solomon, A.K. (1976) J. Gen. Physiol. 67, 45–66.
54 Svelto, M., Curci, S., Micelli, S., Gallucci, E., Storelli, C. and Lippe, C. (1974) in L. Bolis, K. Bloch, S.E. Lurio and F. Lynen (Eds.), Comparative Biochemistry and Physiology, North-Holland, Amsterdam, pp. 367–370.
55 Kaplan, M.A., Hays, L. and Hays, R.M. (1974) Am. J. Physiol. 226, 1327–1332.
56 Hunter, F.R., George, J. and Ospiha, B. (1965) J. Cell. Comp. Physiol. 65, 299–311.
57 Carvounis, C.P., Levine, S.D., Franki, N. and Hays, R.M. (1979) J. Membr. Biol. 48, 269–281.
58 Sha'afi, R.I. and Volpi, M. (1976) Biochim. Biophys. Acta 436, 242–246.
59 Hempling, H.G. (1973) J. Cell. Physiol. 81, 1–10.
60 Owen, J.D. and Solomon, A.K. (1972) Biochim. Biophys. Acta 290, 414–418.

CHAPTER 3

Ion permeability

EDUARDO ROJAS

Department of Biophysics, School of Biological Sciences,
University of East Anglia, Norwich NR4 7TJ, U.K.

1. Introduction

The membranes found both at the surface of all living cells and within them determine compartments. The surface membrane, or plasmalemma, separates the intracellular fluid from the extracellular fluid. The plasmalemma permits the establishment of large concentration differences between these two compartments. This function is also present in the case of intracellular compartments such as the mitochondrion and the endoplasmic reticulum.

The process of transport of ions into and out of these biological compartments is referred to as membrane transport. In general, it involves energy conversion, i.e. energy transfer between the pool of chemical energy [1] and the electrochemical potential stored across most membranes [2].

There are two types of ion transport across membranes. If the transport is such that the ion moves down the electrochemical gradient it is referred to as passive transport. In this case, the energy supplied by the cell metabolism is used to generate and to maintain the electrochemical gradient across the membrane. If, on the other hand, the transport is against the electrochemical gradient, the transport is said to be active. This chapter deals with some aspects of passive transport. We shall consider ion transport produced by two driving forces, namely, the chemical force, which is proportional to the negative value of the chemical potential gradient, and the electrical force, which is proportional to the negative value of the electrical potential. While the ion is in transit through the membrane there are several other forces acting upon it. In general, these forces are of electrical nature but there are examples where frictional forces are thought to be involved. In all cases the transport of ions is diminished or even prevented. When this occurs, one speaks of gating of the ion flux. Obviously, to determine the type of ion permeation involved and to establish its gating mechanism, it is desirable to have detailed information about the membrane structure.

Bonting/de Pont (eds.) Membrane transport
© *Elsevier/North-Holland Biomedical Press, 1981*

2. Theoretical basis for the concept of permeability

(a) The gradient of electrochemical potential as a force

(i) Chemical and electrochemical potential

Thermodynamic quantities which have the same value throughout a homogeneous compartment (temperature, T; pressure, p) are called intensive. Those which are proportional to the amount of component i in that compartment are called extensive (enthalpy, H; entropy, S; free-energy, G). An extensive quantity may be converted to an intensive one by dividing by the amount of component i present. For example, the partial molal free-energy or chemical potential,

$$\mu_i = \left(\frac{\partial G}{\partial n_i}\right)_{T,P,n_j}, i \neq j \tag{1}$$

which makes the total contribution of component i to the free-energy F equal to $n_i\mu_i$, n_i being the number of moles of component i.

At this stage it is convenient to introduce the concept of ideal solution using the same procedure as for gases, i.e. when the intermolecular interactions have been minimized to the limit that the state of the gas is described by

$$P = RT\frac{n}{v}$$

where n is the molar fraction in the volume v. A solution of the solute i is said to be ideal if it obeys van 't Hoff's law, i.e.

$$P_i = RT\, C_i$$

It can be shown that an ideal solution is one in which, for each component i, the total free-energy or chemical potential is

$$\mu_i = \mu_i^S + RT \ln C_i \tag{2}$$

where C_i represents the molarity of the solute i. Real solutions obey Eqn. 2 at the limit of infinite dilution. For general application of Eqn. 2 we introduce a quantity, the activity, so that the form of Eqn. 2 is maintained, i.e. we write,

$$\mu_i = \mu_i^S + RT \ln a_i \tag{3}$$

The activity, a_i, is then the product

$$a_i = \gamma_i C_i,\, 0 < \gamma_i < 1$$

Ion permeability

where γ_i represents the activity coefficient.

Formally, we can write

$$\mu_i^S = RT \ln a_i^S \tag{4}$$

where a_i^S is the activity of the ionic species i in the standard or reference solution.

If the average chemical potential μ_i, of an element of volume, is added to that due to the electrical potential V, one gets the electrochemical potential. Since the electrical potential, V, is defined as the work per unit charge against electrical forces, to bring a mol of ions i from a reference element of volume, where the electrical forces are zero, to that element, it is necessary to multiply its value by $z_i F$ (z_i represents the valence of species i and F the Faraday's constant). Whence, one may write the electrochemical potential η_i as

$$\eta_i = \mu_i + z_i FV \tag{5}$$

To produce a more compact expression, similar to Eqn. 3, we introduce the concept of electrochemical activity as follows.

Let V_0 be the reference electrical potential, then introducing Eqn. 4 into Eqn. 5 one gets:

$$\eta_i = \left(RT \ln a_i^S + z_i FV_0 \right) + \left(RT \ln a_i + z_i FV \right)$$

or

$$\eta = \eta_i^S + RT \ln A_i \tag{6}$$

where

$$A_i = a_i \exp\left(\frac{z_i FV}{RT} \right) \tag{7}$$

or

$$A_i = a_i \xi_i$$

where ξ_i has the meaning of an electrochemical activity coefficient [3].

(ii) The flux equations

The theoretical treatment of diffusion begins with the works of Fourier and Fick.

The fundamental assumption of Fick was that the rate of diffusion across any plane at right angles to the direction of diffusion bears a linear relation to the concentration gradient across the plane in question. That is,

$$\left(\frac{dQ_i}{dt}\right)_x = -D_i A \frac{\partial C_i}{\partial x} \tag{8}$$

where the transport is in the x-direction, D_i is a constant, A is the area of the plane and $\partial C/\partial x$ is the concentration gradient. The theoretical basis for Fick's law was given by Nernst and by Einstein. The rate of diffusion across a plane at right angles to the x-axis per unit area is the flux ϕ_x.

Consider the diffusion system illustrated in Fig. 1A. Assume that the cross section A is a unit area and consider a layer of thickness dx. Taking the concentration in that layer to be C_i, the number of molecules in that element equals $NC_i dx$ where N represents Avogadro's number. Suppose that each surface, A_x, A_{x+dx}, is readily permeable to water, but completely impermeable to the solute. Therefore, a movement of water will occur through A_x and A_{x+dx} from the more dilute to the more concentrated solution, and the layer itself will be moved in the opposite direction until conditions on its two sides are equal. It is possible to stop the water flow by applying an appropriate external pressure to the more concentrated of the two adjacent solutions.

Osmotic pressure may be defined as the pressure which must be applied to a given solution to make the escaping tendency of the solvent which it contains equal to that of the pure solvent under the same conditions.

The osmotic pressure may be taken as equal to p for the solution in immediate contact with the layer on the one side and $p + dp$ for that on the other, A_{x+dx}. Under these conditions water will tend to pass in the positive direction and in consequence to move the layer in the negative direction, the effective pressure available for this purpose being the difference between the two osmotic pressures. As $A_x = A_{x+dx}$ was taken as unity, pressure and force will be numerically equal.

The driving force $F_{x,i}$ acting on a single molecule in the positive direction would then be $-(dp/dx) \cdot (1/NC)$. Such a force would transport the molecules in the element of volume with the same average velocity $\bar{U}_{x,i}$. As the molecules are set in motion at constant velocity, a frictional force $F'_{x,i}$, proportional to the mean velocity and opposing to the motion, will be established. Whence,

$$F'_{x,i} = f_i \bar{U}_{x,i}, \tag{9}$$

where f_i is a frictional coefficient for that ion species. As the velocity is constant,

$$F_{x,i} + F'_{x,i} = 0 \tag{10}$$

The rate of transport per unit time per unit area perpendicular to the x-direction, i.e. the flux $\phi_{x,i}$, equals

$$\phi_{x,i} = C_{x,i} \bar{U}_{x,i} \tag{11}$$

Ion permeability

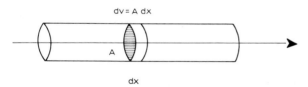

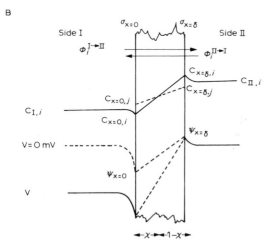

Fig. 1. Diffusion system and membrane profile. A_x is the cross section at a point x and A_{x+dx} at a point $x+dx$. The volume of the element considered is $dV = A\,dx$.

$\sigma_{x=0},\ \sigma_{x=\delta}$ = surface charge density, C/cm².
$\phi_i^{I \to II}$ = efflux of species i, pmol/cm²s.
$\phi_i^{II \to I}$ = influx of species i, pmol/cm²s.
$C_{x=0,j}$ = surface concentration of co-ions.
$C_{x=\delta,j}$ = surface concentration of co-ions.
$C_{x=0,i}$ = surface concentration of counter-ions.
$C_{x=\delta,i}$ = surface concentration of counter-ions.
$C_{I,i},\ C_{II,i}$ = bulk phase concentration of solute.
$\psi_{x=0}$ = surface potential, mV.
$\psi_{x=\delta}$ = surface potential, mV.
X = fraction of the membrane length δ acting on membrane dipoles.

Notice that $V_{II} = 0$. For this reason the difference in potential across the membrane, ΔV, is represented as V.

Assuming that the solution is ideal (i.e. it obeys van 't Hoff's law) then,

$$\phi_{x,i} = -\frac{RT}{Nf_i} \frac{dC_{x,i}}{dx} \tag{12}$$

Comparing Eqn. 8 with Eqn. 12 we have

$$D_i = \frac{RT}{Nf_i} \tag{13}$$

The frictional coefficient can be determined either experimentally or by modelling. Thus, Stokes' law for a sphere of radius r moving in a viscous medium of viscosity η,

$$f = 6\pi\eta r \tag{14}$$

whence,

$$D = \frac{RT}{N}\frac{1}{6\pi\eta r} \tag{15}$$

This important equation is commonly known as the Einstein or the Stokes–Einstein equation, though it was obtained independently by Sutherland.

The Sutherland–Einstein equation has very often been used to obtain information about the radii of the diffusing molecules from the observed values of their diffusion coefficients. In the case of translational free diffusion of neutral molecules the driving force per mole is the negative gradient of the chemical potential. Thus,

$$F_x = -\frac{1}{N}\frac{d\mu_i}{dx} \tag{16}$$

From Eqns. 9, 11, and 16 the flux $\phi_{x,i}$ becomes

$$\phi_{x,i} = -\frac{C_i}{f_i N}\frac{d\mu_i}{dx}. \tag{17}$$

Calculating $d\mu/dx$ from Eqn. 3

$$\frac{d\mu_i}{dx} = RT\frac{1}{a_i}\frac{da_i}{dx}. \tag{18}$$

For ideal solutions

$$\frac{d\mu_i}{dx} = RT\frac{1}{C_i}\frac{dC_i}{dx} \tag{19}$$

and

$$\phi_{x,i} = -\frac{RT}{f_i N}\frac{dC_i}{dx}$$

which is Fick's law.

(b) The Nernst–Planck equation

In the case of a net flow of charges one speaks of an electrical current. This is measured by the rate dQ/dt at which positive charge passes through any specified area A (see Eqn. 8). The current density $i_{x,i}$ at a point x is defined as the vector having the direction of the flow of positive charge and a magnitude equal to the current i per unit area normal to the direction of the flow, A. Defining ρ_x as the charge per unit volume at x, the currrent density equals

$$i_{x,i} = \rho_x \overline{U}_{x,i}. \tag{20}$$

The electrical force F_x acting on a positive charge is proportional to the intensity of the electric field E_x at that point. The electric field, on the other hand, is equal to the negative value of the electrical potential gradient, dV/dx, at that point. For charges moving with a constant mean velocity $\overline{U}_{x,i}$, the size of the electrical force equals that of the frictional force. Defining electrical conductivity of the medium as the ratio between the current density and the electric field E_x at that point,

$$\sigma_x = \frac{i_{x,i}}{E_x},$$

one gets

$$\frac{dV}{dx} = -\frac{1}{\sigma_x A} i_{x,i}.$$

It should be noted that to integrate this equation one must know σ_x.

Assuming that the electrical conductivity σ_x is constant, integration produces the familiar relationship between potential and current,

$$V = Ri$$

where

$$R = \frac{1}{\sigma} \frac{l}{A}.$$

In the case of electrolyte solutions we may write

$$\rho_{x,i} = z_i F C_i \tag{21}$$

where C_i is the molar concentration of ions. On the other hand the mean velocity is proportional to the mobility U_i and to the electric field E_x. Thus,

$$\overline{U}_{x,i} = U_{x,i} E_x \tag{22}$$

As E_x is equal to the negative gradient of V, one may write

$$\bar{U}_{x,i} = -U_{x,i}\frac{dV}{dx}$$

From Eqns. 20, 21, and 22 one gets $\phi_{x,i}$ or current density

$$\phi_{x,i} = -U_{x,i}C_i\frac{d(z_iFV)}{dx} \tag{23}$$

where z_iFV is the electrical potential energy per mole. This last equation is formally identical to Eqn. 17. Both make the flux equal the product of the mobility times the force which is equal to the negative value of the gradient of potential energy.

Consider the system in Fig. 1B. Across the membrane of thickness δ both the chemical and electrical potential gradients act on the positive ions to produce a current density equal to

$$\phi_{x,i} = -z_iFC_{x,i}\left(\frac{1}{f_iN}\frac{d\mu_{x,i}}{dx} + U_i\frac{dV}{dx}\right) \tag{24}$$

Introducing Eqns. 13 and 19 into 24 we get the Nernst–Planck equation,

$$\phi_{x,i} = -z_iF\left(D_i\frac{dC_{x,i}}{dx} + U_iC_{x,i}\frac{dV}{dx}\right) \tag{25}$$

(c) Integration of the Nernst–Planck equation [4]

In order to resolve Eqn. 25, one integrates for $0 < x < \delta$ (see Fig. 1B). It is obvious that one must either know C_x, U_x and V_x or one must assume values as a function of x.

(i) The constant field hypothesis
Assuming that the membrane is homogeneous and neutral, and that the gradient of the electrical potential is constant across the membrane, Eqn. 25 may be directly integrated. Goldman [5] made these assumptions and Hodgkin and Katz [6] used it to analyse results concerning ion flux. The first assumption is used to relate the concentrations at both surfaces

$$C_{x=0,i} \text{ and } C_{x=\delta,i}, \tag{26a}$$

to those in the compartments I and II. Thus,

$$C_{x=0,i} = \beta_i C_{I,i} \tag{26b}$$

$$C_{x=\delta,i} = \beta_i C_{II,i} \tag{26c}$$

Ion permeability

The second assumption reduces Eqn. 25 to non-homogeneous linear differential equation with the following solution:

$$\phi_i = \frac{\beta_i U_i F}{\Delta x} \Delta V \left[\frac{C_I - C_{II} \exp\left(-\frac{z_i F}{RT}\Delta V\right)}{1 - \exp\left(-\frac{z_i F}{RT}\Delta V\right)} \right] \tag{27}$$

where β is the partition coefficient and $\Delta x = \delta$.

Now on for internal concentration and potential read [ ]$_I$ and V_I. For external solution concentration and potential read [ ]$_{II}$ and V_{II}. As for the potential V and the current i, the usual conventions, the electrical potential of compartment II is taken as reference potential (equal to zero) and flow of positive ions is taken as positive current.

(ii) Definition of permeability

Let the permeability coefficient be defined as

$$P_i = \frac{\beta_i}{\delta} D_i. \tag{28}$$

Introducing Eqn. 13 for D_i, Eqn. 27 becomes

$$\phi_i = P_i \frac{-z_i FV}{RT} \frac{[i]_I - [i]_{II} \exp(-z_i FV/RT)}{1 - \exp(-z_i FV/RT)} \tag{29}$$

Another expression often used can be derived from Eqn. 29. As ϕ_i represents a net flux, namely,

$$\phi_i = \phi_i^{II \to I} - \phi_i^{I \to II} \tag{30}$$

then

$$i_i = z_i F \phi_i$$

where ϕ_i is given by Eqn. 29. If instead of ϕ_i in Eqn. 29 one considers the current carried by that ion one may write

$$g_i = P_i \frac{z_i^2 F^2}{RT} \frac{[i]_I - [i]_{II} \exp(-z_i FV/RT)}{1 - \exp(-z_i FV/RT)}$$

where g_i is the conductance of the membrane to that ion.

(iii) Net ion fluxes

Eqn. 27 represents a thermodynamic description of passive transport which has been used to analyse experiments in excitable non-stimulated cells. It states that provided the driving forces are the concentration and the electrical potential gradient, then the net flux for univalent cations can be calculated using either Eqn. 27 or Eqn. 29.

(iv) Unidirectional ion fluxes and the flux ratio

For $[i]_I = 0$, Eqn. 29 describes the influx $\phi_i^{II \to I}$. Conversely, for $[i]_{II} = 0$, it describes the efflux $\phi_i^{I \to II}$.

In any diffusional process, when equilibrium is reached, the net flux is zero. Therefore,

$$\phi_i^{II \to I} = \phi_i^{I \to II}. \tag{31}$$

The dependence of these unidirectional fluxes on transmembrane potential V and ion concentrations has been studied in a variety of cells. However, in order to establish whether or not Eqn. 29 can be used to describe the experimental measurements, it is necessary to control both the extracellular and intracellular concentrations and, ideally, the transmembrane potential.

To calculate a flux ratio [7–9] one writes Eqn. 29 as follows

$$\frac{\phi_i^{II \to I}}{\phi_i^{I \to II}} = \frac{[i]_{II}}{[i]_I} \exp\left(-\frac{z_i F}{RT}\right) V \tag{32}$$

This equation was used by Hodgkin and Keynes [10] to analyse results on ^{42}K fluxes across the plasmalemma of axons from *Sepia officinalis*.

(v) The Goldman equation

Another important consequence of Eqn. 27 is the prediction of a membrane potential value in a system with more than one permeant ion species participating in the equilibrium.

Let us consider three ion species, say Na^+, K^+ and Cl^-, in equilibrium. Whence

$$\phi_{Na^+} + \phi_{K^+} - \phi_{Cl^-} = 0 \tag{33}$$

An equivalent statement of the same condition is to assume that the total current is zero. Thus,

$$i_{Na^+} + i_{K^+} - i_{Cl^-} = 0. \tag{34}$$

Introducing Eqn. 29 for Na^+, K^+ and Cl^- into Eqn. 33 one gets for the membrane potential

$$V = \frac{RT}{F} \ln \frac{P_{K^+}[K^+]_{II} + P_{Na^+}[Na^+]_{II} + P_{Cl^-}[Cl^-]_I}{P_{K^+}[K^+]_I + P_{Na^+}[Na^+]_I + P_{Cl^-}[Cl^-]_{II}} \tag{35}$$

Ion permeability

This equation was derived by Goldman and is called Goldman's equation.

Eqn. 35 has been used in a number of preparations to study the dependence of V on ionic composition and to estimate permeability values in channels of excitable cells [6,11,12].

(vi) Net ionic fluxes and resting membrane conductance

Eqn. 32 was used by A.L. Hodgkin [13] to calculate the membrane conductance to K^+ and to Cl^-.

Suppose that the cell is in equilibrium with

$$\phi_i^{II \to I} = \phi_i^{I \to II}.$$

As the current density through the membrane i_m is the sum of the contributions of potassium i_{K^+}, sodium i_{Na^+} and chloride i_{Cl^-}, if the equilibrium is disturbed by applying small current di then the potential will change by dV. Since,

$$i_i = z_i F \left(\phi_i^{II \to I} - \phi_i^{I \to II} \right)$$

or

$$i_i = z_i F \left(\frac{\phi_i^{II \to I}}{\phi_i^{I \to II}} - 1.0 \right) \phi_i^{I \to II}$$

Then the contribution of the species i to the conductance, i.e. g_i, is

$$\frac{\partial i_i}{\partial V} = z_i F \phi_i^{I \to II} \frac{\partial}{\partial V} \left(\frac{\phi_i^{II \to I}}{\phi_i^{I \to II}} \right) + z_i F \left(\frac{\phi_i^{II \to I}}{\phi_i^{I \to II}} - 1 \right) \frac{\partial \phi_i^{I \to II}}{\partial V}$$

As V approaches the reversal potential for the current the second term on the right hand side of this equation approaches zero and

$$\frac{\partial}{\partial V} \left(\frac{\phi_i^{II \to I}}{\phi_i^{I \to II}} \right)$$

can be calculated from Eqn. 32 to give

$$g_i = \frac{z_i^2 F^2 \phi_i}{RT}. \qquad (36)$$

The total membrane conductance is the sum of the individual ionic conductances, i.e.

$$g_m = \frac{F^2}{RT} \sum_{i=1}^{n} z_i^2 \phi_i. \qquad (37)$$

3. The measurement of ionic permeabilities

The application of the concepts discussed in section 2 of this chapter is best illustrated using the results of experiments on intracellularly perfused giant excitable cells as in these preparations it is possible to control both the ionic composition as well as the membrane potential. To control the membrane potential a feedback amplifier is used. For details on this part of the methods the reader is referred to the original papers.

(a) Cation permeabilities in resting electrically excitable cells

(i) Permeabilities from tracer fluxes
In 1961 two different groups of investigators [14,15] independently developed techniques to replace the axoplasm of squid giant axons by artificial solutions without altering axonal membrane potentials. Since then this preparation has been widely used to study further the origin of these potentials and many experiments have been performed under a variety of intra- and extraaxonal conditions.

For example, to measure unidirectional sodium fluxes, Rojas and Canessa-Fischer [16] used the chamber diagrammatically shown in Fig. 2a.

Sodium fluxes were computed by multiplying the measured flow of radioactivity in cpm/cm^2/s by the reciprocal of the specific activity of the solution measured in pmol, 10^{-12} M/cpm. Fig. 2b shows part of the time course of a typical sodium efflux experiment. Intracellular perfusion was begun with 600 mM KF; after 13 min, this solution was replaced by 500 mM KF, 100 mM NaF and then, after 23 min, by a solution of the same composition, but with ^{22}Na added.

In this figure the vertical axis represents the sodium efflux measured during a 10-min period. Each column represents one determination. It can be seen that the sodium efflux remained almost constant for long periods. This figure also demonstrates that the resting outward movement of sodium is increased by impulse propagation. Determination 14 represents the efflux during external stimulation of the fiber. As there were 50 propagated action potentials per second recorded on the oscilloscope, the average extra efflux per impulse in this particular experiment is seen to be about 6 pmol/cm^2.

Fig. 2c shows part of the time course of a sodium influx experiment. In this case the perfusion was again started with 600 mM KF and 2 mM NaF. After 20 min the external seawater was replaced by ^{22}Na seawater.

It is clear from this figure that the sodium influx again remained almost constant during the internal perfusion with a given sodium concentration. Table 1 summarises the data on resting sodium fluxes obtained by internal perfusion with different concentrations of sodium. As shown in Table 1 the resting sodium efflux is affected by the internal sodium concentration. There is an increase in sodium efflux from 0.1 pmol/cm^2 s to 34 pmol/cm^2 s as the internal sodium concentration is increased from 2 to 200 mM. To calculate resting P_{Na^+} values from the data in Table 1 we

Ion permeability

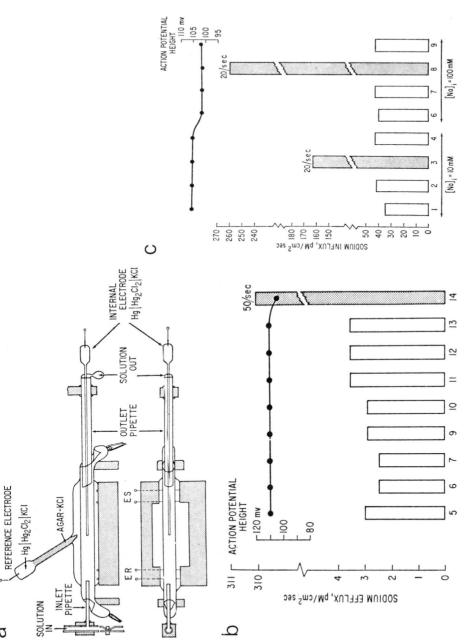

Fig. 2. (a) Simplified diagram of the experimental arrangement used for internal perfusion. The upper diagram represents a section through the perfusion chamber with cannulas and electrodes in position. The lower diagram represents a top view of the chamber. (b) Resting sodium efflux in internally perfused giant axon. Vertical axis is efflux measured in pmol/cm² s. Determinations were made every 10 min. Determination 14 was measured under electrical stimulation pulsing the fibre 50 times/s during 10 min. The recording of the action potential height is shown on top of the columns. Axon diameter 875 μm. (c) Effect of impulse propagation of sodium influx. This figure compares the influx during rest and during activity. Axon diameter 900 μm, area for sodium exchange, 0.64 cm². Experiment performed at room temperature of 17°C. Frequency of stimulation is indicated at the top of each column. (Reproduced with permission from Rojas and Canessa-Fischer [16].)

TABLE 1

Experimental and theoretical flux ratios and permeability values for non-stimulated resting squid giant axons

(a) [Na]$_I$	(b) V	(c) $\phi_{Na}^{I \to II}$	(d) $\phi_{Na}^{II \to I}$	(e) $\phi_{Na} = \phi_{Na}^{II \to I} - \phi_{Na}^{I \to II}$	(f) P_{Na} 10^8 cm/s	(g) $\dfrac{\phi_{Na}^{II \to I}}{\phi_{Na}^{I \to II}}$	(h) Eqn. 32	(i) (h)/(g)	(j)
		pmol/cm^2 s							
2	−53.2	0.09 ± 0.02	42.9 ± 10.5	42.8	4.1	476.7	1898.4	3.8	477
10	−56.0	0.14	40.0	39.8	3.8	285.7	380.4	1.3	247
100	−51.3	5.7 ± 3.6	48.7 ± 6.0	43.0	4.4	8.5	32.9	3.9	31.5
200	−47.0	34.0 ± 9.0	64.5 ± 15.0	30.5	3.0	1.9	13.5	7.1	13.3
Mean value ± Standard deviation								4.0 ± 2.4	

re-write Eqn. 29 as follows

$$P_{Na^+} = \phi_{Na^+} \dfrac{25}{-V} \dfrac{1 - \exp\left(-\dfrac{V}{25}\right)}{[Na^+]_I - [Na^+]_{II} \exp\left(-\dfrac{V}{25}\right)}$$

where the value of RT/zF is taken as 25 mV. P_{Na^+} values obtained are also presented in Table 1. Table 1 presents under column (g) the flux ratios calculated using the data under columns (c) and (d) and in column (h) the values calculated using equation (32). It may be seen that the theoretical values are larger than the measured ratios.

In the case of the K^+ experiments the flux ratios predicted are smaller than the measured ratios. Hodgkin and Keynes assumed that potassium ions cross the membrane in a single file through narrow channels [10]. This mechanism explains why a unidirectional flux in the direction opposite to that of the net flux is reduced. In these experiments V_{K^+} was varied by adjusting $[K^+]_{II}$ and V was controlled by current flow. In the case of the experiments in Table 1 the reduction in V due to an increase in $[Na^+]_I$ from 2 to 200 mM was less than 7.1 mV namely from −53.1 to −46 mV. As for the results in Table 1, column (g), they can not be explained by assuming a single file mechanism as this will increase the theoretical ratio.

Eqn. 32 is also written as

$$\dfrac{\phi_i^{II \to I}}{\phi_i^{I \to II}} = \exp\left(-\dfrac{z_i F}{RT}\right)(V - V_i^*) \tag{38}$$

Ion permeability

where

$$V_i^* = \frac{RT}{z_i F} \ln \frac{[i]_{II}}{[i]_I}$$

Considering V_i^* as a reversal potential, i.e. as the membrane potential at which the net flux ϕ_i is zero, rather than as a Nernst potential, to explain the data on Table 1, column (g), we may write Eqn. 32 as follows.

$$\frac{\phi_i^{II \to I}}{\phi_i^{I \to II}} = \frac{[i]_{II}}{[j]_I \alpha + [i]_I} \exp\left(-\frac{z_i F}{RT}\right) V \tag{39}$$

where i and j are two ion species using the same channel and α is a selectivity coefficient. Thus, column (j) in Table 1 gives the predicted ratios assuming a selectivity coefficient of 0.01.

(ii) Permeabilities from the Goldman equation

To illustrate the use of Eqn. 35 to estimate resting permeability let us consider the data in Table 2 on resting potentials in giant barnacle muscle fibres under intracellular perfusion [12].

First the effects of increasing $[K^+]_{II}$ on the resting potential of the barnacle muscle fibres which were internally perfused with a solution of 200 mM K acetate were measured. From these measurements the following ratios were obtained

$$P_{K^+} : P_{Na^+} : P_{Cl^-} = 1.0 : 0.05 : 0.1.$$

Second, the effects of different cations from the inside were measured. The data is summarised in Table 2.

TABLE 2
Permeability ratios for the membrane of the giant barnacle muscle [12]

i	$[i]_I$	V	$\frac{P_i}{P_{K^+}}$
	mM	mV	
Li^+	200	−29.7	0.48
Na^+	200	−37.6	0.65
K^+	200	−48.4	1.00
Rb^+	200	−35.5	0.60
Cs^+	200	−23.6	0.37
NH_4^+	200	−24.0	0.38
$Tris^+$	200	−31.5	0.51
TEA^+	200	−21.0	0.34

$[Na^+]_{II} = 430$ mM, $[K^+]_{II} = 10$ mM.

The main conclusion to be drawn from Table 2 is that the resting membrane permeabilities follow the sequence

$$P_{K^+} > P_{Na^+} > P_{Rb^+} > P_{Li^+} > P_{Cs^+}.$$

(b) Permeability changes in stimulated electrically excitable cells

(i) Sodium inflow in axons under voltage clamp
The experimental results to be described here were done using techniques whereby a single giant axon of the squid (*Dosidicus gigas* and *Loligo forbesii*) could be internally perfused and voltage clamped. External application of ^{22}Na ions and collection of the perfusate allowed determinations of the sodium inflow. From measurements of the membrane current during, and after, the application of voltage clamp pulses the inward sodium flux as predicted by the equations of Hodgkin and Huxley could be calculated and compared to the measured flux [17–20].

Fig. 3 shows the results of measurements of the total inward tracer sodium flux (open circles) resulting from voltage-clamped depolarising pulses of about 3 ms duration as a function of the absolute membrane potential during the pulse. (Potential is referred to the external solution as ground. Flux is plotted upward but inwardly directed currents would be negative quantities.)

The filled circles are the predicted fluxes as calculated from the membrane current measurements.

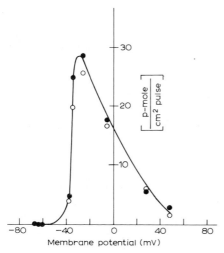

Fig. 3. Comparisons between the measured tracer sodium flux (○) and electrically measured ionic flux during the early transient current following a depolarising voltage-clamp pulse of about 3 ms duration (●). Axons from *Dosidicus gigas* internally perfused with 550 mK KF, pH 7.3 immersed in K-free artificial sea water. Temp. 17°C. Control experiments with tetrodotoxin were used (using the same axon) to subtract current components due to potassium. (Used with permission from Atwater et al. [17].)

The most important conclusion to be drawn from this experiment is that the inward current recorded from a giant axon in response to a controlled displacement of the membrane potential in the depolarizing direction is indeed carried by Na^+ as indirectly shown by Hodgkin and Huxley.

(ii) Timing the flux during a rectangular voltage clamp pulse
The results of Fig. 3 are for the total flux per cycle. In order to compare the flux which occurs *during* the pulse, for different pulse durations it is necessary to subtract that component which occurs during the tail following the pulse. Control experiments demonstrate that this current is indeed sodium and for these experiments TEA was added to the internal perfusion fluid with the result that potassium channel currents were virtually eliminated. The filled circles of Fig. 4 are a plot of the measured inward sodium flux as measured with the radioactive ^{22}Na ions *minus* the tail component as computed from the measured membrane current. The dotted line is the inward flux during the long pulse shown in part A of Fig. 4 as calculated from the current.

(iii) Timing the sodium and potassium permeability changes during an action potential
After timing the sodium flux during a voltage clamp pulse by direct measurement with radioactive tracers it immediately followed that to do something similar during the course of an action potential would be very desirable and a number of possible approaches were considered. The one described here originated with Rojas and Bezanilla and in retrospect seems simplicity itself. Somehow the action potential

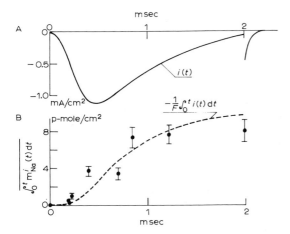

Fig. 4. Extra influx of sodium during the pulse as a function of the pulse duration for experiment TN-55. (A) Membrane current record obtained during the last run in experiment TN-55 corrected for the capacitative transient. This curve was integrated to give the dashed curve in B. (B) Each point represents the extra influx measured by tracers minus the influx during the "tail" calculated from current records. (Used with permission from Bezanilla et al. [18].)

must be stopped, at various times, and measurements of inward flux performed as for the voltage clamp experiments described above. A squid axon is prepared for simultaneous internal perfusion and voltage clamping. With the single pole-double throw switch in the open loop position the total membrane current is zero. Applying a pulse of current through the resistance connected to the internal axial wire results in the production of a non-propagated action potential.

Under these space-clamped conditions the action potential is referred to as a "membrane" action potential. With the switch shown in Fig. 5a in the voltage-clamp position, the membrane potential is under control. The problem was to switch from zero current to voltage-clamp mode quickly at various times during the membrane action potential.

Thus, the potential at which the membrane will be clamped during the control period is preset, an action potential is initiated and then stopped suddenly at the desired moment.

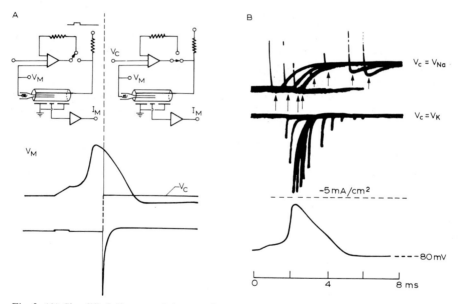

Fig. 5. (A) Simplified diagram of the experimental procedure. Upper part: a simplified diagram of the system used to switch on the membrane potential control system. Lower part: the recorded membrane potential and membrane current. On the left side of Fig. 5A the membrane is in open loop current clamp condition because the switch at the output of the control amplifier is connected to the auxiliary feed-back loop. The action potential is excited by a short pulse of voltage through a very high resistor connected to the axial wire. On the right side of Fig. 5A the switch connects the output of the control amplifier to the axial wire and the membrane is under voltage-clamp control. The free course of the action potential is interrupted and the recorded potential is equal to the command potential. Simultaneously membrane currents are recorded as shown in the bottom of the Fig. 5A. (B) The superimposed traces shown in the upper part were obtained with V_c set at 60 mV. For the records in the middle V_c was set at -80 mV. (Used with permission from F. Bezanilla et al. [19].)

Ion permeability 79

Measuring the membrane current at a very short time (after the voltage clamp has charged the membrane capacitance) for various potentials during the control period yields a current-voltage curve for the membrane at that time. It was found that these current-voltage curves were approximately linear so that the total membrane conductance is a meaningful, and simple, quantity which we could measure as a function of time during the membrane action potential. The curve of membrane conductance vs. time during the action potential was found to have a time course similar to that found by Cole and Curtis using impedance measurements at 20 KHz.

If the membrane potential during the control period is close to the reversal potential for potassium channels (see above) then no net current should be flowing through these channels and the initial current just following the interruption would be proportional to the conductance of the sodium channels. These transient currents following interruption are shown just above the action potential in Fig. 5b.

The upper set of curves in Fig. 5b are the currents following interruption of the action potential to a membrane potential close to the reversal potential for the sodium channels. Similar considerations apply. If the instantaneous-current voltage curves are reasonably linear then, with corrections for series resistance and leakage, the points indicated by the arrows give the time course of the potassium channel conductance during the action potential.

(iv) Timing the sodium flux during an action potential

All of the methods mentioned above were combined by Atwater et al. [17] to get an independent measure of the net flux during the course of a "membrane" action potential.

Stimulating the perfused axon ten times/s and collecting the perfusate about every 200 s the action potential was interrupted at various times. The membrane current during the period of voltage control after the interruption was recorded and the component of the flux during the "tail" subtracted from the net measured flux.

For details the original papers must be consulted. In Fig. 6 the results of several experiments are presented, showing the extra sodium influx as a function of the time at which the action potential was interrupted compared to the predictions as calculated from the equations of Hodgkin and Huxley. To get the time course of the sodium entry during the action potential, one may differentiate the curve in Fig. 6.

There are no longer any experimental or logical reasons to doubt that under ordinary conditions the action potential of the squid axon membrane does indeed arise, as suggested by Hodgkin, Huxley and Katz in 1949, from the membrane conductance changes to sodium and potassium ions and the inward flow of sodium ions occurs with a time course as calculated from the equation of Hodgkin and Huxley in 1952.

(c) Selectivity ratios for monovalent cations

(i) Permeability ratios from the reversal potential

This section deals with the cationic selectivity of the sodium channel of the squid

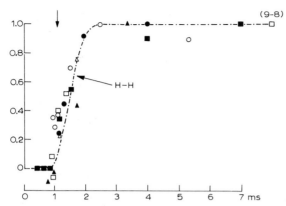

Fig. 6. Normalised integrated sodium influx during the course of a membrane action potential determined using radioactive sodium and action potential interruption. The curve is not drawn through the points but was calculated using the HH equations.

axon membranes. This is measured as a permeability ratio.

Chandler and Meves applied Eqn. 35 using the reversal potential for the early currents measured under voltage clamp conditions and calculated the following [21]

$$P_{Li} : P_{Na} : P_K : P_{Rb} : P_{Cs} = 1.1 : 1.0 : 1/12 : 1/40 : 1/61$$

This work provided evidence suggesting that the Na-channel can be characterised by a very specific sequence of alkali cation selectivity which, as we shall see later on, is different than that of the K-channel.

Similar voltage-clamp studies have been carried out in myelinated nerve fibres and the permeability sequence determined is given in Table 3 [23]. It should be pointed out here that Hille extended the series of cations to include some organic species. It was pointed out that in the nodal membrane of myelinated nerve fibres the permeabilities to Cs^+, Rb^+, Ca^{2+} and Mg^{2+} were too small to measure.

In order to compare the selectivity properties of the Na-channels let us examine unpublished results obtained using the giant axon of squid and some of the critical cations in Table 3.

There are several reports which indicate that small organic cations can maintain excitability of squid giant axons in sodium-free solutions [24]. Binstock and Lecar [25] carried out voltage-clamp experiments using NH_4^+ and found that NH_4^+ permeates both Na and K channels.

Fig. 7 illustrates the results of a typical experiment. Fig. 7A shows a family of voltage-clamp current records for an intact fibre bathed in 230 mM Na-saline. Fig. 7B shows a family of membrane current records when the fibre was bathed in 230 mM hydrazine-saline. Fig. 7C shows a family of membrane current records when the fibre was bathed in the 230 hydrazine-saline after the addition of 20 mM TTX. Fig.

TABLE 3
Permeability ratios for monovalent cations in the sodium channel of nerve

Node of Ranvier [23]		Giant axon of squid	
Ion	P_{ion}/P_{Na}	Ion	P_{ion}/P_{Na}
Sodium	1.000	Hydrazine	1.400
Hydroxylamine	0.940	Lithium	1.100
Lithium	0.930	Sodium	1.000
Hydrazine	0.590	Hydroxylamine	0.500
Thallium	0.330	Guanidine	0.32
Ammonium	0.160	Potassium	0.04
Formamidine	0.140	Aminoguanidine	0.04
Guanidine	0.130	Caesium	0.01
Hydroxyguanidine	0.120	Methylamine	<0.001
Potassium	0.086	Rubidium	<0.001
Aminoguanidine	0.060	Phenylhydrazine	<0.001
Methylamine	<0.007		

P_{ion}/P_{Na} calculated from reversal potential and Eqn. 35.

7D shows a family of membrane action potentials obtained in the same hydrazine solution but after removal of the TTX and for different depolarising and hyperpolarising current clamp pulses. There is a TTX-sensitive inward current. Neither the kinetics nor the maximum value of the outward currents are affected by hydrazine in a noticeable manner. These results suggest that hydrazine permeates the axolemma through the sodium system.

Fig. 8 shows the current-voltage relations (calculated from families of current records shown in Fig. 7) for the early conductance change and the delayed conductance change. It can be seen that in Fig. 8 the I–V curves for the potassium system are not considerably affected by hydrazine whereas the I–V curves for the early permeability change are modified. The size of the maximum inward current is reduced and the reversal potential is decreased. The slope conductance for large depolarisations (when the early outward current is mainly carried by internal sodium) is slightly reduced suggesting that it is harder for the Na ions to wash away the hydrazine ions that are filling the channel.

This comparison of the amplitude of the currents in a test and in a control solution can be used to calculate ionic selectivity of the sodium channel provided that the sodium outward currents are not affected by the presence of the test cation in the external solution. However, these results (amplitude of the outward currents) suggest that the sodium outward currents are somehow affected by hydrazine. These effects have not been observed in the Ranvier node.

Another way to estimate the selectivity from voltage-clamp measurements is to use the "reversal potential" (zero early current) [21,22] and calculate the ratio P_i/P_{Na} of the Goldman equation which would account for the change in reversal potential induced by the organic cation. The ratio calculated in this last case is

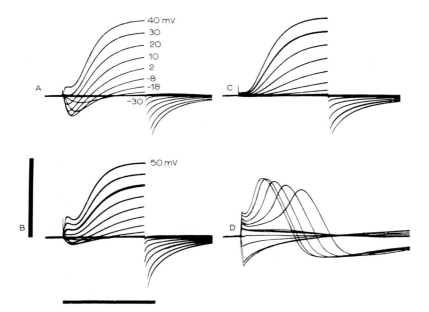

Fig. 7. Experimental protocol to measure selectivity ratios in the giant axon of squid. (A) Set of superimposed membrane current records in response to depolarising pulses to −60, −50, −40, −30, −18, −8, 2, 10, 20, 30 and 40 mV with the fibre in 230 mM Na$^+$-sea water. (B) Set of superimposed membrane current records in response to depolarising pulses to −40, −30, −18, −8, 2, 10, 20, 30, 40 and 50 mV with the fibre in 103 mM hydrazinium$^+$-sea water. (C) Set of current records after the addition of 50 mM TTX to the hydrazinium$^+$-sea water. (D) Superimposed membrane action potentials and passive membrane response for depolarising and hyperpolarising current pulses of increasing size, with the fibre in hydrazinium$^+$-sea water after the removal of TTX. Calibrations: Vertical represents 2.75 mA/cm^2 for A, B, C and 100 mV for D. The horizontal calibration represents 10 ms for A, B, C and D. Henderson and Hasselbach's equation for a proton acceptor of the type R-NH$_2$,

$$\text{R-NH}_3^+ \rightleftharpoons \text{H}^+ + \text{R-NH}_2,$$

$$K_a^1 = \frac{\text{R-NH}_2\,\text{H}^+}{\text{R-NH}_3^+},$$

as

$$\text{pH} = \text{p}K_a^1 + \log\frac{\text{R-NH}_2}{\text{R-NH}_3^+}$$

is used to calculate the concentration of hydrazinium (assuming a pK_a^1=7.97 at 3.5°C) as 103 mM. (Unpublished records from E. Rojas and R.E. Taylor.)

independent of the fraction of the sodium channels that are open.

We have used the reversal potential method as follows:

$$V_i^* = RT/F \ln \left(P_i[i]_{\text{II}}/(\cdot/\cdot)\right)(P_{\text{Na}}[\text{Na}]_{\text{I}} + P_{\text{K}}[\text{K}]_{\text{I}})$$

Ion permeability

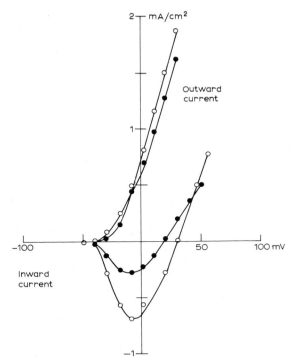

Fig. 8. Current-voltage relationships in Na$^+$-sea water (○) and in hydrazinium$^+$-sea water (●).

$V^*_{Na} = 30$ mV for $[Na^+]_{II} = 230$ mM;

$V^*_{Hy^+} = 19$ mV for a calculated $[Hy^+] = 103$ mM

$P_{Hy^+}/P_{Na} = 1.4$ in the sodium channel. Holding potential was set at -71 mV. Maximum Na$^+$ conductance 28.6 m mho/cm^2. Maximum hydrazinium$^+$ conductance 15.9 m mho/cm^2. (Unpublished data from E. Rojas and R.E. Taylor.)

where V^*_i is the reversal potential measured with the test solution.

It is always possible to compare the reversal potentials for two consecutive runs with two different solutions to obtain the following relationship:

$$V^*_{Na} - V^*_i = RT/F \ln \left(P_{Na}[Na]_{II}/P_i[i]_{II} \right)$$

where V^*_{Na}, V^*_i, T, $[Na]_{II}$ and $[i]_{II}$ are quantities experimentally measured. One can calculate the ratio P_i/P_{Na} which is a measure of the selectivity of the channel. Table 3 summarises data for the giant axon.

Table 4 gives the permeability ratios for various cations in the potassium channel of the node of Ranvier [23]. These values should be compared with those in Table 3

TABLE 4
Permeability ratios for monovalent cations in potassium channel calculated from reversal potentials [23]

Ion	P_{ion}/P_K	Ionic radii Å
Lithium	<0.018	0.60
Sodium	<0.010	0.95
Potassium	1.00	1.33
Thallium	2.3	–
Rubidium	0.91	1.48
Ammonium	0.13	–
Hydrazine	<0.029	–
Hydroxylamine	<0.025	–
Caesium	<0.077	1.69
Methylamine	<0.021	–
Formamidine	<0.020	–
Guanidine	<0.013	–

measured in the same preparation. It is clear that the permeability to Na⁺ in the potassium channel is very small. It should be noted that ammonium ions can permeate the nodal membrane using both channels. This is also the case for the membrane of the giant axon of the squid.

(ii) Permeability ratios from tracer fluxes [78]
Fig. 9 presents the results of an experiment designed to determine the voltage dependence of the permeability ratio. In this case we measured the voltage dependence of the sodium net extra flux and the unidirectional fluxes of calcium, $\Delta\phi_{Ca}^{II\rightarrow I}$, and sodium $\Delta\phi_{Na}^{II\rightarrow I}$. Vertical axis represents the extra influx in pmol/cm² pulse for sodium either measured with ^{22}Na or calculated by integration of the inward currents. The ordinate for the extra calcium influx is proportional to the extra sodium influx as follows

$$\Delta\phi_{Ca}^{II\rightarrow I} = 30.3 \cdot 10^{-4} \Delta\phi_{Na}^{II\rightarrow I}.$$

The horizontal axis represents the membrane potential during the pulses.

The results of measurements of calcium influx using ^{45}Ca in those experiments for which we could estimate the sodium influx from ^{22}Na simultaneously, indicate that the quantity

$$\alpha^* = \frac{2[Ca]_{II}\Delta\phi_{Na}^{II\rightarrow I}}{[Na]_{II}\Delta\phi_{Ca}^{II\rightarrow I}}$$

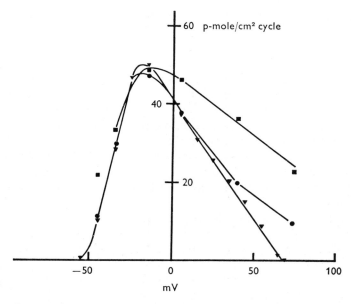

Fig. 9. Voltage dependence of the selectivity ratio P_{Na}/P_{Ca}. $[Na]_{II}=460$ mM. $[K]_I=275$ mM. $V^*_{Na}=70$ mV giving a selectivity ratio $P_{Na}/P_K=11$. ▼, estimated extra sodium entry; ●, measured extra sodium entry; ■, measured extra calcium entry. (Reproduced from Rojas and Taylor [78].)

is close to 100. Thus, the selectivity ratio calculated with

$$P_{Na}/P_{Ca} = \alpha^*/\left[\exp(\psi_{x=\delta}F/RT) + \exp(\psi_{x=0}+V)F/RT\right]$$

where $V=V_p$ in Fig. 9 is nearly 200.

In the model proposed by Hille (to be discussed in Section 4b (i)) to explain some of the properties of the selectivity filter of the sodium channel, sodium ions permeate the channel together with one water molecule forming a permeating unit with its largest diameter close to 5.5 Å. The penetration of the selectivity filter takes place in such a way that an oxygen within the channel acts as a hydrogen bond acceptor to the water molecule of the unit, 5 Å away from a negative charge at the entrance of the filter (see Fig. 14) which attracts the Na^+ part of the unit. It is not difficult to show that a calcium ion with one water molecule could also constitute a permeating unit since its crystal radius is 0.99 Å and with one water molecule would barely fit the narrow part of the selectivity filter. However, this argument should also be applicable to Mg^{2+} with an even smaller crystal radius of 0.65 Å. Accordingly, Mg^{2+} should permeate the sodium channel but the evidence indicates the contrary. It could be argued, however, that the high energy required to partially remove the water molecules of hydration from the magnesium in solution would constitute the impediment against Mg^{2+}.

4. Mechanisms

(a) The concept of ionic channel

The development of the concept of ionic channel started with the realisation by Bernstein that cellular excitability was a property of the membrane. The starting point at the experimental level was the observation by Cole and Curtis that, concomitant with a propagated electrical impulse (manifestation of cellular electrical excitability) in the squid giant nerve fibre, a decrease in the electrical resistance took place with no detectable change in the membrane capacitance. This result lent strong support to Bernstein's concept and clearly indicated that the most plastic components of the axolemma, the proteins, underwent structural transitions leading to a transient increase in ionic fluxes.

The problem of molecular organisation of electrically and chemically excitable membranes is of fundamental importance in explaining Cole and Curtis's observation and understanding the various mechanisms possible that could generate conductance changes in cellular membranes.

For example, it is known that proteins are important membrane constituents; the ratio by weight of proteins to lipids varies between 1.5:1 and 4:1 for different membranes. Since the bulk of the membrane capacitance is associated with the lipid phase, the constancy of the membrane capacitance during the passage of an electrical impulse may be interpreted assuming that only a minute fraction of these proteins (a few molecules scattered throughout within the nerve membrane) underwent transitions inducing dramatic changes in conductivity.

(i) Hodgkin–Huxley channels [26–29]

The first experimental results which suggest the existence of two different membrane structures controlling the movements of sodium and potassium during a nervous impulse was produced by Hodgkin et al. [26]. They observed that if a squid giant axon is subjected to sudden displacements of the membrane potential in the positive direction, the resulting membrane current i_m could be analysed as the sum of three components,

$$i_m = i_{Na} + i_K + i_L$$

where i_{Na} represents a transient inward component, i_K a sustained outward component and i_L a small leakage current. Hodgkin and Huxley [29] described the kinetic properties of these components using the following equations.

$$i_{Na} = \bar{g}_{Na} m^3 h (V - V^*_{Na}), \tag{40}$$

$$i_K = \bar{g}_K n^4 (V - V^*_K), \tag{41}$$

$$i_L = \bar{g}_L (V - V^*_L). \tag{42}$$

Ion permeability

Hodgkin and Huxley postulated that m, h, n are first-order processes. Thus,

$$\frac{dj}{dt} = \alpha_j(1-j) - \beta_j j \tag{43}$$

for $j = n$, m, h. It was also assumed that the rate constants α_j and β_j for the transitions depend on the absolute value of the membrane potential during the rectangular pulse. Therefore, the kinetic characteristics of the ionic currents, namely, i_{Na} and i_K, are different. Following Eqn. 40 i_{Na} is the product of two variables. Thus, i_{Na} is a second-order process (activation followed by inactivation) and i_K is a first-order process (activation).

In the description of the conductance change, Hodgkin and Huxley [27,29] assumed that the sodium and potassium systems were two independent and non-interacting molecular processes.

Fig. 10 illustrates the results of one experiment that lent strong support to the hypothesis that activation and inactivation are independent membrane transitions. A giant axon was intracellularly perfused with a caesium fluoride solution and i_{Na} for various depolarising voltage-clamp pulses was recorded. Parts A, B and C present three superimposed i_{Na} records made 1, 2 and 5 min after the addition of alkaline

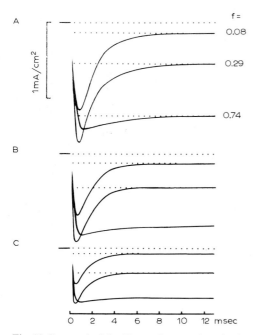

Fig. 10. Removal of the Na conductance inactivation by internal application of alkaline proteinase b. (A) Potential during the pulses was 0 mV. f represents the fraction of the total number of Na-channels with the inactivation removed. (B) Potential during the pulses 20 mV. (C) Potential during the pulses 40 mV. (Unpublished records from E. Rojas and B. Rudy.)

proteinase b to the internal solution. The fraction of i_{Na} with the inactivation removed is shown next to the i_{Na} record. It may be seen that after 2 min of internal application of the proteinase i_{Na} is larger than the control. The observed increase is predicted by the Hodgkin–Huxley model [29].

The controversy of whether i_{Na} and i_K pass through the membranes in separate structures or through a single set of pathways which conduct first i_{Na} and then i_K [31] was resolved by showing that the total membrane current could be made to exceed the maximum value of i_{Na} or i_K, as a single pathway could not simultaneously conduct i_{Na} and i_K [32,33].

The sodium and potassium systems were proven to be non-interacting molecular processes by the experimental observation that i_{Na} could be blocked reversibly by external application of the drug tetrodotoxin without affecting either the kinetic or the steady-state properties of i_K [34]; the equilibrium dissociation constant for this effect was close to 3 nM [30]. Further support for this concept of non-interacting systems was obtained from the demonstration of reversible blockage of i_K induced by tetraethylammonium, without noticeable effects on i_{Na} [35,36].

The direct measurement of the time course of the influx of Na^+ during a depolarising rectangular voltage-clamp pulse, and during an action potential discussed in Section 3b, resolved the question of specificity of the voltage-clamp measurements, and the concept of two independent kinetic components in the membrane currents, each representing an ionic pathway with different ionic selectivity, was established without doubt.

The first objective of our work was to redetermine the kinetics and steady-state properties of g_{Na} and compare them with those of the gating currents. For each fibre, we measured the sodium currents [37,38] during turn-on and turn-off of g_{Na} and calculated the voltage dependence of the time constant for the activation of the sodium conductance.

τ_m values obtained according to the procedure outlined elsewhere [38] are plotted in Fig. 11A. The τ_m values from -140 to -50 mV were obtained from the i_{Na} transients during turn-off of g_{Na} with repolarisation by plotting on a semilogarithmic paper as a function of time the digitised values of i_{Na} smaller than 0.6 mA/cm^2. In this case, the trace was digitised every 10 μs. Each i_{Na} tail decayed with a single exponential time course and regardless of the value of V_h, always intercepted the time axis at time zero. This time constant was multiplied by three and after allowing for temperature differences (assuming $Q_{10} = 3$), was plotted in Fig. 11A.

The continuous curve drawn in Fig. 11A represents the least-squares fit of the experimental values. A computer program provided estimates for the coefficients in the empirical relationship

$$\tau_m = \frac{1}{A \exp BV + C \exp DV} \tag{44}$$

The following coefficients were obtained from the curve fitting shown in Fig. 11A: $A = (2.60 \pm 0.095)/\text{ms}$; $B = (25.39 \pm 1.610)/\text{V}$; $C = (0.40 \pm 0.055)/\text{ms}$; and $D =$

Ion permeability

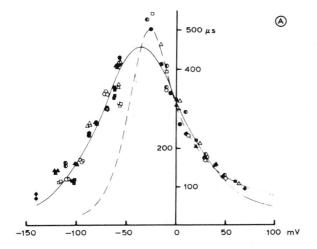

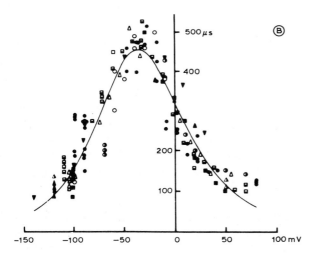

Fig. 11. (A) Voltage dependence of τ_m for the activation of g_{Na}. For $V_p < -50$ mV, the time constants were evaluated from "tail" current records. The different symbols represent separate experiments. (B) Voltage dependence of the relaxation time constants of the gating currents. All results were normalised to a standard temperature of 6.3°C assuming $Q_{10} = 3$. For $V_p > -50$ mV, $\tau(V)$ was measured during the pulses; for $V_p < -50$ mV, it was measured from the "tail" of the gating current and V_h was varied. (○, ■, ▲, ▼, △, ◐) in fibres perfused with high Cs$^+$ (the different symbols represent different experiments); (▲, ▬, ●) in fibres perfused with low Cs$^+$ (50 mM CsF plus 900 mM sucrose), plotted with membrane potentials shifted 9 mV in a negative direction ($\tau(V)$ values from Table 3 in [41]). The lines were computed to give a least-squares best fit of the points in A or in B (for parameters see text). (Adapted from Keynes and Rojas [38].)

$(-27.0 \pm 1.524)/V$. The dashed curve in Fig. 11 represents the values of τ_m at 6.3°C calculated from the Hodgkin and Huxley equations [29] (Eqns. 20 and 21, resting potential taken as -57 mV; see Table 3 in [38]).

In each experiment before the gating current runs, we measured the $i_{Na} - V$ curve in 1/4 Na-SW, and after correcting for the effects of the uncompensated fraction of the measured resistance in series (for further details, see Keynes and Rojas [38]), we measured the reversal potential for the sodium currents, V_{Na}, and estimated the limiting value of $g_{Na,max}$.

In Fig. 12A we illustrate some of the $m_\infty - V$ curves obtained with perfused fibers in 1/4 Na-SW (m_∞ values were calculated as described in [38]). The solid curve shown in this figure was computed to give a least-squares fit with the points for potentials more negative than -20 mV in the form

$$m_\infty = \frac{1}{1 + C/A \exp(D-B)V} \tag{45}$$

with $C/A = (0.150 \pm 0.009)$ and $D - B = (-55.5 \pm 6.21)/V$. The dashed curve was computed to give the least-squares fit with all the points, with $C/A = (0.080 \pm 0.007)$ and $D - B = (80.1 \pm 9.10)/V$. In the perfused axon, however, the state of the membrane was better described by the constant field equation than by g_{Na}. Recalculation of m_∞ from P_{Na} shifted the curve about 10 mV in a positive direction [38].

(ii) Molecular transitions associated with the activation of channels

From the data discussed in the preceding section, there is no doubt about the applicability of the Hodgkin–Huxley kinetics to describe i_{Na}, namely m^3h kinetics for the squid axon when the holding potential before the test pulse is set near the physiological value of the resting potential (that is to say -70 mV). At high negative holding potentials, excellent fitting of the sodium current using Hodgkin–Huxley kinetics is preserved after introducing an initial delay in the start of the sodium current [38].

Regardless of the permeation mechanism involved, it seems unavoidable that one must postulate the existence of charges (either attached to carriers or to "gates" present near or in the channels) that are displaced with undetermined efficiency by changes in the electrical field across the membrane.

The main difficulty in proposing a definite test for this hypothesis was the lack of a method to measure directly the voltage-dependent rearrangements of charges preceding ionic conductance changes [39,40]. To be able to detect these molecular rearrangements it was necessary to eliminate the much larger sodium and potassium currents. This was done by various ionic substitutions of permeant ions by impermeant ones (on the outside, $Tris^+$ in place of Na^+ and K^+, isethionate$^-$ in place of Cl^-; on the inside Cs^+ in place of K^+ and F^- in place of intracellular free anions) [38,41].

Ion permeability

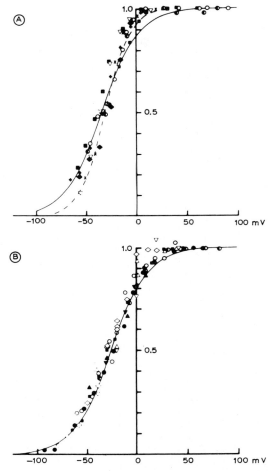

Fig. 12. (A) m_∞-V relationship for fibres perfused with high Cs$^+$ in 1/4 Na-SW. ◐, ■, ○, fibres perfused with high Cs$^+$ and bathed in 1/4 Na-SW made with chloride; +, ×, ◇, ▽, fibres perfused with high Cs$^+$ in 1/4 Na-SW made with isethionate, with 15 mM CaCl$_2$ (the calcium activity in this solution was identical to that of the 1/4 Na-SW made with chloride). (B) Voltage dependence of the steady-state distribution of charges. ◐, ■, ○, ▲, ▼, △, ▽, fibres internally perfused with high Cs$^+$ (the different symbols represent separate experiments); ◇, ●, fibres internally perfused with low Cs$^+$ and all experimental points shifted toward more negative values by 9 mV. All experiments were carried out in an Na- and K-free saline made with isethionate in place of chloride. Curve was computed by the least-squares fit with the experimental points (see text). (Adapted from Keynes and Rojas [38].)

A typical record of displacement currents, under the experimental conditions given above is depicted in Fig. 13A [42].

The basic properties of the remaining asymmetrical currents shown in Fig. 13 (middle records) may be summarised as follows:

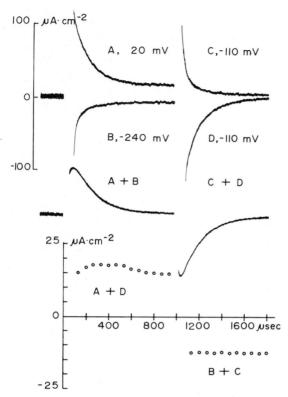

Fig. 13. Single sweep displacement current records. Top: superimposed membrane current records for 0.9 ms pulses from: −110 to 20 mV (transient A during pulse, transient D after pulse) and to −240 mV (transient B during pulse, transient C after the pulse). Middle: records shown in the upper part after subtraction of the linear capacitative and leakage currents (A+B and C+D). Fibre perfused with 300 mM CsF, 400 mM sucrose, 5 mM Tris–HCl at pH 7.3 in potassium and sodium-free saline at 6.5°C. Bottom: summing A+D and B+C reveals the asymmetry in the transients A and D and the perfect linearity in the transients B and C.

The on-response, during a pulse taking the membrane potential from a negative towards a positive value, is an outward current (outward movement of positive charge or inward movement of negative charge) characterised by a fast rise to a peak value (time to peak less than 50 μs in the giant axon and too small to measure in the node of Ranvier) [43,44], followed by a monotonically declining current. Immediately after the membrane potential is switched back to a negative holding potential, the off-response obtained is an inward current characterised by an exponential decay towards the base line. For brief pulses, the size of the charge transferred in one direction during the pulses is exactly the same as that transferred in the opposite direction after the pulse. A single-exponential function is used to fit the on response and the relaxation time constant, τ_{on}, is voltage-dependent. The off response can be fitted with a single exponential and if the τ_{on} and τ_{off} values obtained are lumped together in one single distribution, the voltage dependency obtained may be represented by a symmetrical bell-shaped function, $\tau(V)$, of membrane potential, with a maximum at about -35 mV in both types of fibres. The data for the squid giant axon are shown in Fig. 11B.

The steady-state distribution of the charge transferred, during positive potential steps, from a negative holding potential of -100 mV, is a sigmoidal curve, $Q_{on}(V)$, saturating at positive potentials exceeding 25 mV with a midpoint in the neighbourhood of -25 mV. The Boltzman's factor for the distribution is 19 mV in the squid giant axon and 15 mV in the node of Ranvier. The data for the squid giant axon are shown in Fig. 12B [38].

If we now compare $\tau(V)$ for the asymmetry current with $\tau_m(V)$, as defined by Hodgkin–Huxley [29] namely, $(\alpha + \beta)^{-1}$, both curves are found to lie on a very similar bell-shaped curve, with a peak of just under 500 μs at 6°C in the case of the giant axon [38] and 200 μs at 10°C in the case of the node of Ranvier [44].

Detailed comparisons between the steady-state properties of the sodium conductance and the steady-state charge distribution have been made only in the case of the squid giant axon, these are shown in Fig. 12 [38]. The midpoint of the $m_\infty(V)$ curve lays at -35 mV if calculated directly with the conductance data obtained from sodium-current data and, at -25 mV if calculated using permeability data calculated from I_{Na} using the constant field model for the electrical profile across the membrane. Therefore, both steady-state distributions have exactly the same position in the potential axis. The $m_\infty(V)$ curve may be fitted, using Boltzman's law, with a kT/a factor equal to 19 mV (kT/a, where k is the Boltzman constant and a is the effective valence of the gating particle), again in perfect agreement with the slope for the $Q_{on}(V)$ curve.

The essential points of disagreement are circumscribed to the questions of whether Q_{on} remains equal to Q_{off} for a pulse lasting longer than the time to peak of i_{Na} and whether the on response obeys first- or higher-order kinetics.

For the squid giant axon some preliminary data presented by Armstrong and Bezanilla showed that for pulses exceeding the time to peak of i_{Na} corresponding to that pulse, Q_{off} decreases with the duration of the pulse from Q_{on} to a third of Q_{on} [45,46]. A similar observation has been reported in a more extensive study using myelinated nerve fibres [44]. This reduction in Q_{off} with pulse length gave rise to the notion that the gating currents were inactivated by the positive potential during the test pulse [44].

The question of whether the on response obeys first- or higher-order kinetics has been studied systematically only in the case of myelinated nerve fibres [43,44]. The results showed that the net charge transfer during and after pulses of equal size can be analysed in terms of a single exponential component. It should be pointed out here that with the data-acquisition system used in these experiments the maximum error in the determination of the charge is about 8% of the average maximum charge displaced in the myelinated nerve fibre. Therefore, the maximum contribution of any other non-exponential component to the charge $Q(t)$ will be 8% of the maximum charge [43,44]. This conclusion supports the notion that the charge movement is of first-order transition and bears important implications as to the mechanism.

(b) Mechanisms for permselectivity

Table 4 presents a list of the ionic radii for some of the monovalent cations used in the experiments described in this chapter. Baylis [47] stated that

> "if one ion be larger than the other, there might be only a small number of pores permeable to the larger ion, so that for a considerable time an electromotive force might exist."

Conway related the hydrated radius of the ion to the permeability coefficient. He concluded that the critical size is at a hydrated radius of 4 Å [48].

Mullins recognised that the barrier to movement of a heavily hydrated ion into a narrow channel is the energy required to dehydrate the ion. The size of the energy barrier would decrease if, as water molecules are shed from the ion, they are replaced by groups from the wall of the channel. Mullins suggested cylindrical pores of 3.65 Å radius for Na^+ and 4.05 Å for K^+, sizes equal to the crystal radius given in Table 4 plus the diameter of one water molecule (2.72 Å) [49].

These three examples of theoretical explanations of the selectivity are based on ion and channel dimensions and on ion–channel interactions.

Hille's model puts together these two basic concepts and explains a great number of experimental results [23,50–52].

(i) Hille's selectivity filter

The essential experimental observations supporting this model are as follows: TTX block sodium channels of nerve and muscle membranes at nanomolar concentrations [22,30,34,53,54]. TTX acts only from outside and not when perfused inside axons [55] or muscle. The measurements of the rates of blocking and unblocking of i_{Na} by TTX together with measurements of the equilibrium blockage of i_{Na} by TTX show that TTX action involves a reversible one-to-one interaction with an external receptor to block Na-channels [53,56].

It has been proposed that a guanidinium group found in TTX forms the blocking complex by entering the Na-channel and becoming stuck there because the remaining ring structure of TTX is too wide to pass [57].

Hille extended this hypothesis, proposing that the part of the channel where TTX binds is the part of the channel conferring the selectivity properties [51]. The proposed structure included a 3 by 5 Å constriction of the Na-channel formed by a ring of six oxygen atoms. One group in the ring was supposed to be an ionised carboxylic group.

This hypothesis was extended to account for the results obtained with saxitoxin, STX, another potent Na-channel blocking toxin with two guanidinium groups in its structure [51].

Fig. 14 shows a diagrammatical representation of Hille's model. The hypothetical receptor interaction shown in Fig. 14 is drawn to show points of contact between

Ion permeability 95

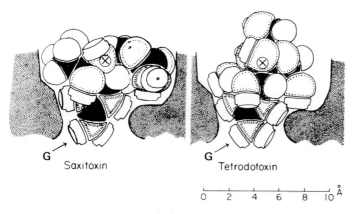

Fig. 14. Saxitoxin and tetrodotoxin on their receptors. Drawn from CPK models. The stippled areas represent the receptor in sagital section with narrow selectivity filter below. Most of the receptor is hydrogen bond accepting and there is a negative charge associated with the selectivity filter. A circled X has been drawn in the same position with respect to the receptor in the two cases. The X falls on an −OH group attached to an unusually electropositive carbon. G, guanidinium group. (Reproduced with minor modifications from [57].)

TTX and the filter and possible hydrogen bonding structures both in the TTX molecule and in the filter. To accommodate both toxins the external opening of the Na-channel must have a 9 by 10 Å cross section before narrowing to the 3 by 5 Å filter where the guanidinium group binds.

(c) Gating mechanisms

Let us propose that in series with the selectivity filter of the Na channel there is a system of charged gates located in the low dielectric region of the membrane [58]. These dipoles are arranged in two possible configurations. In one of them, the positive charge of the dipole induces a positive electric field in the Na channel. This electric field is an energy barrier for the flow of Na ions, and therefore in this state, the Na channel is closed. The level of the energy barrier in the Na channel depends on the number of gates remaining in the closed state. A decrease in the level of the energy barrier is produced by the removal of the positive end of the dipole from the blocking position to a membrane region from where it can no longer influence the Na channel. Assuming that Na ions can cross the channel only after all the hypothetical dipoles undergo a transition from state 1 to state 2 (state 1 is identified with blocking position), it is possible to estimate the number of gating particles per channel by direct application of Boltzman's law to both the steady-state sodium conductance curve and the steady-state distribution of gating charges. The minimum membrane potential change that produces an e-fold change in sodium conductance is 6.5 mV, and the minimum potential change required to produce an e-fold change in charge displacement is given in Table 5B as about 19 mV. The ratio therefore should be independent of any factor relating the efficiency with which the membrane potential displaces individual charges in the membrane and should give the coordination number involved in the process of activation of each sodium channel. Thus $19/6.5 = 2.9$ is consistent with the original Hodgkin–Huxley formulation, which is expressed in terms of an exponent equal to 3 for the m-process [29].

In order to confer specific properties on the gated sodium net flux (sodium current) using this gating mechanism, it is necessary to propose a specific physical model for the movement of the gates. In the case of rotary diffusion, in which the dipoles undergo a first-order transition over a single energy barrier the Na channels exist either in the fully conducting state or in the nonconducting state. This results from the postulated property that the gates, free of mutual interactions, are arranged in only two possible configurations.

(i) The two-state transition model [37,58]

During a potential step, an intramembrane charge movement takes place, which continues in time until a new equilibrium distribution is achieved. Taken together the results strongly suggest that the on-response in Fig. 13 follows an exponential time course, thus indicating that the charge movement is a first-order process. Assuming that the charges have to cross a single energy barrier, it can be shown that

TABLE 5
(A) Parameters for single-energy-barrier model calculated from kinetic and steady-state data [58]

Type of parameter	τ_m	τ	m_∞	$q(V_p)/Q_{max}$	Unit
V_0	−35.82	−41.57	−34.18	−26.05	mV
kT/a	−19.08	−19.49	−18.02	−18.52	mV
a	−1.27	−1.24	−1.27	−1.31	charges
x	0.52	0.56	–	–	
$\delta\epsilon/kT$	8.8	9.2	–	–	
$\alpha_0 = \beta_0 = 1$					

$kT = 24.2$ meV

(B) Comparison of the gating systems present in the nodal and giant nerve fibre membranes

Parameter	Unit	Node of Ranvier [44]	Squid axon [38]
Q_{max}	$e/\mu m^2$	17200	1882
Temperature coefficient			
for Q_{max}		1.0	1.0
for $\tau(V_p)$		2.4	2.5
V_0	mV	−33.7	−26.0
a	electronic charges	−1.65	−1.3
kT/a	mV	14.9	18.5

Ion permeability

the rate constants are exponential functions of the energy difference between the top of the barrier and the level from which the charges are being displaced. Thus,

$$\alpha = \alpha_0 \exp((\epsilon_1 - \epsilon_b)/kT), \tag{46a}$$

$$\beta = \beta_0 \exp((\epsilon_2 - \epsilon_b)/kT), \tag{46b}$$

where ϵ_b is the size of the energy barrier (see Eqn. 50) and k is the Boltzman constant. The relaxation time constant is therefore

$$\tau = 1/(\alpha + \beta). \tag{47}$$

The relaxation time constant of the asymmetrical displacement currents has a rather large temperature coefficient. Such a large temperature effect might be accounted for by introducing the temperature dependence in the energy barrier ϵ_b. However, it is possible to show that the coefficients α_0 and β_0 contain a dynamic frictional coefficient as a multiplying factor, and this coefficient certainly depends on temperature. In line with this interpretation, α_0 and β_0 can be expressed in terms of an empirical relationship, making α_0 and β_0 proportional to $\exp(2(t - 6.3)/kT)$ where t is in °C; this agrees fairly well with the temperature dependence of $\tau(V)$ measured experimentally [41,58].

Fig. 11B represents 132 values of $\tau(V)$ measured in 25 different fibres. All results were normalised to a standard temperature of 6.3°C. The empirical curve drawn through the points is of the form of Eqn. 5 with $A = (2.76 \pm 0.145)$/ms; $B = (22.445 \pm 2.140)$/V; $C = (0.342 \pm 0.084)$/ms; and $D = (-28.860 \pm 2.660)$/V.

In previous publications, we have presented a formal treatment of the steady-state distribution of charges in terms of Boltzman's law with two allowed energy states. It was shown that for $V_h = -100$ mV, the energy difference between these states is given by

$$\epsilon_1 - \epsilon_2 = kT[q(V_p, \infty)/(Q_{max} - q(V_p, \infty))],$$

where Q_{max} is as has been defined for Fig. 12B, and $q(V_p, \infty)$ is the charge transferred during a membrane potential step from V_h to V_p. It was also shown that the energy difference for the steady-state distribution, $\epsilon_1 - \epsilon_2$, is a linear function of V_p (see Fig. 9a in [38]).

Fig. 12B shows the voltage dependence of the normalised steady-state distribution of charges, that is to say the $q(V_p, \infty)/Q_{max} - V$ curve. It may be seen that above $+40$ mV, there was no further asymmetrical transfer of charge, so that saturation of the system of mobile charges had indeed been achieved. The curve in Fig. 12B was computed by the least-squares best fit with the experimental points (as for Fig. 12A), with $C/A = (0.2450 \pm 0.0088)$ and $D - B = (54.0 \pm 1.80)$/V.

As previously shown (see Fig. 15 in [41]), the value of V_p which makes $\epsilon_1 - \epsilon_2$ equal to zero, designated V_0, depends on experimental conditions. Thus, lowering the ionic strength of the internal solution unshields surface charges on the membrane, which is equivalent to adding to the energy level ϵ_1 an electrostatic energy equal to $\psi_i \cdot \sigma_i$, where ψ_i is the inner surface potential and σ_i is the inner surface charge density. Therefore the fact that $V_0 \neq 0$ indicates the presence of an asymmetry in the surface potentials.

From Eqns. 7a and 7b we obtain

$$\frac{\alpha}{\beta} = \frac{\alpha_0}{\beta_0} \exp[(\epsilon_1 - \epsilon_2)/kT]. \tag{48}$$

The relaxation time constant defined by Eqn. 47 and the steady-state distribution of charges $q(V_p, \infty)/Q_{max}$ (which is equal to $\alpha/(\alpha + \beta)$) represent two expressions for α and β in terms of measurable experimental quantities. Therefore it is possible to evaluate the rate constants at any given potential V_p. The results showed that the rate constants are exponential functions of V_p, and that the energy differences governing the rate constants are, to a first approximation, linearly dependent on the membrane potential.

Assuming that the space charge density within the low dielectric constant (ξ_m) region of the membrane is close to zero, integration of the Poisson equation

$$\frac{d^2\psi}{dx^2} = -\frac{4\pi}{\xi_m}\rho \tag{49}$$

leads to a constant gradient of potential or constant electric field. If we neglect image forces, the observation that not only $\epsilon_1 - \epsilon_2$, but also $\epsilon_1 - \epsilon_b$ and $\epsilon_2 - \epsilon_b$ are linear functions of V_p can be used to define the energy barrier as

$$\epsilon_b = a(\chi(V + \psi_1) + (1-\chi)\psi_0) + \delta\epsilon, \tag{50}$$

where ψ_0 is the outer surface potential, χ is a reaction coordinate such that $0 < \chi < 1$, and $\delta\epsilon$ represents the nonelectrostatic contribution to the energy barrier. Then, writing

$$\epsilon_1 = a(V + \psi_1), \tag{51a}$$

and

$$\epsilon_2 = a\psi_0, \tag{51b}$$

we obtain

$$\epsilon_1 - \epsilon_2 = a[V - (\psi_0 - \psi_1)] \tag{52a}$$

or

$$\epsilon_1 - \epsilon_2 = a(V - V_0). \tag{52b}$$

Ion permeability

Also,

$$\epsilon_1 - \epsilon_b = a(1-\chi)(V-V_0) - \delta\epsilon \qquad (53a)$$

and

$$\epsilon_2 - \epsilon_b = a[-\chi(V-V_0)] - \delta\epsilon. \qquad (53b)$$

Eqns. 46a and 46b may be written in terms of Eqns. 53a and 53b as

$$\alpha = \alpha_0 \exp[(a(1-\chi)(V-V_0) - \delta\epsilon)/kT] \qquad (54a)$$

and

$$\beta = \beta_0 \exp[(-a\chi(V-V_0) - \delta\epsilon)/kT] \qquad (54b)$$

and therefore

$$\alpha/\beta = (\alpha_0/\beta_0) \exp[a(V-V_0)/kT]. \qquad (55)$$

It can easily be shown that the coefficients A, B, C and D used in empirical Eqns. 44 and 45 for Figs. 11 and 12 are:

$$\alpha = a_0 \exp[(-a(1-\chi)V_0 - \delta\epsilon)/kT], \qquad (56a)$$

$$B = a(1-\chi)/kT, \qquad (56b)$$

$$C = \beta_0 \exp[(a\chi V_0 - \delta\epsilon)/kT], \qquad (56c)$$

$$D = -a\chi/kT, \qquad (56d)$$

$$D - B = -a/kT, \qquad (56e)$$

$$A/C = \alpha_0/\beta_c \exp(-aV_0/kT). \qquad (56f)$$

Figs. 11 and 12 were used to calculate the various parameters in Eqns. 56a through 56f. The results of these calculations are summarised in Table 5A.

Table 5B summarises the data for comparison [43,44]. In order to calculate current densities for the experiments on the nodal membrane, the area was assumed to be 50 μm^2 [43,44]. It may be seen that (1) Q_{max} in the nodal membrane is about ten times larger than in the squid axon membrane; (2) the temperature coefficients for Q_{max} and for V_p are very similar for both membranes; (3) V_0 in the node is 7.7 mV more negative; (4) the effective charge of the dipoles, a, is somewhat larger in the nodal membrane, and (5) the least change in membrane potential to produce an e-fold change in the steady-state distribution of charges (kT/a) is 14.9 mV in the node and 18.5 mV in the squid axon [58].

These results suggest that the kinetic and steady-state properties are qualitatively similar in both membranes [38,41,43,44]. The steady-state distribution curve is steeper in the node ($kT/a = 14.9$ mV) than in the case of the squid giant axon ($kT/a = 18.5$ mV). However, the $P_{Na} - V_p$ curve is less steep in the node (7.1 mV for an e-fold change) than the $g_{Na} - V_p$ curve is in the giant fibre (6.5 mV for an e-fold change), giving a ratio of 2.1 (14.9/7.1), which is in excellent agreement with the kinetic data for the P_{Na} change in the node. The similarity seems too good to be merely a coincidence and strongly supports the idea that each channel incorporates a number of gating dipoles in its structure (three in the case of the giant fibre and two in the case of the nodal membrane), and that the sigmoid increase of g_{Na} during a depolarising pulse arises from this structural property of the channel, and not from cooperativity. It should not be forgotten that for giant axons perfused with caesium fluoride solutions, the m_∞ curves should be calculated from permeability data using constant field assumptions and not from sodium conductance data.

If, for the activation of each Na channel in the node two particles with an effective charge of 1.65 electron charges have to be displaced from blocking to open position, $Q_{max}/3.3$ equals 5212 Na channels/μm^2. In the case of the giant fibre, we have estimated the number of Na channels as 483 channels/μm^2. Therefore the density of Na channels is 12.5 times greater in the nodal membrane. However, the maximum sodium conductance in the node is about 15 nmho/μm^2 and in the giant axon about 1.2 nmho/μm^2. Therefore the ratio, $15/1.2 = 12.5$, is very similar to the ratio of densities, suggesting that the conductance of a single Na channel is the same in both preparations, namely, 2–3 pmho.

(ii) The aggregation-field effect model of Rojas [58]

We propose that sodium channels are not permanent structures but functional aggregates of integral proteins (proteins completely or partially embedded in the lipid matrix of the membrane, and held in the membrane mainly by hydrophobic interactions). We postulate that channel formation involves the following components: (1) Selectivity filter or amphipathic proteins (integral proteins with two parts, one hydrophilic which protrudes from the membrane, and the other hydrophobic, which is embedded in the lipid matrix located on the outside of the membrane and which serves as organisers of the sodium channel (see Fig. 15A). We identify this protein with the tetrodotoxin-binding protein of Levinson and Ellory with a molecular weight of 200 000. (2) Gating molecules or hydrophobic proteins (Fig. 15B) with a large dipole moment. From the relaxation time constant data for gating currents we estimate the molecular weight of 100 000 for these subunits.

Once the sodium channel is formed (Fig. 15C) the size of the aggregate is 500 000. The energy barriers within the pore formed between the segment of the organiser protruding into the hydrophobic part of the membrane and the gating units will determine whether Na^+ could go through, the driving force to set this motion being the electrochemical potential gradient for Na^+ (Fig. 15D).

It is possible to show that a positive electric field in the channel is sufficient to block the sodium flux. Thus, using the theory of rate processes [59] and assuming an

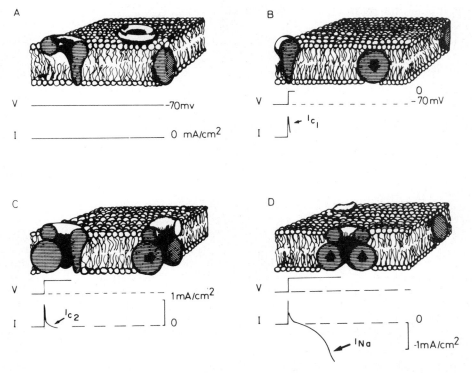

Fig. 15. "Photographs" of four stages during the transitions underlying sodium channel formation. (A) Rectangular portion of a nerve membrane at rest. A section through the hypothetical selectivity filter is shown. (B) Within the first 10 μs after a sudden decrease in membrane potential, Cole's capacity component charges generating Ic. (C) Rotation of gating dipoles starts immediately after changing V but the transition leading to the organisation of the sodium channel has already occurred. (D) I_{Na} starts when the channel has been formed and the charges on their gates do not represent an energy barrier. Reproduced from Rojas and Bergman [42].

arbitrary potential profile with a single barrier located at the unidirectional Na fluxes are equal to

$$\phi_{Na}^{I \to II} = |Na|_{II} \frac{kT}{h} l \exp\left(-\frac{U_0}{kT} - \frac{U_1}{kT} - \chi \frac{aV}{kT}\right) \tag{57a}$$

$$\phi_{Na}^{II \to I} = |Na|_{I} \frac{kT}{h} l \exp\left(-\frac{U_0}{kT} - \frac{U_1}{kT} + (1-\chi) \frac{aV}{kT}\right) \tag{57b}$$

where χ gives the location of the barrier (see Fig. 1), kT/h is the frequency factor, l is the jump length in front of the energy barrier, U_0 is the energy for the Na ion entering or leaving the pore, and U_1 is the height of the energy barrier in the pore.

Assuming that the partition coefficient equals

$$\beta = \exp(-U_0/kT) \tag{58}$$

and diffusion coefficient,

$$D = kT/hld \tag{59}$$

where d is the thickness of the membrane, the net stationary flux equals

$$\phi_{Na} = \frac{I_{Na}(V_p, \infty)}{F}$$

$$= \beta D/d \, \exp(-U_1/kT) \exp(-\chi aV/kT)\{|Na|_i \exp(aV/kT) - |Na|_0\}. \tag{60}$$

It may be seen that increasing U_1, the energy barrier, decreases and therefore the sodium current is gated. It can be shown that if U_1 is negative, the $\phi_{Na}(V_p, \infty)$ is not gated. For negative values of U_1,

$$I_{Na} = \frac{\beta \cdot D \cdot F}{d}[|Na|_i \exp(aV/kT) - |Na|_0]/[\exp(aV/kT) + 1], \tag{61}$$

that is to say the stationary sodium flux is not a function of U_1. The results remain valid for electrical profiles other than the constant electrical field.

(iii) Channel counting and single channel conductance

Two transport mechanisms have been proposed: by carriers and through pores. Hodgkin suggested that "sodium crosses the membrane by combining with a carrier molecule which bears a negative charge equivalent to 3 or 4 electrons." Carriers transporting ions across a membrane are limited by translational diffusion. In the case of valinomycin, for example, the maximum rate of transport has been estimated at $2 \cdot 10^4$ ions/s [60,61]. On the other hand, ion transport through a pore is limited by, among other factors, the access resistance. In a pore with an access resistance of $2 \cdot 10^8 \, \Omega$ (corresponding to a circular entrance with a radius of about 3 Å), the rate of entry of ions into the pore has been estimated at $5 \cdot 10^8$ ions/s.

Without information about the density of conductance units (carriers or pores), it is impossible to calculate the maximum rate of transport in the sodium channels.

The notion that a discrete number of conductance units could be measured in nerve membranes originated from an observation in myelinated nerve fibres i.e., the discreteness in the fluctuations of threshold voltage responses [62].

The first measurements were made by titration of the sites associated with the conductance units by the molecules of TTX, which block the sodium conductance change. Using TTX and employing voltage-clamp methods, we were able to estimate

Ion permeability

the density of the sodium conductance units in the squid giant nerve fiber to be between 300 and 600 μm^2 [53].

The possibility of measuring the small fluctuations in i_{Na} and i_K occurring at nerve membrane level led to the realisation that such phenomena could provide insight into the microscopic mechanisms of ionic permeability changes. This technique had been successfully employed to examine membrane potential fluctuations and membrane current fluctuations due to the chemically mediated open-closed kinetics of ionic channels [63–68].

To estimate single-channel conductance Stevens proposed the following equation

$$\gamma_i = \frac{\sigma_i}{\mu_i(1-f)(V-V_i^*)} \tag{62}$$

where σ is the variance of the fluctuating current, μ is the mean value of the current, V is the membrane potential, V_i^* is the equilibrium potential and f is the fraction of ionic channels activated [69].

Table 6 gives estimates of Na-channel density for various excitable cells. As in the two state model the maximum conductance g_{max} is related to the conductance of a single channel γ by the formula

$$g_{max,i} = A_i \gamma_i$$

where A is the channel density.

In the node of Ranvier and in the giant axon of squid it is possible to calculate

TABLE 6
Estimation of the density of Na-channels in nerves

Preparation	Method	Channels/μm^2	Ref.
Squid giant axon	Rate of TTX blockage	300–600	53,54
Squid giant axon	TTX binding	350	73
Crab nerve	TTX equilibrium blockage	49	72
Lobster nerve	TTX equilibrium blockage	36	72
Rabbit vagus nerve	TTX equilibrium blockage	75	72
Squid giant axon	Gating currents	483	41
Squid giant axon	Gating currents	350	74
Squid giant axon	Sodium current fluctuations	330	70
Node of Ranvier	Gating current	5212	43

TABLE 7
The conductance of a single sodium channel

Method	Preparation	Parameters and values used	κ Ωcm	γ_{Na} pS	Ref.
Rate of TTX blockage and measurement of g_{Na}	Squid giant axon (*Loligo*)	$A = 5 \; 10^{10}/cm^2$ $g_{Na} = 120 \; mS/cm^2$	33	2.4	54
	Skeletal muscle fibres from *Rana temporaria*	$A = 3 \; 10^{10}/cm^2$ $g_{Na} = 120 \; mS/cm^2$	110	4.0	75
Q_{max} from gating current and mV for e-fold changes in g_{Na} and Q_{on}	Squid giant axon (*Loligo*)	$Q_{max} = 30 \; 10^{-9} \; C/cm^2$ g_{Na}/e with 6 mV Q_{on}/e with 19 mV	33	2.5	41
	Node of Ranvier (*Rana esculenta*)	$Q_{max} = 274 \; 10^{-9} \; C/cm^2$ g_{Na}/e with 7 mV Q_{on}/e with 15 mV	110	2.9	43
Sodium current fluctuations	Squid giant axon (*Loligo*)	Formula 62	33	4.0	70
	Node of Ranvier (*Rana esculenta*)	Formula 62	110	7.9	71
Q_{max} from gating current and $g_{Na,max}$	Skeletal muscle fibres from *Rana temporaria*	$Q_{max} = 28 \cdot 10^{-12} \; C$ $g_{Na} = 124 \; mS/cm^2$	110	0.75	76
	Giant axon from crab (*Carcinus maenas*)	$Q_{max} = 27 \; 10^{-9} \; C/cm^2$ $g_{Na} = 130 \; mS/cm^2$	33	2.4	77

the conductance of Na and K channels. Table 7 summarises the results obtained using different methods. The conductivity of the ionic medium in which the measurements were done is also given in Table 7 to normalise the values to a standard conductivity. It may be seen that the conductance per channel is a constant value for preparations from various species.

The resistance of a single unit is near $2000 \cdot 10^8 \; \Omega$. A potential difference of 100 mV across this resistance could drive ions at a rate of $500 \cdot 10^4$ ions/s, a value greater than the maximum rate of transport estimated for certain carriers.

Another interesting observation to be made from the data on single-channel conductance in Table 7, is the remarkable constancy of its value even though there is a considerable distance (in the evolutionary sense) between the various species used. This suggests that the channel design has been preserved along the millions of years of evolution.

References

1 Lehninger, A.L. (1971) Bioenergetics, 2nd ed., Benjamin, Menlo Park, CA.
2 Mitchell, P. (1966) Chemiosmotic Coupling and Energy Transduction. Glynn Research Laboratories. Bodmin, Cornwall, England.

3 Tandford, C. (1961) Physical Chemistry of Macromolecules. Wiley, New York.
4 Schögl, R. (1966) Ber. Burnsenges. Phys. Chem. 70, 400–447.
5 Goldman, D.E. (1943) J. Gen. Physiol. 27, 37–60.
6 Hodgkin, A.L. and Katz, B. (1949) J. Physiol. 108, 37–77.
7 Behn, U. (1897) Ann. Phys. Chem. Neue Folge 62, 54.
8 Ussing, H.H. (1949) Acta Physiol. Scand. 19, 43–56.
9 Teorell, T. (1949) Annu. Rev. Physiol. 11, 545–564.
10 Hodgkin, A.L. and Keynes, R.D. (1955) J. Physiol. 128, 61–88.
11 Hodgkin, A.L. and Horowicz, P. (1959) J. Physiol. 148, 127–160.
12 Lakshminarayanaiah, N. and Rojas, E. (1973) J. Physiol. 233, 613–634.
13 Hodgkin, A.L. (1951) Biol. Revs. 26, 339–409.
14 Baker, P.F., Hodgkin, A.L. and Shaw, T.I. (1961) J. Physiol. 157, 25P.
15 Oikawa, T., Spyropoulos, C., Tasaki, S. and Teorell, T. (1961) Acta Physiol. Scand. 52, 195–199.
16 Rojas, E. and Canessa-Fischer, M. (1968) J. Gen. Physiol. 52, 240–257.
17 Atwater, I., Bezanilla, F. and Rojas, E. (1969) J. Physiol. 201, 657–664.
18 Bezanilla, F., Rojas, E. and Taylor, R.E. (1970) J. Physiol. 207, 151–164.
19 Bezanilla, F., Rojas, E. and Taylor, R.E. (1970) J. Physiol. 211, 729–751.
20 Atwater, E., Bezanilla, F. and Rojas, E. (1970) J. Physiol. 211, 753–765.
21 Chandler, W.K. and Meves, H. (1965) J. Physiol. 180, 788–820.
22 Rojas, E. and Atwater, I. (1967) Proc. Natl. Acad. Sci. USA 57, 1350–1355.
23 Hille, B. (1975) Ionic selectivity of Na and K channels of nerve membranes. in Eisenman (Ed.), Membranes: a Series of Advances, Vol. 3, Marcel Dekker, New York, pp. 255–323.
24 Tasaki, I. (1963) J. Gen. Physiol. 46, 755–772.
25 Binstock, L. and Lecar, H. (1969) J. Gen. Physiol. 53, 342–361.
26 Hodgkin, A.L., Huxley, A.F. and Katz, B. (1952) J. Physiol. 116, 424–448.
27 Hodgkin, A.L. and Huxley, A.F. (1952) J. Physiol. 116, 449–472.
28 Hodgkin, A.L. and Huxley, A.F. (1952) J. Physiol. 116, 473–496.
29 Hodgkin, A.L. and Huxley, A.F. (1952) J. Physiol. 117, 500–544.
30 Hille, B. (1970) Progr. Biophys. Mol. Biol. 21, 3–32.
31 Mullins, L.J. (1968) J. Gen. Physiol. 52, 550–556.
32 Rojas, E. and Armstrong, C.M. (1971) Nature New Biol. 229, 177–178.
33 Armstrong, C.M., Bezanilla, F. and Rojas, E. (1973) J. Gen. Physiol. 62, 375–391.
34 Narahashi, T., Moore, J.W. and Scott, W.R. (1964) J. Gen. Physiol. 47, 965–974.
35 Armstrong, C.M. and Binstock, L.J. (1965) Gen. Physiol. 48, 859–872.
36 Hille, B. (1968) J. Gen. Physiol. 51, 199–219.
37 Rojas, E. and Keynes, R.D (1975) Phil. Trans. Roy. Soc. Lond. B 270, 459–482.
38 Keynes, R.D. and Rojas, E. (1976) J. Physiol. 255, 157–189.
39 Keynes, R.D. and Rojas, E. (1973) J. Physiol. 233, 28–30P.
40 Armstrong, C.M. and Bezanilla, F. (1973) Nature 242, 459–461.
41 Keynes, R.D. and Rojas, E. (1974) J. Physiol. 239, 393–434.
42 Rojas, E. and Bergman, C. (1977) TIBS 1, 6–9.
43 Nonner, W., Rojas, E. and Stämpfli, R. (1975) Pflügers Arch. 354, 1–18.
44 Nonner, W., Rojas, E. and Stämpfli, R. (1978) Pflügers Arch. 375, 75–85.
45 Armstrong, C.M. and Bezanilla, F. (1977) J. Gen. Physiol. 70, 567–590.
46 Armstrong, C.M. and Bezanilla, F. (1975) Ann. N.Y. Acad. Sci. 264, 265–277.
47 Baylis, W.M. (1915) Principles of General Physiology, 1st ed., Longmans, Green, London.
48 Conway, E.J. (1954) Symp. Soc. Exp. Biol. VIII, 297.
49 Mullins, L.J. (1959) J. Gen. Physiol. 42, 817–829.
50 Hille, B. (1971) Proc. Natl. Acad. Sci. USA 68, 280–282.
51 Hille, B. (1975) Biophys. J. 15, 615–619.
52 Hille, B. (1975) J. Gen. Physiol. 66, 535–560.
53 Keynes, R.D., Bezanilla, F., Rojas, E. and Taylor, R.E. (1975) Phil. Trans. Roy. Soc. Lond. B 270, 365–375.

54 Rojas, E. (1973) Acta Physiol. Lat. Am. 23, 90-92.
55 Narahashi, T., Anderson, N.C. and Moore, J.W. (1966) Science 153, 765-767.
56 Schwarz, J.R., Ulbricht, W., Wagner, H.H. (1973) J. Physiol. 233, 167-194.
57 Kao, C.Y. and Nishiyama, A. (1965) J. Physiol. 180, 50-66.
58 Rojas, E. (1976) Cold Spring Harbor Symp. Quant. Biol. XL, 305-320.
59 Glasstone, S., Laidler, K.J. and Eyring, H. (1941) The Theory of Rate Processes. McGraw-Hill, New York.
60 Stark, G., Ketterer, B., Benz, R. and Läuger, P. (1971) Biophys. J. 11, 981-989.
61 Stein, W.D. (1968) Nature 218, 570-576.
62 Verveen, A.A. and Derksen, H.E. (1969) Acta Physiol. Pharmacol. Neerl. 15, 353-379.
63 Katz, B. and Miledi, R. (1972) J. Physiol. 224, 665-699.
64 Anderson, C.R. and Stevens, C.F. (1973) J. Physiol. 235, 655-691.
65 Colquhoun, D., Dionne, V.E., Steinbach, J.H. and Stevens, C.F. (1975) Nature 253, 204-206.
66 Katz, B. and Miledi, R. (1973) J. Physiol. 230, 707-717.
67 Neher, E. and Sackman, B. (1976) J. Physiol. 258, 705-729.
68 Neher, E. and Sackman, B. (1975) Proc. Natl. Acad. Sci. USA 72, 2140-2144.
69 Stevens, F. (1972) Biophys. J. 12, 1028-1047.
70 Conti, F., De Felice, L.J. and Wanke, E. (1975) 248, 45-82.
71 Conti, F., Hille, B., Neumcke, B., Nonner, W. and Stämpfli, R. (1976) J. Physiol. 262, 699-727.
72 Keynes, R.D., Ritchie, J.M. and Rojas, E. (1971) J. Physiol. 223, 235-254.
73 Levinson, S.R. and Meves, H. (1975) Phil. Trans. R. Soc. Lond. B. 270, 349-352.
74 Armstrong, C.M. and Bezanilla, F. (1974) J. Gen. Physiol. 63, 533-52.
75 Allen, T.J.A., Rojas, E. and Suarez-Isla, B. (1980) J. Physiol. 308, 100-101P..
76 Rojas, E. and Suarez-Isla, B. (1980) J. Physiol. 301, 46-47P.
77 Arispe, N., Quinta Ferreira, E. and Rojas, E. (1979) J. Physiol. 295, 11-12P.
78 Rojas, E. and Taylor, R.E. (1975) J. Physiol. 252, 1-27.

CHAPTER 4

Channels and carriers in lipid bilayers

JAMES E. HALL

Department of Physiology and Biophysics, University of California, Irvine, CA 92717, U.S.A.

1. Perspectives

Carriers and channels both provide a similar service: they reduce the activation energy of translocation of an ion across the low dielectric constant membrane. To a first approximation, the energy of translocation of a free ion is nearly equal to the Born charging energy for a sphere having a radius equal to that of the ion in question. The Born energy depends on the reciprocal of ionic radius and thus delocalization or smearing out of the ionic charge reduces the energy of translocation substantially. The Born charging energy also depends on the reciprocal of the dielectric constant of the membrane, so an increase in the effective dielectric constant of the membrane interior reduces the activation energy of translocation. Most channels and carriers probably delocalize the charge of their substrates and increase the local dielectric constant. Also each channel or carrier likely reduces translocation energy by specific interactions unique to itself, such as specially charged sites to enhance selectivity. Fundamentally carriers and channels both reduce energy of translocation and are therefore likely to have certain features in common.

(a) Terminology

The term "channel" has a fairly specific meaning in the literature of excitable membranes and membrane transport [1]. It means a simple site of permeation with too high a conductance to be a carrier [2–4]. A pure "carrier" combines with its substrate on one side of the membrane. The site of combination moves to the other side of the membrane and the carrier dissociates from its substrate. This definition of carrier allows for both rotational carriers and diffusional carriers. Channels are usually thought of as pores or fixed passages in the membrane through which the ions (or channel substrates) move.

(b) How to tell a channel from a carrier

Clearly many ion-transport mechanisms can be imagined having features of both carriers and channels. The experimental problem is one of making measurements of

Bonting/de Pont (eds.) Membrane transport
© *Elsevier/North-Holland Biomedical Press, 1981*

sufficient refinement to allow construction of a detailed model of a given mechanism. At present, experimental techniques allow distinction between channels and carriers under some conditions.

Since ion carriers and ion channels promote the passage of electric current through membranes, electrical measurements are essential to determine their properties. The fundamental instruments are a voltmeter to measure the voltage across the membrane, an ammeter to measure the current and a power source to prescibe the current, the voltage or the charge applied to the membrane.

A very powerful technique of increasingly great importance is fluctuation analysis [5,6]. Usually, the voltage across the membrane is clamped by the external circuitry, and fluctuations in the current are recorded. Analysis gives estimates of the magnitude and time course of the single conduction sites in the membrane. Most importantly, ion channels and ion carriers show *qualitatively* different kinds of fluctuations. Under voltage-clamp conditions, the current due to ion carriers shows a decrease of noise per unit bandwidth at low frequencies. The current noise due to ion channels turning on and off shows a decrease at high frequencies. Fig. 1 shows a comparison of channel noise exemplified by alamethicin [8] with carrier noise exemplified by valinomycin [7].

Fundamental differences in mechanism of conduction are responsible for this. A carrier which has just carried an ion across the membrane must go back or reset before it can carry another ion in the same direction. A fixed channel on the other hand, if it is open, is equally available to carry ions from both sides of the membrane. Thus, for a carrier, large, long-lasting fluctuations (low-frequency noise) are damped out because of the nature of the carrier mechanism, which must reset.

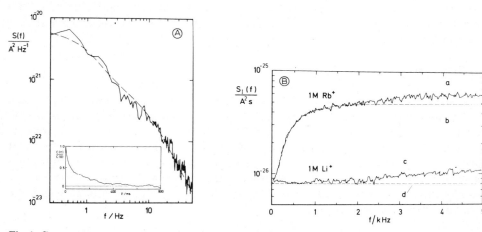

Fig. 1. Comparison of the current noise due to a carrier with that due to a pore. (A) Current fluctuations in the presence of valinomycin with lithium (which is poorly transported) and the rubidium (which is well transported) as the cation. Noise increases at high frequencies because the empty carrier must return after carrying one ion across the membrane. (From [8].) (B) Current fluctuations in the presence of alamethicin. Noise decreases at high frequencies because the channels open and close at a limited rate. (From [7].)

For a channel, high-frequency noise is damped out at frequencies greater than the opening or closing frequency of the channel. Differences in the fluctuation spectrum, if experimentally accessible, can thus distinguish channels from carriers.

Another very important way of identifying channels is the observation of unit conductance steps. Such steps were first seen in lipid bilayers doped with excitability-inducing material (EIM) [11]. Steps due to many other substances have now been seen in lipid bilayers and very excitingly in biological membranes [12]. To observe steps, the voltage is clamped at a fixed value, and the total number of open channels in the membrane limited in some way to a small number. In the bilayer, the amount of channel-forming substance added to the system is reduced so that each membrane has very few active channels. In biological membranes, the area of the membrane observed is minimized so that the area under observation contains only a minimum number of channels. Fig. 2 shows examples of single-channel recordings from both bilayers and muscle.

In both cases, single-step currents occur with amplitudes ranging from about 0.2 pA to 1 nA, depending on the channel-former and the experimental conditions, especially the conductance of the membrane bathing solution. The smallest currents correspond to translocation of about $10^7 - 10^8$ ions/s. A carrier diffusing back and

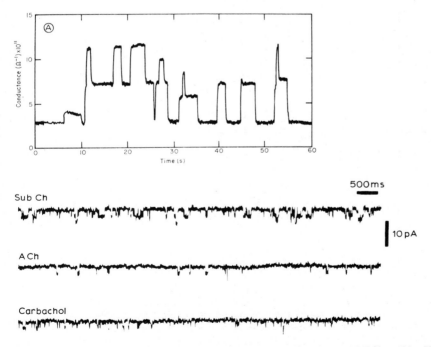

Fig. 2. Conductance steps for two channels. (A) Valine gramicidin A in 1 M KCl at 100 mV. (From [19].) (B) Current recordings from frog muscle showing acetylcholine channel steps in the presence of different cholinergic agonists. (From [12].)

forth across the membrane with a diffusion constant of 10^{-5} cm^2/s (the diffusion constant of a very fast small ion in free aqueous solution) could just barely make 10^8 trips/s with no time for loading and unloading its substrate. We also know that charged ions cannot diffuse anywhere near that fast across the membrane because of the electrostatic energy barrier. Thus, a mechanism capable of transporting ions at

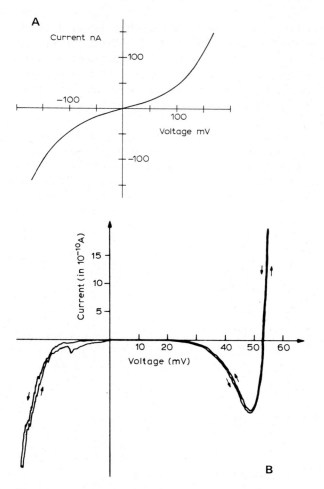

Fig. 3. Current-voltage curves for a carrier and a voltage-dependent channel. (A) Current-voltage curve of nonactin, a carrier selective for K$^+$ in 0.1 M KCl. The current increases more than linearly with the voltage showing that transport is limited by a translocation reaction rather than a surface reaction. (Hall, J.E., unpublished). (B) Current-voltage curve of alamethicin in a salt gradient (0.005 M KCl on one side of the membrane and 0.5 M KCl on the other). The number of channels increases with voltage in the same way in both branches of the curve, but between +10 and +53 mV, the salt gradient causes the current flow through the open channels to be negative. This gives rise to the negative resistance region. (From [34].)

such a large rate is almost certainly not a carrier, but something like a channel or pore.

The dependence of the conductance on voltage is a particularly important characteristic of both carriers and channels. If the voltage is varied slowly enough so that the current reaches steady-state at each voltage, a current-voltage (I–V) curve can be constructed. An "ohmic" conductance mechanism has a linear current-voltage curve, but channels and carriers have for the most part non-linear I–V curves. Fig. 3 shows two current-voltage curves for: A carrier, nonactin; and channel in a black lipid film (alamethicin). These curves are all non-linear, but channel I–V curves generally show much steeper dependence of the current on voltage than do carrier I–V curves. While not all channels have conductances as strongly voltage-dependent as those illustrated, no carrier shows a voltage-dependence as strong as that exhibited by the voltage-dependent channels.

(c) Ion selectivity and its consequences

The ability of biological membranes to allow the passage of particular ions, excluding all others, has long been a puzzle of some concern. Studies of carriers and channels in lipid bilayers have much increased our understanding of selectivity and its structural basis, particularly for nonactin and valinomycin as carriers and for gramicidin as a channel.

The working out of the details of selectivity is often quite complicated, but relatively simple concepts can be very useful. In a particularly useful type of model, selectivity is calculated by taking the difference between energy increase due to dehydration of the ion and energy decrease due to an attractive site of a given charge and size [13]. Taking field strength of the site as a parameter and using known energies of hydration generates several selectivity sequences for the alkali cations. One of these is the experimental sequence of selectivity for nonactin and valinomycin (both of which are very selective for K^+ over Na^+). Other ionophores are selective for different ions. Channels can have two or more selectivity sites, a fact which complicates the analysis, but does not change the fundamental concept.

One consequence of ion selectivity in some voltage-dependent channels is negative resistance: For example, the negative resistance region in curve b of Fig. 3 (from 20 mV to 50 mV). Because of the salt gradient across the membrane (100:1) and the slight selectivity of the alamethicin channel for potassium over chloride, no current flows through an open alamethicin channel at a voltage of about 53 mV, the channel's *reversal potential.* Above 53 mV, the current through an open channel will be positive (upward in Fig. 3), below negative (downward in Fig. 3). But the number of open channels also depends on voltage. An empirical formula for the number of open alamethicin channels, n, (2) as a function of voltage is

$$n = (C_{ala})^9 (C_{salt})^4 \exp(V/V_0)$$

(C_{ala} = alamethicin concentration, C_{salt} = salt concentration, V = membrane potential, V_0 = voltage which causes e-fold conductance charge.)

$$I_{single} = \gamma_{single}(V - V_R)$$

(I = current, γ_{single} = average single-channel conductance, V_R = channel-reversal potential.)

The membrane current is then

$$I_{total} = n\gamma(V - V_R)$$
$$= (C_{ala})^9 (C_{salt})^4 \exp(V/V_0)(V - V_R)$$

(n = number of channels.)

Because V_0 is only about 4 mV, the magnitude of the current (for fixed alamethicin and salt concentrations) is determined primarily by the value of $\exp(V/V_0)$, but the sign of the current is that of $(V - V_R)$. As V increases from zero toward V_R (53 mV in Fig. 3), channels begin to turn on, but the sign of $V - V_R$ is negative. As V approaches very close to V_R (closer than V_0), the current will become less negative until it reaches zero at V_R. All the while the number of open channels continues to increase with voltage.

This negative resistance is a very important consequence of selectivity, and the negative resistance in the peak-current-versus-voltage curve for the sodium channel in nerve arises in exactly the same way as that for alamethicin. Monazomycin, another voltage-dependent channel, shows very similar negative resistance, and has been well described by Muller and Finkelstein [14].

We will now discuss a few specific examples of carriers and channels illustrating the methods and concepts briefly outlined above.

2. Carriers and matters unique to them

(a) The carrier model

A model for carrier conduction, which is applicable to many different carriers, has been worked out [3,4,15]. Fig. 4 shows this model in detail. There are two types of reactions which occur: Homogeneous reactions, where the metal ion (M) combines with the carrier (S) in the same phase and heterogeneous reactions, where the metal ion combines with a carrier in a different phase. The model considers three phases: The aqueous phase, the membrane surface, and the interior of the membrane. An ion can thus find its way across the membrane by two distinct pathways. It can combine with the carrier in the aqueous phase (homogeneous reaction), remain complexed long enough to enter the surface of the membrane, and then move across the membrane to the opposite surface and then to the opposite aqueous phase where it dissociates from the carrier. Or it can combine with a carrier at the membrane

Channels and carriers

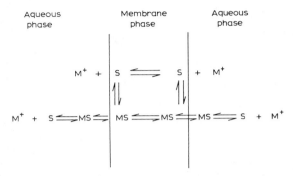

Fig. 4. The principal reactions of the carrier model. M$^+$ denotes a metal ion. S denotes the uncomplexed carrier. MS devotes the complex. The reactions in the top line are *heterogeneous*: An ion in the aqueous phase joins or leaves a carrier in the membrane phase. The reactions in the lower line are homogeneous: Complexation or dissociation takes place between a carrier and an ion in the same phase. (Ions cannot exist in the membrane phase because of high Born charging energy.)

surface (heterogeneous reaction), move complexed across the membrane and be released by the carrier at the opposite surface. In both cases, the free carrier has to diffuse back to get another ion.

The second pathway is the most efficient and has been extensively studied. It has proven, in fact, possible to measure most of the rate constants in the reaction scheme for some carriers by applying voltage pulses and observing the current relaxation [3]. The voltage is pulsed from zero to the chosen value, and the current jumps to an initial value and then relaxes with time constants characteristic of the rate constants of the carrier in question (see Fig. 5 for an example).

The current per unit area is

$$J = F\{K'_{ms}[MS^+]'_m - K''_{ms}[MS^+]''_m\}$$

where K'_{ms} and K''_{ms} are the voltage-dependent rates of translocation, $[MS^+]'_m$ and $[MS^+]''_m$ are the concentrations of carrier ion complex on the left and right sides of the membrane and F is the Faraday. The complex translocation rate constants depend on voltage approximately as

$$K'_{ms} = K_{ms} \exp(nFV/RT)$$

$$K''_{ms} = K_{ms} \exp(-nFV/RT)$$

where n is a factor of 0.5 or less.

At 0 Volts, $K'_{ms} = K''_{ms}$, and the surface concentrations of carrier ion complex are given by the equilibrium values determined by the surface reaction rate constants:

$$[MS^+]_m = \frac{K_R}{K_D}[M]_{aq}[S]_m$$

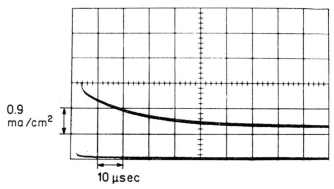

Fig. 5. Current relaxation of valinomycin following a 38 mV voltage pulse (10^7 M valinomycin in 1 M KCl). (From [35].)

where K_R is the association rate, K_D the dissociation rate, $[M]_{aq}$ the ion concentration in the aqueous solution and $[S]_m$ the free carrier concentration in the membrane.

Just after a voltage is applied to the membrane, the initial values of carrier complex surface concentrations determine the current. But because the translocation rates change on application of the voltage, the surface concentrations of carrier complex change with time, and the current relaxes to a new initial value. The rate constants of the various reactions will determine both the final value of the current (the amplitude of the relaxation) and the time course by which the current attains its final value.

(b) Special limiting cases

Several limiting cases are of particular interest. If the surface reaction rates and the translocation rate of free carrier of complex are very fast compared to translocation, the concentrations of complex at the surfaces will remain unchanged when voltage is applied. Under these conditions, the current will be proportional to the hyperbolic sine of the voltage:

$$J = 2FK_{ms}[S][M]\frac{K_R}{K_D}\sinh\left(\frac{nFV}{RT}\right)$$

This case is realized by nonactin at potassium concentrations less than 100 mM in bacterial phosphatidyl ethanolamine membranes.

If on the other hand, either the rate constants for complexation (K_R) or that for dissociation (K_D) is comparable to the rate of translocation, the change in complex concentration at the interface will be appreciable. This case is realized for valinomycin, and the rate constants are all sufficiently close to one another to make measurement of all rate constants relatively simple. Readers interested in details should consult Läuger [3] for a relatively elementary treatment, Hladky [15] for a very complete treatment.

An analog of valinomycin, slightly altered in structure by substituting a D-proline for D-hydroxyvaline and L-proline for L-lactic acid is called PV (for proline valinomycin). This compound has such a strong affinity for potassium that K_D is nearly zero, and PV remains complexed to potassium essentially permanently, even in solution. Addition of PV to an aqueous solution containing potassium leads to very large relaxation and currents which are essentially independent of potassium concentration. PV in the presence of excess potassium thus forms a light complex which behaves almost like a lipophilic ion. The maximum current density which can flow under these conditions is given by

$$J_{max} = \frac{FD}{\delta}[PV]_{aq}$$

where D is the diffusion coefficient of $PV\text{-}K^+$, δ the thickness of the unstirred layer and $[PV]_{aq}$ the aqueous concentration of PV added.

A second structural alteration of valinomycin, substitution of only L-proline for L-lactic acid, yields PV lac. Its conductance properties realize yet another limiting case, that of current limited by back flux of the neutral carrier. This case can be distinguished from the case of solution complexation (illustrated by PV), because the time constant for relaxation depends on both the concentration of carrier and the concentration of ion.

3. Channels and matters unique to them

Channels almost all exhibit step-wise, discrete changes in conductance. Existence of these steps can be taken as prima facie evidence of a channel mechanism. The macroscopic conductance of a membrane containing many channels acting independently should be predictable from the properties of a single channel, in fact, the single-channel properties satisfactorily account for the multi-channel properties of nearly all known channels. In the following paragraphs, we will consider three classes of channels, show how their single-channel properties explain their many-channel properties, and discuss how channel structures relate to channel function for those channels where sufficient information is available.

(a) Gramicidin—the most studied channel

Gramicidin A is a linear polypeptide containing fifteen amino acids, alternately D and L. Hladky and Haydon [16] first showed that gramicidin A induces discrete steps of current in lipid bilayers. Tosteson et al. [17] first proposed that the gramicidin channel is a dimer, because the conductance appeared to depend on the square of the gramicidin concentration. However, because the solubility of gramicidin in water is very low, this kind of data is not very reproducible and other ways of confirming the dimer model were sought. Bamberg and Läuger [18] proposed a model for the action of gramicidin based on dimerization. In some membrane systems, the gramicidin conductance is voltage-dependent, although not nearly so much as that of the strongly voltage-dependent channels we will discuss next. When a voltage-pulse is applied, the current of a membrane doped with gramicidin rises rapidly to an initial value and then increases with a single time constant to a final value at long time (shown in Fig. 6).

Bamberg and Läuger [18] formulated a kinetic model assuming that the rate constants for formation and break up of the channel were voltage-dependent. This model predicted that the time constant for relaxation of the current should depend on the square root of the mean 0 V membrane conductance if the channel was a dimer. This dependence was found.

The dimer model was further supported by the work of Veatch et al. [19]. They used a fluorescent analogue of gramicidin to measure directly the concentration of gramicidin in the membrane, thus avoiding the solubility problems which had caused trouble in earlier work. Their results also strongly supported the dimer model. Finally, use of a negatively charged analogue of gramicidin (O-pyromellityl gramicidin), much more water-soluble than gramicidin A, confirms the dependence of conductance on the square of gramicidin concentration. This analogue must also be added to *both sides* of the membrane to be effective. This result shows what had been previously suspected, that each side of the membrane contributes one monomer to the channel dimer.

There is also a wealth of structural information on how the actual conducting dimer is arranged in the membrane. It now seems almost certain that gramicidin forms a dimer in which the N-formyl groups are in the center of the membrane and

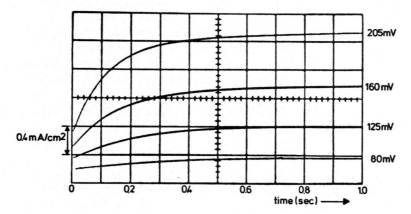

Fig. 6. Kinetic response of gramicidin current to voltage pulses of various magnitudes (cf. Fig. 5). (From [18].)

join together two separate helices by hydrogen bonding. A particular helical model embodying this notion was first proposed by Urry [20], but many of the experiments designed to test this model benefited from consideration of an alternate model devised by Veatch and colleagues in which the two monomers formed an intertwined double helix [21].

NMR measurements show that water-soluble shift reagents alter chemical shifts of the O-terminus of gramicidin more than those of the N-terminus [22]. Synthetic derivatives with charged N-terminus work only when the N-terminus can be protonated, but charged O-terminus derivatives are all active if added to both sides of the membrane simultaneously. The N-terminus cannot tolerate a charged group, but the O-terminus can [23]. Finally, dimerized gramicidins, linked N-terminus to N-terminus, form active channels with single-channel properties similar to those of natural gramicidin.

The single-channel properties of gramicidin have been investigated in more detail than those of other channel formers. It is even possible to make some attempts to correlate single-channel properties directly with structure. In fact, the X-ray structure of gramicidin is consistent with a two-site ion binding model [24] (to be discussed shortly). Gramicidin single-channel conductance depends on the activity of the cationic species in the solutions bathing the membrane. At low concentrations, the conductance is proportional to activity of permeant species, but as the activity is increased, the single-channel conductance increases less per unit activity increase than at low concentrations. At still higher concentrations, the conductance for some ions (especially cesium) can even decrease with increasing concentrations.

These facts can be explained quantitatively by a fairly simple model. If one assumes that only a single ion can pass through the pore at one time, the low concentration data predict proportionality between single-channel conductance and activity. Assuming that the channel has an ion binding site near its mouth predicts that increasing ion concentration will decrease the number of free channels available. This will in turn lead to saturation of conductance with increasing ion concentration. Finally, because the channel is symmetrical, the possibility that the two sites on opposite sides of the membrane may be occupied simultaneously exists. The translocation rate of ions is thus concentration-dependent, and if the binding sites are inside of the membrane far enough so they are at different electrical potential, than the aqueous phase, their occupancy and rate constants of occupation and dissociation will also be voltage-dependent. Different ions also have different affinities for the binding sites. Thallium in particular binds strongly, but passes through the channel poorly (relative to cesium or potassium). Thallium can thus act as a blocking ion and reduce the conductance due to other ions, even when present itself only in small amounts. Finally, the "two-site model" predicts that the channel permeabilities measured by bionic potential measurements based on the Goldman–Hodgkin–Katz equation will be concentration-dependent because of differences in ion affinity for the sites.

(b) Voltage-dependent channels

Voltage-dependent channels are physiologically important in nerve and muscle. They provide the basis for the action potential and are thus fundamental to the operation of our muscles and our minds. Nonetheless, though kinetic models of voltage-dependent channels abound, models based on physical principles and known structural properties are relatively few. Studies of voltage-dependent channels in thin lipid membranes were originally initiated in hopes that they might supply this lack. To some extent, they have done so, but whether or not voltage-dependent channels in nerve and muscle have any features in common with channels studied in vitro remains to be seen.

(i) Channels which turn on with voltage

Channels like alamethicin and monazomycin which turn on with voltage probably have a similar structural basis for their mode of action. These channels have steep current-voltage curves and strong power dependence of conductance on concentration. They do not turn on immediately with the application of a voltage, but show time-delayed turn on characteristics, suggestive of one or more intermediate states between the closed and open state. The structure of the alamethicin-monazomycin type channel is very different from the structure of the gramicidin channel. The molecules are rod-like and not cylindrical or helical like gramicidin. They thus probably form channels by aggregating together like barrel staves, rather than forming a tube through the membrane. The aggregate is loose and may change shape or gain or loose molecules [25,26]. Single-channel recordings support this picture. Unlike the steps recorded for gramicidin (see Fig. 2), the alamethicin step sizes are not uniform. The steps size is smallest for the lowest level and increases progressively up to about the sixth level. This suggests that low-level changes in conductance level reflect changes in conformation of a single unit, rather than turning on unrelated units widely distributed in the membrane.

This conclusion is confirmed by raising the voltage slightly. Then new levels appear whose conductances are sums of those of original single-unit levels. Voltage thus turns on new conducting units. Each unit then fluctuates among the structurally allowed range of conducting states.

Other indirect, but interesting experiments, support the idea of a barrel-like structure for the alamethicin channel. Donavan and Latorre [26,27] found that very large charged molecules with a hydrophobic moiety go through the alamethicin channel much faster than expected on the basis of single-channel conductance. I have found that the phospholipid molecules making up the membrane can cross the membrane when alamethicin channels are open, but not when they are closed. If the channel is barrel-like, the long hydrophobic parts of the molecules could stay in the hydrocarbon region membrane and the hydrophobic parts of the molecule could slip into the lumen of the channel between the barrel staves.

Finally, amphotericin and nystatin also appear to form channels by an aggregation mechanism, but in a voltage-independent way.

(ii) Channels which turn off with voltage

EIM [11], hemocyanin [30], voltage-dependent anion conductance (VDAC) [31], colicins [32], and some proteinaceous materials extracted from biological membranes form channels which turn off with voltage. Fig. 7a shows the current-voltage curve for EIM as an example of this class of channel former. At low concentrations of EIM, membranes sometimes contain only one channel. As the voltage is made positive, the current begins to decrease. The single-channel recordings (Fig. 7b) show this as an increased probability of the lower conductance level, and the multi-channel membrane current-voltage curve shows it as a decrease in average conductance. Thus, increasing voltage reduces the lifetimes of the higher conductance level of a single EIM channel and increases the lifetimes of the lower conductance levels. This dependence of channel lifetime on voltage exactly explains the voltage dependence of the conductance of a many-channel membrane. The details of lifetime dependence on voltage and the precise conductance changes which occur vary considerably from substance to substance, but all of the "EIM-like" channels have in common voltage-current curves which are explained by the properties of many single channels acting in a statistically independent way.

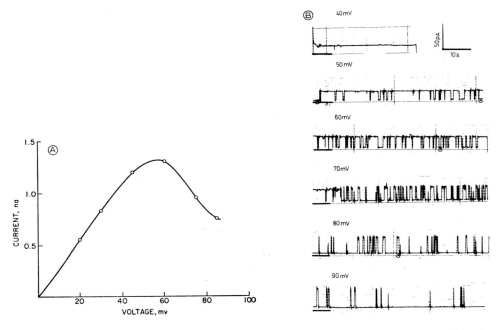

Fig. 7. The single-channel properties of EIM (and other channels by similar reasoning) explain the macroscopic conductance properties. (A) Current-voltage curve for EIM showing how EIM turns off with increasing voltage. (B) Series of EIM single-channel recordings at different voltages. As the voltage increases, the channel spends less time in the on-conducting state and more time in the off-state. In a many-channeled membrane, this means that fewer channels are likely to be on at high voltages. This can quantitatively explain the shape of the current-voltage curve (From [30].)

4. New directions

The incorporation of native conductance mechanisms isolated from real membranes reliably into planar lipid membranes has been a long-standing goal. It seems, however, that the day may be coming soon when this will be accomplished. Several groups now report reconstitution of acetylcholine receptors in planar membranes. The isolation of VDAC [31] from mitochondria and K^+ channels from sarcoplasmic reticulum [33] shows that channels not yet known to have physiological significance exist in biological membranes, or alternatively channel properties may be altered by the environment of the planar film.

The reconstitution of voltage-dependent channels from nerve and muscle in planar bilayers will very likely be an early accomplishment of the eighties, and this will enable us to begin to test the ideas of channel structure and function developed from the study of channel-forming antibodies in the seventies.

We can expect to gain greater understanding of the role of lipid in altering conductances by using channels and carriers as probes of membrane structure. This effort is actually well under way at present. The interested reader is referred to Benz et al. [36] for an introduction to this problem of using carriers as probes.

Acknowledgements

I would like to thank Mary Ann Tacha, Karen Jessup and the staff of the Department of Physiology and Biophysics for their help in preparing the manuscript. Some of the work reported here was supported by a grant from the United States National Institutes of Health HL23183 and the author is a recipient of an NIH Research Career Development Award.

References

1 Hille, B. (1970) in J.A.V. Butler and D. Noble (Eds.), Progress in Biophysics and Molecular Biology, Vol. 21, Pergamon, pp. 1–31.
2 Hall, J.E. (1978) in G. Giebisch, D.C. Tosteson and H.H. Ussing (Eds.), Membrane Transport in Biology, Vol. I, Concepts and Models, Springer, Berlin, pp. 475–531.
3 Läuger, P. (1972) Science 178, 24–30.
4 Stark, G. (1978) in G. Giebisch, D.C. Tosteson and H.H. Ussing (Eds.), Membrane Transport in Biology, Vol. I, Concepts and Models, Springer, Berlin, pp. 448–473.
5 Stevens, C.F. (1972) Biophys. J. 12, 1028–1047.
6 Stevens, C.F. (1975) Fed. Proc. 5, 1364–1369.
7 Kolb, H.-A. and Läuger, P. (1978) J. Memb. Biol. 1, 167–187.
8 Kolb, H.-A. and Boheim, G. (1978) J. Memb. Biol. 38, 151–191.
9 Alvarez, O. and Latorre, R. (1981) Physiol. Rev. in press.
10 Haydon, D.A. and Hladky, S.B. (1972) Quart. Rev. Biophys. 5, 187–282.
11 Ehrenstein, G., Lecar, H. and Nossal, R. (1970) J. Gen. Physiol. 55, 119–133.
12 Neher, E. and Sakmann, B. (1976) Nature 260, 799–802.

13 Eisenman, G. and Krasne, S.J. (1973) in C.F. Fox (Ed.), MTP International Review of Science, Biochemistry of Cell Walls and Membranes, Biochemistry Series One, Vol. 2, Butterworth, London, pp. 27–59.
14 Muller, R. and Finkelstein, A. (1972) J. Gen. Physiol. 60, 263–284.
15 Hladky, S. (1979) in F. Bronner and A. Kleinzeller (Eds.), Current Topics in Membranes and Transport, Vol. 12, Academic Press, New York, pp. 53–164.
16 Hladky, S.B. and Haydon, D.A. (1972) Biochim. Biophys. Acta 274, 294–312.
17 Tosteson, D.C., Andreoli, T., Tiffenberg, M. and Cook, P. (1968) J. Gen. Physiol. 51, 3735–3845.
18 Bamberg, G. and Läuger, P. (1973) J. Memb. Biol. 11, 177–194.
19 Veatch, W., Mathies, Eisenberg, M. and Stryer, L. (1975) J. Mol. Biol. 99, 75–92.
20 Urry, D.W. (1972) Proc. Natl. Acad. Sci. USA 69, 1610–1614.
21 Veatch, W.R., Fossel, E.T. and Blout, E.R. (1974) Biochemistry 13, 4756–4770.
22 Weinstein, R., Wallace, B.A., Blout, E.R., Morrow, J.S. and Veatch, W. (1979) Proc. Natl. Acad. Sci. USA 76, 4230–4234.
23 Bamberg, E., Alpes, H.-J., Apell, H.-J., Bradley, R., Harter, B., Quelle, M.-J. and Urry, D.W. (1979) J. Memb. Biol. 50, 257–270.
24 Koeppe, R.E. II, Berg, J.M., Hodgson, K.O. and Stryer, L.S. (1979) Nature 279, 723–725.
25 Boheim, G. (1979) J. Memb. Biol. 14, 277–303.
26 Baumann, G. and Mueller, P. (1974) J. Supramol. Struct. 2, 538–557.
27 Neher, E. (1975) Biochim. Biophys. Acta 401, 540–544.
28 Donovan, J. and Latorre, R. (1979) J. Gen. Physiol. 73, 425–451.
29 Holz, R. and Finkelstein, A. (1976) J. Gen. Physiol. 67, 703–729.
30 Latorre, R., Alvarez, O., Ehrenstein, G., Espinoza, M. and Reyes, J. (1975) J. Memb. Biol. 25, 163–182.
31 Schein, S.J., Kagan, B.L. and Finkelstein, A. (1978) Nature 276, 159–163.
32 Schein, S.J., Colombini, M. and Finkelstein, A. (1976) J. Memb. Biol. 30, 99–120.
33 Miller, C. and Rosenberg, R. (1979) J. Gen. Physiol. 74, 457–478.
34 Eisenberg, M., Hall, J.E. and Mead, C.A. (1973) J. Memb. Biol. 14, 143–176.
35 Stark, G., Ketterer, B., Benz, R. and Läuger, P. (1971) Biophys. J. 11, 981–994.
35 Benz, C., Frölich, O. and Länger, P. (1977) Biochim. Biophys. Acta 464, 465–481.

CHAPTER 5

Concepts of mediated transport

W.D. STEIN

Department of Biological Chemistry, Institute of Life Sciences, Hebrew University, Jerusalem, Israel

1. Introduction

We saw in Chapter 1 that the movement of many types of substances across cell membranes could be accounted for by the process of their dissolution in the lipid phase of the membrane and their subsequent diffusion through this lipid phase to leave the membrane at the opposite face. Yet there were many substances which could not be included within such a simple model. For these substances, permeation was much faster than the simple model would have predicted, permeation was often not a linear function of substrate concentration, and was often inhibitable by analogues of the substrate or by reagents capable of attacking proteins. For all such deviant substances one might suggest that there exist membrane components, probably proteins, which mediate the movement across the membrane of these substances. Such mediated transport may be a net movement of substrate down its electrochemical gradient, in which case the process is often termed facilitated diffusion. In other cases, the net movement of substrate may be against the prevailing electrochemical gradient, when the phenomenon is termed active transport. We proceed to discuss in turn both these forms of mediated transport, emphasising the very strong links between them. Our approach will be to start with the very simplest model for mediation, to see what its predictions are, and then to introduce successively more complex models until most of the concepts of mediated transport are laid bare.

2. The kinetic analysis of facilitated diffusion

A very natural approach to the study of the mechanism of action of these mediated transport systems is through kinetic analysis. Here one measures the rate of movement of a particular substrate across the cell membrane and sees how this rate is affected by varying, for example, the concentration of the substrate at one or other face of the membrane. One might instead vary the electrical potential across the membrane, or the temperature or measure the rate of substrate movement in the presence of inhibitors of transport. Such studies are valuable insofar as they lead to the construction of models for the transport system, to the exact analysis of the predictions of such models and to the comparison of these predictions with experimental data. We shall see how one can, step by step, propose progressively more complex conceptual models for a transport system, rejecting these models when they conflict with the experimental data, until one reaches a model which is not yet rejectable by the available data. Insofar as such models are physically plausible they will in general suggest non-kinetic experiments designed to test them and in this way a closer and closer approach to an understanding of the mechanism of transport at the molecular level can be attained.

(a) Description of the experimental procedures

We will first describe certain experimental situations which lend themselves well to kinetic analysis. (We confine our discussion to non-charged substrates. The effect of charge has been considered elsewhere [1,2].) There are four of these of principal importance:

(i) The zero trans *procedure*
This is conceptually the simplest procedure. Here one sets the substrate concentration to zero at one face of the membrane (the *trans* face), while measuring the rate of transport of substrate from the opposite face (the *cis* face), when this face is exposed to various concentrations of substrate. Clearly there are two zero *trans* experiments, one (efflux) in which the *trans* face is bathed by the medium external to the cell, the other (influx) in which it is the inner face of the membrane which is *trans*. Examples of the data obtained in zero *trans* experiments are given in Fig. 1. Typically, the rate of transfer of substrate reaches a limiting maximum velocity as the substrate

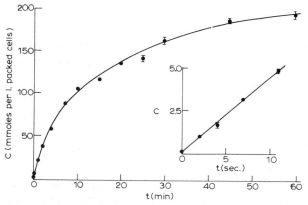

Fig. 1. Zero *trans* uptake of galactose by human red blood cells at 20°C. Time course of uptake from a solution containing 250 mM galactose in a salt medium to a total osmolarity of 560 mosM/l. The inset shows the very early portion of the uptake curve. The overall curve is linear at such short times but becomes more and more curved as galactose accumulates within the cell. (Data taken, with permission, from [33].)

concentration is increased (Fig. 2). This limiting velocity we can denote by V_{12}^{zt}, for the experiment where the *cis* face is face 1 of the membrane. The substrate concentration at which one half this maximum velocity is reached we denote as K_{12}^{zt} in this experiment. V_{21}^{zt} and K_{21}^{zt} denote the corresponding quantities for the experiment where face 2 of the membrane is the *cis* face. It is often found that the hyperbola

$$v_{12}^{zt} = \frac{S_1 V_{12}^{zt}}{K_{12}^{zt} + S_1} \qquad (1)$$

is a good description of how the rate of transport in the 1 to 2 direction, V_{12}^{zt}, depends on the substrate concentration, S_1, at face 1 of the membrane. A corresponding equation which describes transport in the 2 to 1 direction is obtained by interchanging the subscripts 1 and 2 in Eqn. 1.

Eqn. 1 has the form of the Michaelis–Menten equation, well-known to students of enzyme kinetics. The various ways in which these equations and the similar forms that we will come across later in this chapter can be manipulated so as to yield values of V_{12}^{zt} and K_{12}^{zt}, given experimental determination of v_{12}^{zt} at various values of S_1, are described in the classic texts of enzymology (see [3] and [4], and for transport studies [5]).

(ii) The equilibrium exchange procedure
In this procedure one first allows the cell to come into equilibrium with the substrate at a particular concentration. That is, the cell is first loaded with the substrate. If the substrate is uncharged or if there is no electrical potential gradient across the membrane, the substrate concentrations at the two faces of the membrane will be

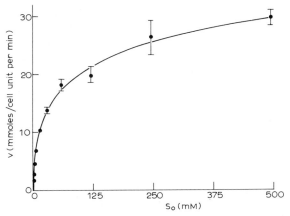

Fig. 2. Initial slopes of such uptake curves as a function of external galactose concentration. For this system there appears to be two parallel components of uptake, one of high affinity for sugar and the other of lower affinity. (Data taken, with permission, from [33].)

equal at equilibrium. After this equilibrium is reached, a quantity of radioactively labelled substrate is added to one face of the membrane, in concentrations so low as to not to add materially to the substrate concentration at that face. The rate of flow of substrate across the membrane is measured. The experiment is repeated at a set of different equilibrium concentrations and results similar to that of Fig. 2 are typically obtained. As in the zero *trans* procedure, a limiting velocity, which we can denote as V^{ee} is typically found and a half-saturation concentration K^{ee} can be defined. An equation of the form

$$v^{ee} = \frac{SV^{ee}}{K^{ee} + S} \qquad (2)$$

is often a good description of how the unidirectional flow of substrate, V^{ee}, in the 1 to 2 direction depends on the substrate concentration, S (the same at face 1 as at face 2). The unidirectional flux in the 2 to 1 direction is of necessity equal to that in the 2 to 1 direction at equilibrium. As before, V^{ee} and the appropriate term in K^{ee} can be found from measurements of v^{ee} at different values of the equilibrium substrate concentration.

(iii) The infinite trans *procedure*
Here the concentration of substrate at one face of the membrane (the *trans* face) is set at a limitingly high value, i.e. so high that measurements to be described in what follows are not significantly affected if this substrate concentration is further increased. (We will see later that this condition may not be able to be fulfilled for certain transport systems. Whether it can be fulfilled or not is an important clue as to the appropriateness or otherwise of certain models for the transport systems.) The substrate concentration at the *cis* face is set at various values and the unidirectional flux (in the *cis* to *trans* direction) of labelled substrate is measured at each *cis* substrate concentration. As the substrate concentration at the *cis* face is increased, the unidirectional flux will approach a limit given by V^{ee}. V^{ee} is identically that measured in the equilibrium exchange procedure since in both cases the concentration at both faces of the membrane is limitingly high. One can define a term K^{it}_{12} as the substrate concentration at face 1 at which one-half the limiting value of the flux is reached. An equation of the form

$$v^{it}_{12} = \frac{S_1 V^{ee}}{K^{it}_{12} + S_1} \qquad (3)$$

is often appropriate, with a corresponding form for the flux in the 2 to 1 direction (where now face 1 is the *trans* face and has the limiting high concentration of unlabelled substrate). Values of the terms in V^{ee} and K^{it} can be found from the determination of v^{it} at various values of S.

This experiment requires the same technical resources as the equilibrium exchange procedure and a comparison of the two is, as we shall see later, extremely instructive.

Concepts of mediated transport

(iv) The infinite cis procedure
For this procedure one, once again, sets the substrate concentration at one face (now the *cis* face) of the membrane at a limitingly high level—as defined in the preceding subsection (if the system is such that this condition can be fulfilled). One now measures the net *cis* to *trans* flow of substrate when the substrate concentration at the *trans* face is set at various levels. In the complete absence of *trans* substrate, the net flow of substrate has a maximal value and is of course equal to the maximal velocity of the corresponding zero *trans* experiment, i.e., V_{12}^{zt} or V_{21}^{zt}. As the substrate concentration at the *trans* face is increased, the net *cis* to *trans* flow is reduced, since the substrate can, in general, move in the *trans* to *cis* direction. One can define K_{21}^{ic} and K_{12}^{ic} as the half-saturation concentration of substrate at the *trans* face, which is just sufficient to reduce the net flow of substrate in the *cis* to *trans* direction to one-half of the maximal value.

An equation of the form

$$v_{12}^{ic} = \frac{K_{12}^{ic} V_{12}^{zt}}{K_{12}^{ic} + S_2} \tag{4}$$

with a corresponding equation for the net flow in the 2 to 1 direction is often appropriate. Fig. 3 depicts data obtained in such infinite *cis* experiments.

This experiment requires the same technical resources as does the zero *trans* procedure and once again a comparison of the data obtained by the two procedures can be very useful.

(b) Some general considerations

This simple, formal description of these experimental procedures conceals a number of formidable technical difficulties in their application. The foremost of these arises from the nature of the transport event itself. Of necessity transport involves the movement of substrates from one membrane face to another, a process which in general alters those substrate concentrations which must themselves be fixed during the experiment. Measurements of the initial rate of the transport event (that is, the rate before any significant change in substrate concentration has occurred), are, by definition, sufficient to allow meaningful data to be accumulated. But the small changes that must occur in such cases are often difficult to measure sensitively. Integration with respect to time of the initial rate equations allows use of the technique where the change of substrate concentration with time is followed, and leads to the obtaining of a maximum amount of information. Another approach is to use very large volumes of bathing solution as the extracellular phase. Changes in concentration at the extracellular face are thus minimised. Detailed descriptions of the experimental procedures used can be found in Eilam and Stein [6] and in the original papers cited in the legends to Figs. 1 to 3.

The presence of a significant unstirred layer at one or both faces of the membrane

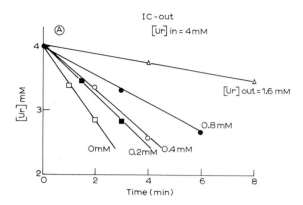

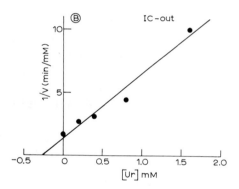

Fig. 3. Infinite *cis* efflux from human red blood cells at 25°C. Upper curve, efflux of uridine as a function of time, into various concentrations of external uridine, as indicated. Internal uridine was 4 mM. Lower curve, slopes of the regression lines here, plotted reciprocally as a function of the external uridine concentration. The intercept on the abscissa gives the negative value of the external K^{ic}, which was here 0.25 mM. (Data taken, with permission, from [8].)

will make it impossible to set the conditions for certain procedures. Thus, for example, an unstirred layer will, in general, ensure that a rigorous zero *trans* condition cannot be set up. The concentration of the substrate in the bulk solution (the *trans* face) can be set indistinguishably different from zero, yet within the unstirred layer at the *trans* face there is, by definition, a gradient of substrate so that the transport system sees a definite, non-zero concentration of substrate at the *trans* face. The treatment of kinetic data for systems in which a significant unstirred layer is present is dealt with by Lieb and Stein [1,2].

3. First model for facilitated diffusion—the simple pore

(a) Kinetic analysis

The fact that so many transport systems (cf. Figs. 1–3) display the same characteristic form for the dependence of velocity on substrate concentration as do enzyme systems strongly suggests that a formal analysis of transport kinetics along the lines of that of enzyme kinetics might be valuable. The simple Michaelis–Menten or hyperbolic velocity vs. substrate concentration curve for enzymes has traditionally been interpreted as arising from the combination between enzyme and substrate, with the subsequent breakdown of this complex to product and free enzyme. One writes

$$E + S \underset{b_1}{\overset{f_1}{\rightleftharpoons}} ES \underset{f_2}{\overset{b_2}{\rightleftharpoons}} E + P \tag{5}$$

where E is the enzyme, S the substrate, P the product and ES is the enzyme-substrate complex (which is also the enzyme-product complex for the back reaction). The coefficients in b represent the rate of $b$reakdown of the complex, the coefficient in f, the rate of its $f$ormation. When the concentration of P is vanishingly small, i.e., during the very early stages of the enzymic reaction, the rate of the reaction will be b_2 times the concentration of ES. At low concentrations of S, the concentration of ES will be proportional to S; at high S, the concentration of ES and hence the rate of enzymic reaction will reach a limiting value, independent of S, and the classic hyperbolic form of the Michaelis–Menten equation can be derived.

If, now, we are to consider the interaction between a transport system and its substrate we will naturally proceed on the same lines. The symbol E will represent a site present in the membrane which interacts with the substrate S. ES is the complex between site and substrate resulting from this interaction—and the "reaction" here results in the appearance of substrate at the opposite face of the membrane. We will write S_1 as the substrate present at face I of the membrane, while S_2 is the substrate present at face II. Then we can write, as the simplest possible formulation

$$E + S_1 \underset{b_1}{\overset{f_1}{\rightleftharpoons}} ES \underset{f_2}{\overset{b_2}{\rightleftharpoons}} E + S_2 \tag{6}$$

We do not at this stage make any assumptions as to what may be the details of the transport event. We say simply that transport occurs after, and as a result of, the interaction of the substrate S with the site E. Diagramatically we can represent this model as in Fig. 4. From the figure we can see that this model looks like and is in fact, a simple pore. We can, indeed, formally define the simple pore as a system which behaves according to the formalism of Eqn. 6. Again the rate constants in b are for the breakdown of the complex; f, those for its formation. Subscripts 1 refer to processes originating at face I, those with 2, to processes originating in face II.

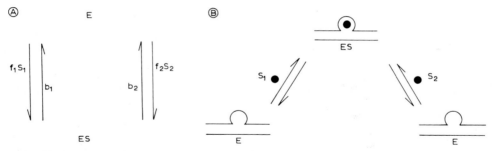

Fig. 4. The simple pore. (A) formal representation; (B) cartoon diagram. Rate constants f_1 and f_2 describe the formation, b_1 and b_2 the breakdown of the pore-substrate complex at sides 1 and 2 of the membrane, respectively. The solid circle represents the substrate molecule. E is the unoccupied pore, ES is the pore occupied by the substrate S.

How can we work out the predictions of this simple model? We note first that the unidirectional flow of substrate from side I to side II is given by the rate at which substrate from side I is transformed into the form ES (i.e. by $f_1 S_1 E$) multiplied by a factor which allows for the fact that a part only of ES breaks down in the direction towards side II, while a part breaks down once again at side I. The required factor is clearly $b_2/b_1 + b_2$. Thus the unidirectional flux in the direction 1 to 2 is given by

$$f_{12} = \frac{b_2}{b_1 + b_2} \cdot f_1 S_1 E \tag{7}$$

Thus if we know the concentration of E, we can calculate the unidirectional flux — and if this is known all else is known, since any net flux is always composed of corresponding unidirectional fluxes. To work out the concentration of E in this simple case, we make use of what is known as the *steady-state assumption*. This assumption is that the concentration of intermediate forms (here ES) in such a reaction reaches a steady state within a time well below the resolution of the conventional kinetic analyses. That is, the formation of ES balances its breakdown (at fixed values of S_1 and S_2) within a fraction of a second. With this assumption (well-proven for enzyme kinetics) we may write:

Formation of ES = breakdown of ES (8)

Now ES is formed by the interaction of free E with S_1 or S_2 at rates governed by the relevant rate constants f_1 and f_2 and it breaks down at a rate given by concentration ES and the rate constants b_1 and b_2. Thus

$$f_1 S_1 E + f_2 S_2 E = (b_1 + b_2) \, ES \tag{9}$$

The total concentration of site can be written as Tot where

$$E + ES = \text{Tot} \tag{10}$$

This equation is known as the conservation equation since it arises from the fact that the total amount of site, in free or combined form, has to be conserved, however the relative amount of E and ES varies.

Eliminating ES from the simultaneous Eqns. 9 and 10 and simplifying we get:

$$E = \frac{b_1 + b_2}{b_1 + b_2 + f_1 S_1 + f_2 S_2} \cdot \text{Tot} \tag{11}$$

so that the unidirectional flux in the direction 1 to 2 is

$$f_{12} = \frac{b_2 f_1 S_1 \text{Tot}}{b_1 + b_2 + f_1 S_1 + f_2 S_2} \tag{12}$$

This, now, is the desired result. We have an Eqn. 12 which describes how the unidirectional flux f_{12} varies with the concentrations of substrate at the two sides of the membrane S_1 and S_2, in terms of the rate constants of the model. Clearly, for the corresponding flux in the 2 to 1 direction, we need merely interchange the subscripts 1 and 2 in the result given by Eqn. 12. We can now see whether the model correctly predicts the four types of experimental procedure that we defined above.

For the subsequent discussion it is convenient to transform Eqn. 12 into a form involving experimentally accessible parameters. We multiply each term in the numerator and denominator of the right-hand side of Eqn. 12 by $(1/b_1 + 1/b_2)/f_1 f_2$. On making the substitutions given in columns 1 and 2 of Table 1 and putting $b_1 f_2 = b_2 f_1$ (see below), Eqn. 12 transforms into:

$$f_{12} = \frac{S_1}{Q + R_{12} S_1 + R_{21} S_2} \tag{13}$$

A corresponding equation for the unidirectional flux f_{21} will be obtained if the subscripts 1 and 2 are interchanged in Eqn. 13.

TABLE 1
Interpretation of fundamental transport parameters in terms of rate constants for the simple pore and for the complex pore

Parameter	Simple-pore interpretation	Complex-pore interpretation
R_{12}	$\dfrac{1}{b_2}$	$\dfrac{1}{b_2} + \dfrac{1}{g_1}\left(\dfrac{b_2 + g_2}{b_2}\right)$
R_{21}	$\dfrac{1}{b_1}$	$\dfrac{1}{b_1} + \dfrac{1}{g_2}\left(\dfrac{b_1 + g_1}{b_1}\right)$
Q	$\dfrac{1}{f_1} + \dfrac{1}{f_2}$	$\dfrac{1}{f_1} + \dfrac{1}{f_2} + \dfrac{b_1}{f_1 g_1}$
		$= \dfrac{1}{f_1} + \dfrac{1}{f_2} + \dfrac{b_2}{f_2 g_2}$
Constraint:	$b_1 f_2 = b_2 f_1$	$b_1 f_2 g_2 = b_2 f_1 g_1$

(i) Zero trans *procedure on the simple pore*
We take side 2 as the *trans* side. Since the *trans* concentration is zero we put $S_2 = 0$ in Eqn. 13. We obtain, on dividing through by Q,

$$f_{12} = \frac{S_1}{Q + R_{12}S_1} \tag{14}$$

Now this is precisely the form of Eqn. 1, if we put $V_{12}^{zt} = 1/R_{12}$ and $K_{12}^{zt} = Q/R_{12}$. Furthermore, the ratio V_{12}^{zt}/K_{12}^{zt}, the limiting permeability, is given by $1/Q$ (Table 2). Clearly the simple-pore model can account for a Michaelis–Menten type dependence of transport velocity on substrate concentration, in the zero *trans* procedure.

(ii) Equilibrium exchange on the simple pore
We substitute $S_1 = S_2 = S$ in Eqn. 13 and obtain:

$$f_{12} = \frac{S}{Q + (R_{12} + R_{21})S} \tag{15}$$

Once again, this is of the form of Eqn. 2, if we put $V^{ee} = 1/(R_{12} + R_{21})$ and $K^{ee} = Q/(R_{12} + R_{21})$ while $V^{ee}/K^{ee} = 1/Q$, with the corresponding forms for the 2 to 1 flux, with the subscript 1 and 2 interchanged. We get a Michaelis–Menten form, once again.

The permeability at limitingly low substrate concentration for the equilibrium exchange procedure in a particular direction is the same as that for the zero *trans* procedure in that direction (since the *trans* concentration is, in the limit, zero in both cases). At equilibrium, these unidirectional flows are numerically equal, so from Eqn. 12, since S_1 and S_2 are negligible in comparison with b_1 and b_2, we have:

$$b_2 f_1 S_1 = b_1 f_2 S_2$$

but $S_1 = S_2$, therefore

$$b_1 f_2 = b_2 f_1 \tag{16}$$

This relation between the rate constants in b and f must hold for a system which is not linked to any source of energy input and which has therefore no possibility of a net flow of substrate under equilibrium conditions.

(iii) The infinite trans *procedure on the simple pore*
For this procedure, we take S_2 as the concentration at the *trans* face and let S_2 become limitingly high in Eqn. 12 or 13. Then the surprising result is obtained that the unidirectional flux in the *cis*-to-*trans* direction is zero at all finite values of S_1. An equation of the form of Eqn. 3 cannot be obtained. There is no maximum velocity in the infinite *trans* procedure at accessible substrate concentrations and a finite value for the half-saturation concentration K_{12}^{it} cannot be obtained. The conclusion follows:

If the infinite *trans* procedure yields a finite value for the half-saturation concentration, the simple-pore model of Fig. 5 cannot be an adequate description of the system and the simple-pore model is to be rejected.

(iv) The infinite cis procedure on the simple pore
For this net transport experiment, the net flow of substrate in the 1 to 2 direction is obtained using Eqn. 13 and its analogue with the subscripts 1 and 2 interchanged. We have

$$\text{Net}_{12}^{ic} = \frac{(S_1 - S_2)}{Q + R_{12}S_1 + R_{21}S_2} \qquad (17)$$

If now the *cis* side concentration, S_1, is set at a limitingly high value the net flow of substrate is equal to $1/R_{12}$. This is the maximum zero *trans* velocity and this value will not change as S_2 is increased to any finite concentration. (This is so by definition since with S_1 limitingly high, no finite value of S_2 can affect the value of the numerator or denominator of Eqn. 17.) An equation of the form of Eqn. 4 with a finite half-saturation concentration at the *trans* face K_{12}^{ic} cannot be obtained for a simple pore. Once again, if the infinite *cis* experiment can be performed so that a finite value of the half-saturation concentration K_{12}^{ic} is measurable, the simple-pore model must be rejected.

(b) Some further tests for the simple pore

We have seen that if the infinite *trans* and infinite *cis* experiments give finite values for their respective half-saturation procedures, then the simple pore model cannot hold. But if these experiments do not yield finite values for such parameters, then the simple pore is temporarily a satisfactory model and must be further tested. We consider now some such tests.

From Eqn. 14 and the analogous form with 1 and 2 interchanged, $1/V_{12}^{zt} = R_{12}$ and $1/V_{21}^{zt} = R_{21}$. Similarly from Eqn. 15,

$$1/V^{ee} = R_{12} + R_{21} \qquad (18)$$

Hence we have

$$1/V^{ee} = 1/V_{12}^{zt} + 1/V_{21}^{zt} \qquad (19)$$

This result must hold if the system behaves as a simple pore. V^{ee}, V_{12}^{zt}, and V_{21}^{zt} can be measured and the applicability or otherwise of Eqn. 19 can be used as a rejection criterion for the simple pore. Eqn. 19 itself implies that V^{ee} is smaller than, or equal to, both V_{12}^{zt} and V_{21}^{zt}. Any other result rejects the simple pore as defined in Eqn. 6. (Where such a test has been made for transport systems, the most commonly found result is indeed that V^{ee} is bigger than at least one of the V^{zt} terms. Such systems cannot be simple pores.)

TABLE 2

Interpretation of experimentally measurable transport parameters in terms of fundamental transport parameters for the "pore"

Experimental parameter	Fundamental interpretation
Maximal velocities:	
Zero *trans*:	
V_{12}^{zt}	$\dfrac{1}{R_{12}}$
V_{21}^{zt}	$\dfrac{1}{R_{21}}$
Equilibrium exchange:	
V^{ee}	$\dfrac{1}{R_{12}+R_{21}}$
Michaelis (half-saturation) parameters:	
Zero *trans*:	
K_{12}^{zt}	$\dfrac{Q}{R_{12}}$
K_{21}^{zt}	$\dfrac{Q}{R_{21}}$
Equilibrium exchange:	
K^{ee}	$\dfrac{Q}{R_{12}+R_{21}}$
Permeability $\left(\dfrac{\text{velocity at limitingly low substrate concentration}}{\text{substrate concentration}}\right)$	$\dfrac{V_{12}^{zt}}{K_{12}^{zt}} = \dfrac{V_{21}^{zt}}{K_{21}^{zt}} = \dfrac{V^{ee}}{K^{ee}} = \dfrac{1}{Q}$

If we multiply Eqn. 19 through by V^{ee} and make use of the fact that the ratio of maximum velocities to half-saturation concentrations (V/K) is the same for a measurement in a particular direction (see above) we obtain

$$\frac{K^{ee}}{K_{12}^{zt}} + \frac{K^{ee}}{K_{21}^{zt}} = 1 \tag{20}$$

This is another useful rejection criterion and one which implies furthermore that

$$K^{ee} < K_{12}^{zt}$$

and

$$K^{ee} < K_{21}^{zt}$$

predictions which are often contradicted by experimental results for transport systems.

4. Second model for facilitated diffusion: the complex pore

When the experimental data are sufficient to reject the simple pore, defined in Eqn. 6 and depicted in Fig. 4, it becomes necessary to propose and test a more complex model. We recall that the simple pore was one with merely a single binding site accessible to substrate from both sides of the membrane. We might now proceed to what is perhaps a more realistic model, one which takes into account the finite thickness of the membrane. We assume that there are now two binding sites along the pore. Substrate combines with the site at one end of the pore, is transferred within the pore to the second binding site (at rates governed by the rate constants g_1 and g_2) from which it can emerge at the opposite face. A formal picture of this model is given in Fig. 5. One assumes that only one molecule of substrate can be combined with the pore at any one time (a singly accepting pore).

It is easy to show that this model gives precisely the same formal kinetic predictions as does that of the simple pore. The meaning of the parameters in Q and R of Eqn. 13 in terms of rate constants is different for the two models (the relevant results are collected in Table 1), but the prediction in terms of *experimentally determinable parameters* are identical. Making the pore model more complex in this particular way does not save it, so that if the simple pore is rejectable so is this more complex pore.

This result is to be expected on the analogy between transport kinetics and enzyme kinetics. It is a well-known result in this latter discipline [4] that the introduction of intermediate forms in a kinetic scheme will not affect the steady-state predictions of that scheme, if these intermediate forms are not able to combine with a substrate or product species (or some modifier). Now, the transition between ES_1 and ES_2 in Fig. 5 is just such a transition between forms which do not *combine* with substrate or product, and hence this step cannot be seen by steady-state methods. The simple pore of Fig. 4 is thus kineticaly equivalent at the steady-state level to the more complex pore of Fig. 5 and indeed to any more complex pore involving an indefinite number of such intermediate transitional forms between ES_1 and ES_2.

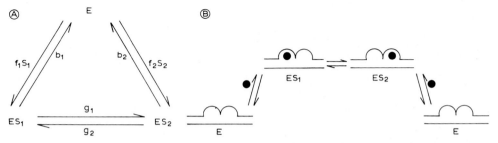

Fig. 5. The complex pore. (A) formal representation; (B) cartoon diagram. Symbols as in Fig. 4, except that rate constants g_1 and g_2 describe the interconversion of the two forms of the pore-substrate complex, ES_1 and ES_2.

5. Third model for facilitated diffusion: the simple carrier [2]

(a) Introduction

To accommodate the existence of determinable infinite *trans* and infinite *cis* kinetic parameters we will have to make the model of Fig. 5 more complex in a way other than the assumption of intermediate ES forms. Now, the essential feature of the infinite *trans* and *cis* experiments is this: although there is a limitingly high concentration of substrate at one face of the membrane, substrate at the opposite face at concentrations within an accessible range can influence the transport kinetics. The interaction of substrate and system at one face of the membrane is shielded from the events at the opposite face. Another way of looking at this is: substrate at one face of the membrane is not in competition with that at the other face. To account for this in terms of a model, one must erect some sort of barrier across the membrane to separate the two sets of substrate-site interactions. The most relevant type of barrier is an energy barrier. We modify the simple pore, therefore, by defining two forms of the free site E_1 and E_2 accessible exclusively to sides 1 and 2 of the membrane, respectively, and separated by an activation-energy barrier. The kinetic scheme of Fig. 6 represents such a model. The rate constants k_1 and k_2 describe the rates of interconversion of the two forms E_1 and E_2, these rate constants being, of course, determined by the heights of the activation energy barrier postulated between these two forms.

We note that the transition between the forms E_1 and E_2 is between forms which interact with different species of the substrate (S_1 and S_2, respectively). It is to be expected, therefore, that such transitions can be revealed by kinetic analysis at the steady-state level. We will see later that the model of Fig. 6 contains all the kinetic features of the classical carrier model. Thus we define a system which behaves according to the kinetic scheme of Fig. 6 as "the simple carrier" [2].

To calculate the unidirectional flux from, say, side 1 on this scheme we proceed as before, where we considered the simple pore. The flux is given by the rate at which S_1 enters the system, that is, by the term $f_1 S_1 E_1$ (the rate constant for the particular

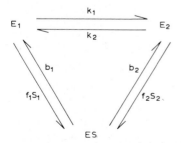

Fig. 6. The simple carrier. Formal representation of kinetic scheme. Symbols as in Fig. 3, except that k_1 and k_2 describe the rates of interconversion of the unoccupied carrier between its two forms, E_1 and E_2.

Concepts of mediated transport

process, multiplied by concentrations of those species taking part in the reaction), multiplied by a factor $b_2/b_1 + b_2$ which accounts for the fact that part of this substrate is released into side II while part returns to side I. We need to find the concentration of E_1. We set up the steady-state condition for the intermediate form as follows. Steady-state for ES:

$$f_1 S_1 E_1 + f_2 S_2 E_2 = (b_1 + b_2) ES \tag{21}$$

steady state for E_1:

$$k_2 E_2 + b_1 ES = (k_1 + f_1 S_1) E_1 \tag{22}$$

steady-state for E_2:

$$k_1 E_1 + b_2 ES = (k_2 + f_2 S_2) E_2 \tag{23}$$

We also have the relevant conservation equation:

$$E_1 + E_2 + ES = \text{Tot } E \tag{24}$$

Of these four equations, one of the steady-state equations is redundant since it can be derived from the conservation equation and two of the three steady-state equations. We have three independent simultaneous equations in the three unkowns E_1, E_2 and ES. These can be solved in the conventional manner (for example, by determinants) and one obtains that:

$$\frac{E_1}{\text{Tot } E} = \frac{b_1 f_2 S_2 + k_2 (b_1 + b_2)}{\Sigma} \tag{26}$$

$$\frac{E_2}{\text{Tot } E} = \frac{b_2 f_1 S_1 + k_1 (b_1 + b_2)}{\Sigma} \tag{27}$$

$$\frac{ES}{\text{Tot } E} = \frac{f_1 k_2 S_1 + f_2 k_1 S_2 + f_1 f_2 S_1 S_2}{\Sigma} \tag{28}$$

where Σ is the sum of the numerators of the right-hand sides of Eqns. 26–28.

From these equations, and using the argument in the paragraph above, the unidirectional flux from side I to side II becomes, after simplifying

$$f_{12} = \frac{\left(b_2 f_1 k_2 S_1 + \left(\dfrac{b_1 b_2}{b_1 + b_2} \right) f_1 f_2 S_1 S_2 \right) \text{Tot } E}{(k_1 + k_2)(b_1 + b_2) + f_1 (b_2 + k_2) S_1 + f_2 (b_1 + k_1) S_2 + f_1 f_2 S_1 S_2} \tag{29}$$

By analogy with Eqn. 16, $b_2 f_1 k_2 = b_1 f_2 k_1$, for no active transport. To transform Eqn. 29 into a form involving measurable parameters only, we multiply each term on the right-hand side of the equation by the factor $(b_1 + b_2)/b_1 b_2 f_1 f_2$, and then make the substitution listed in columns 1 and 2 of Table 3. We obtain:

$$f_{12} = \frac{KS_1 + S_1 S_2}{K^2 R_{00} + KR_{12} S_1 + KR_{21} S_2 + R_{ee} S_1 S_2} \tag{30}$$

where $R_{00} + R_{ee} = R_{12} + R_{21}$.

This equation gives the unidirectional flux of substrate in the direction 1 to 2. A corresponding equation with the subscripts 1 and 2 interchanged gives the flux in the direction 2 to 1. The fluxes are in terms of the concentrations S_1 and S_2 and measurable parameters as we proceed to show.

(b) The zero trans *and equilibrium exchange procedures on the simple carrier*

Precisely as we proceeded in the discussion of the simple pore, we can derive the predictions of the simple carrier for the procedures of zero *trans* and equilibrium exchange. Simple Michaelis–Menten forms are obtained with the maximum velocity and half-saturation concentrations listed in Table 4.

(c) The infinite trans *procedure on the simple carrier*

If we take side 2 as the *trans* solution and let the value of S_2 in Eqn. 30 become limitingly high, we obtain after simplification:

$$v_{12}^{it} = \frac{S_1}{KR_{21} + R_{ee} S_1} \tag{31}$$

TABLE 3

Interpretation of fundamental transport parameters in terms of rate constants for the simple carrier and for the conventional carrier

Parameter	Simple-carrier interpretation	Conventional-carrier interpretation
R_{12}	$\frac{1}{b_2} + \frac{1}{k_2}$	$\frac{1}{b_2} + \frac{1}{k_2} + \frac{1}{g_1} \frac{b_2 + g_2}{b_2}$
R_{21}	$\frac{1}{b_1} + \frac{1}{k_1}$	$\frac{1}{b_1} + \frac{1}{k_1} + \frac{1}{g_2} \frac{b_1 + g_1}{b_1}$
R_{00}	$\frac{1}{k_1} + \frac{1}{k_2}$	$\frac{1}{k_1} + \frac{1}{k_2}$
R_{ee}	$\frac{1}{b_1} + \frac{1}{b_2}$	$\frac{1}{b_1} + \frac{1}{b_2} + \frac{1}{g_1} + \frac{1}{g_2} + \frac{g_2}{g_1 b_2} + \frac{g_1}{g_2 b_1}$
K	$\frac{k_1}{f_1} + \frac{k_2}{f_2}$	$\frac{k_1}{f_1} + \frac{k_2}{f_2} + \frac{b_1 k_1}{g_1 f_1} \left(= \frac{b_2 k_2}{g_2 f_2} \right)$
Constraint:	$b_1 f_2 k_1 = b_2 f_1 k_2$	$b_1 f_2 g_2 k_1 = b_2 f_1 g_1 k_2$

Concepts of mediated transport

TABLE 4

Interpretation of experimentally measurable transport parameters in terms of fundamental transport parameters for the carrier [a]

Procedure	Maximum velocity	Michaelis parameter	"Permeability"
Zero *trans*			
$1 \to 2$	$V_{12}^{zt} = \dfrac{1}{R_{12}}$	$K_{12}^{zt} = K \dfrac{R_{00}}{R_{12}}$	$\dfrac{V_{12}^{zt}}{K_{12}^{zt}} = KR_{00}$
$2 \to 1$	$V_{21}^{zt} = \dfrac{1}{R_{21}}$	$K_{21}^{zt} = K \dfrac{R_{00}}{R_{21}}$	$\dfrac{V_{21}^{zt}}{K_{21}^{zt}} = KR_{00}$
Infinite *cis*			
$1 \to 2$	as zero *trans*	$K_{12}^{ic} = K \dfrac{R_{12}}{R_{ee}}$	no meaning
$2 \to 1$	as zero *trans*	$K_{21}^{ic} = K \dfrac{R_{21}}{R_{ee}}$	no meaning
Equilibrium exchange			
$1 \to 2 = 2 \to 1$	$V^{ee} = V_{12}^{ee} = V_{21}^{ee}$ $= \dfrac{1}{R_{ee}}$	$K^{ee} = K_{12}^{ee} = K_{21}^{ee}$ $= K\dfrac{R_{00}}{R_{ee}}$	$\dfrac{V^{ee}}{K^{ee}} = KR_{00}$
Infinite *trans*			
$1 \to 2$	$V_{12}^{it} = V^{ee} = \dfrac{1}{R_{ee}}$	$K_{12}^{it} = K \dfrac{R_{21}}{R_{ee}}$	$\dfrac{V_{12}^{it}}{K_{12}^{it}} = KR_{21}$
$2 \to 1$	$V_{21}^{it} = V^{ee} = \dfrac{1}{R_{ee}}$	$K_{21}^{it} = K \dfrac{R_{12}}{R_{ee}}$	$\dfrac{V_{21}^{it}}{K_{21}^{it}} = KR_{12}$

[a] "Permeability" is defined as the velocity of transport divided by the substrate concentration from the side transport is being measured, as the substrate concentration is reduced to limitingly low values.

This is clearly a simple Michaelis–Menten form with a maximum velocity of $v_{12}^{it} = 1/R_{ee}$ and a half-saturation concentration $K_{12}^{zt} = KR_{21}/R_{ee}$. The infinite *trans* procedure will thus give, in general, a half-saturation concentration in an accessible concentration range on the simple carrier model of Fig. 6.

(d) The infinite cis procedure on the simple carrier

The net flow of substrate is given from Eqn. 30 (and its analogue with subscripts 1 and 2 interchanged) on letting the substrate concentration at the *cis* side, taken as side 1, reach limitingly high levels. The net flow in the 1 to 2 direction is, after simplifying,

$$v_{12}^{ic} = \frac{K}{KR_{12} + R_{ee}S_2} \tag{32}$$

When S_2, the concentration at the *trans* side, is zero, the net flow is given by $1/R_{12}$

— as it should be since this is the maximum velocity of the zero *trans* experiment in the 1 to 2 direction. But as S_2 increases, the net flux decreases until at $S_2 = K(R_{12}/R_{ee})$, the net flow is one-half that in the absence of substrate at the *trans* side. The system behaves formally as Eqn. 4, so that for a system which behaves as a simple carrier, the infinite *cis* procedure is performable and a half-saturation concentration K_{12}^{ic} is measurable in an accessible concentration range. Its value is exactly the same as that of K_{21}^{it}, if the carrier model holds.

(e) The simple pore and simple carrier compared

It is important to understand clearly what is the molecular basis of the difference between the simple pore and the simple carrier. Formally the difference is expressed in the difference between the mathematical equation describing the kinetic predictions of the model (Eqns. 13 and 30, respectively). Only the latter contains terms involving the product S_1S_2, such terms allowing for the mutual interaction of the substrate species at the two membrane faces. But why should such terms not appear in the equation for the simple pore? If in Eqn. 29 for the simple carrier, we let the rate constants k_1 and k_2 become very large, only terms containing these constants will remain in the equation so that, after simplifying, we will obtain:

$$f_{12} = \frac{b_2 f_1 S_1}{\frac{k_1+k_2}{k_2}(b_1+b_2)+f_1S_1+f_2\frac{k_1}{k_2}S_2} \tag{33}$$

which is identical in form with the equation for the simple pore.

Thus if the rate constants for the interconversion of the two forms E_1 and E_2 of the carrier become exceedingly large—that is, there is no detectable activation energy barrier between these two forms—the simple carrier reverts to the simple pore. Conversely, if the terms in k are not infinitely great, that is, if a non-zero value for R_{00} can be measured or if the infinite *cis* and infinite *trans* experiments give measurable values for half-saturation parameters, there is a detectable energy barrier between the states E_1 and E_2 of the free carrier. It is important to emphasize that none of this analysis requires that the carrier must move bodily across the membrane during the transition from E_1 to form E_2. All that is required of the model is that *some* conformation change of the site must occur during the transition such that a binding site originally accessible to substrate at face I of the membrane is, after the conformation change, shielded from face I but accessible at face II. If in a real system this transition occurs at a finite rate, terms in Eqn. 29 which do not contain the rate constant for this transition—and these are the cross terms involving S_1 *and* S_2—cannot be neglected and the predictions of the simple pore will be able to be rejected.

TABLE 5
Characterising and testing the simple carrier

(i) *Consistency tests:* Using experimentally determined maximum velocities of zero *trans*, infinite *cis*, equilibrium exchange and infinite *trans* procedures, obtain R_{12}, R_{21} and R_{ee} from Table 4. From the "permeability" for the infinite *trans* experiments and R_{12} and R_{21}, obtain K, using Table 4. Now, using K and the "permeability" for the two zero *trans* experiments and the equilibrium exchange (check that the measured permeabilities are indeed identical!), find R_{00}. All fundamental parameters are now determined and any further determination becomes a test of the model. Thus, any Michaelis parameter is now calculatable using Table 4 and the derived fundamental parameters. Thus by determining K_{12}^{zt} or K_{12}^{ic} and so on, derived and experimental values can be compared and the model tested for consistency. If the model passes all such tests, the system can be characterised in terms of the five fundamental parameters, the four R values and K. It must also be the case that $R_{12} + R_{21} = R_{ee} + R_{00}$.

(ii) *Testing relations of identity:* There are a number of identities which can be derived mathematically and which relate measurable kinetic parameters. These include (with corresponding relations on interchanging subscripts 1 and 2 throughout)

(a) $K_{12}^{it} = K_{21}^{ic}$

(b) $\dfrac{K_{12}^{zt}}{K_{21}^{zt}} = \dfrac{V_{12}^{zt}}{V_{21}^{zt}} = \dfrac{K_{12}^{it}}{K_{21}^{it}} = \dfrac{K_{21}^{ic}}{K_{12}^{ic}} = A$, the asymmetry factor.

(c) $\dfrac{V_{12}^{zt}}{K_{12}^{zt}} = \dfrac{V_{21}^{zt}}{K_{21}^{zt}} = \dfrac{V^{ee}}{K^{ee}} =$ "Permeability of zero *trans*"

(d) $\dfrac{V_{12}^{it}}{K_{12}^{it}} = \dfrac{V_{21}^{it}}{K_{21}^{it}} =$ "Permeability of infinite *trans*"

(e) Using the asymmetry factor A, as defined in (b) above:
$$A + 1 = \dfrac{K^{ee}}{K_{12}^{ic}} + \dfrac{K_{12}^{zt}}{K^{ee}}$$

(f) $(A+1)^2 \geqslant 4 \dfrac{K_{12}^{zt}}{K_{12}^{ic}}$

(g) $\dfrac{1}{K_{12}^{zt}} + \dfrac{1}{K_{21}^{zt}} = \dfrac{1}{K^{ee}} + \dfrac{1}{K}$ $\left(= \dfrac{K^{ee}}{K_{12}^{zt} K_{12}^{ic}} \right)$

(K can also be derived directly from a countertransport experiment, by determining the maximum concentration ratio reached during counterflow [2].)

(iii) *Using irreversible inhibition data.* An irreversible inhibitor of a transport system will often interact preferentially with one form of the carrier—E_1, E_2 or ES. Then, depending on whether substrate is present on both sides of the membrane, one side only or on either side, the rate of inhibition will be different. The effect of substrate on increasing (or decreasing) the rate of inactivation by the inhibitor can be considered as if it were affecting the rate of an enzyme reaction and a maximum effect of substrate and a half-saturation concentration of substrate to reach this maximum effect can be defined. Using the symbols ΔQ for the maximum effect of substrate and ΔK for the half-saturation concentration, with appropriate sub- and superscripts we can write [36]

(h) $\dfrac{\Delta Q_{12}^{zt}}{V_{12}^{zt}} + \dfrac{\Delta Q_{21}^{zt}}{V_{21}^{zt}} = \dfrac{\Delta Q^{ee}}{V^{ee}}$

(i) $\Delta K_{12}^{zt} = K_{12}^{zt}$

(j) $\Delta K_{21}^{zt} = K_{21}^{zt}$

(k) $\Delta K^{ee} = K^{ee}$

(f) Some further tests for the simple carrier [2]

The fact that the infinite *cis* and *trans* experiments can be performed and yield finite values of the respective half-saturation concentration leads, as we have seen, to the rejection of the simple-pore model (and its more complex form). The simple carrier can then temporarily be considered acceptable for such systems as yield finite half-saturation concentrations for these procedures. But the actual value of these parameters may or may not be consistent with the simple carrier and hence one can develop rejection criteria for the simple carrier in terms of the experimentally measurable parameters. The point of such an analysis is the following: For a system which behaves as a simple carrier, the unidirectional flux Eqn. 30 is appropriate and will serve to account for all steady-state experiments involving the single substrate S. Yet Eqn. 30 contains only four independently variable parameters—one form in K and three forms in R (since the forms are connected by $R_{00} + R_{ee} = R_{12} + R_{21}$). Yet there are many more than four experimentally determinable descriptors of the transport process. These include (i) the three maximum velocities for the two zero *trans* procedures and the equilibrium exchange procedure and (ii) the seven half-saturation concentrations (two each for the zero *trans*, infinite *trans* and infinite *cis* procedures and one for the equilibrium exchange procedure). The experimentally determinable descriptors must therefore be mutually interrelated and tests of the applicability of such interrelationships provide rejection criteria for the simple carrier. Some useful rejection criteria are collected in Table 5. The application of some of these criteria to specific systems is discussed in [8] and [9].

6. Fourth model for facilitated diffusion: the conventional carrier

The simple carrier of Fig. 6 is the simplest model which can account for the range of experimental data commonly found for transport systems. Yet surprisingly, it is not the model that is conventionally used in transport studies. The most commonly used model is some or other form of Fig. 7. In contrast to the simple carrier, the model of Fig. 7, the conventional carrier, assumes that there exist two forms of the carrier-substrate complex, ES_1 and ES_2, and that these can interconvert by the transitions with rate constants g_1 and g_2. Now, our experience with the simple- and complex-pore models should lead to an awareness of the problems in making such an assumption. The transition between ES_1 and ES_2 is precisely such a transition as cannot be identified by steady-state experiments, if the carrier can complex with only one species of transportable substrate. Lieb and Stein [2] have worked out the full kinetic analysis of the conventional carrier model. The derived unidirectional flux equation is exactly equivalent to that derived for the simple carrier Eqn. 30, although the experimentally determinable parameters involving K and R terms have different meanings in terms of the rate constants (the b, f, g and k terms). The appropriate values for the K and R terms in terms of the rate constants are listed in column 3 of Table 3. Thus the simple carrier and the conventional carrier behave identically in

Concepts of mediated transport

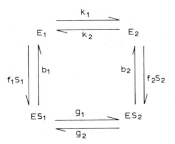

Fig. 7. The conventional carrier. Formal representation of the kinetic scheme. Symbols as in Figs. 5 and 6.

steady-state transport experiments, so that for the description of such experiments the simple carrier is to be preferred (so long as the model cannot be rejected by the available data).

People who have worked many years in the transport field have strong intuitive feelings for the correctness of the conventional carrier model. It seems perfectly natural to assume that the substrate-carrier complex formed at one face of the membrane does undergo a transition to a complex which breaks down at the opposite face. Nevertheless, steady-state kinetic experiments cannot justify this assumption and the simpler model yields decidedly simpler kinetic expressions and to a heightened awareness as to the molecular interpretation of the measurable kinetic parameters, as we proceed to discuss.

The use of the conventional carrier model (as opposed to the simple carrier) has led a number of researchers to consider the following question: can one decide whether or not the carrier is in equilibrium with substrate at a particular face of the membrane—in other words, in the diagram of Fig. 8 is the rate of breakdown of the carrier-substrate complex faster than the rate of transition between the two carrier-substrate complexes, or is the latter process rate-limiting? (Only if the transition between the two carrier-substrate complex forms is rate-limiting, can one say that carrier and substrate are in equilibrium.) A number of tests purporting to make a distinction between these possibilities have been suggested [10,11]. We have seen, however, that steady-state kinetics measurements cannot distinguish between even the most general form of the conventional carrier and the simple carrier and that in the case of the simple carrier the hypothetical steps involving breakdown of the complex or its transformation are not distinguishable. Hence steady-state measurements cannot enable a decision to be made as to whether or not the carrier and substrate are in equilibrium. The problem is not resolvable at the steady-state level and need not be further discussed.

7. A molecular interpretation of the transport parameters

Transport systems which can be characterised as simple pores or as simple carriers yield determined values for the parameters Q or K and the respective values in R. A value of Q and two values of R suffice to characterise the simple pore; a value of K and three values of R determine the simple carrier. It is valuable to consider, for such systems, the molecular interpretation of the experimentally derived parameters.

(a) R_{ij}—The resistance parameters

The terms in R are best considered as resistances. If we refer to Tables 1 and 3 we see that for the simple pore or carrier, these values of R are determined by reciprocals of rate constants. Now, a rate constant measures the rate at which a particular reaction occurs, i.e., the ease at which a reaction takes place along this path. Hence its reciprocal is the resistance (per unit site) for reaction along this path. Resistances will add algebraically if they are present in series. Thus the values of R are resistances or sums of resistances for reactions in series leading to a particular end-result. R_{12} is the resistance obtained from the zero *trans* experiment in the 1 to 2 direction. Since the concentration of S_1 is by definition limitingly high, what now determines the value of the maximum velocity is the overall resistance for the return journey, given by the resistance experienced by the complex on breaking down (for the simple pore)—or by this resistance and that for the return of free carrier (for the simple carrier). A similar argument follows from a consideration of either the equilibrium exchange or the infinite *trans* experiments, where, for both, the maximum velocity is determined by R_{ee}. R_{ee} gives the overall resistance of the breakdown of the carrier at one side of the membrane and then at the other, both processes being necessary in a complete cycle of substrate transfer. The term R_{00} is the total resistance experienced by the empty carrier as it undergoes a complete cycle of movement from one face of the membrane to the opposite face and then back to the initial face. (The more complicated expressions for the terms in R for the complex pore and conventional carrier models can be derived directly from the kinetic diagrams if the laws of the combination of resistances in parallel as well as in series are taken into account.)

(b) K—The intrinsic dissociation constant

The significance of the parameter K can best be appreciated by the following argument: We note that for the two zero *trans* procedures and for the equilibrium exchange procedure the product of the relevant half-saturation concentration term and its appropriate resistance term is given by KR_{00} that is,

$$K_{12}^{zt} R_{12} = K_{21}^{zt} R_{21} = K^{ee} R_{ee} = KR_{00} \tag{34}$$

The first term in Eqn. 34 is the resistance when substrate is on one side of the membrane only, multiplied by the appropriate half-saturation constant; the second term is for the situation where substrate is present only at the second face of the membrane while the third term is for the case where the substrate is present at both faces of the membrane. Now the fourth term contains the resistance when substrate

Concepts of mediated transport

is present at neither side of the membrane. It is natural to think of the term K as expressing the "half-saturation concentration" appropriate to the hypothetical procedure when substrate is present at neither side of the membrane! It expresses, therefore, the half-saturation concentration for the system when the free carrier distribution across the membrane is unperturbed by substrate. It is, thus, the reciprocal of the intrinsic affinity of the substrate for the carrier, being determined by the relative tendency of the free carrier to combine with substrate and move across the membrane compared with its tendency to move as the free carrier (terms in k).

Depending on the values of the resistance terms R, one may get very different values for the measured half-saturation concentrations for the various transport procedures. Nevertheless, all these are derivable (if the carrier model holds) from the intrinsic dissociation constant K and the appropriate pair of resistance terms.

The meaning of Q for the simple pore can be similarly derived and similarly considered.

(c) The asymmetry parameter – R_{21}/R_{12}

In general, one will not be surprised to find a marked asymmetry in transport parameters. The transport rate constants can have any values subject only to the constraint that in the absence of an external source of energy there is no net movement of substrate when the concentration at each face of the membrane is the same. This implies that $b_1 f_2 = b_2 f_1$ for the simple pore or that $b_1 f_2 k_1 = b_2 f_1 k_2$ for the simple carrier. It is the value of the transport resistance R_{12} and R_{21} that will determine whether or not the system will behave asymmetrically. This can be seen by taking the ratio of the derivable half-saturation concentration and maximum velocities as follows:

$$\frac{R_{21}}{R_{12}} = \frac{K^{zt}_{12}}{K^{zt}_{21}} = \frac{V^{zt}_{12}}{V^{zt}_{21}} = \frac{K^{ic}_{21}}{K^{ic}_{12}} = \frac{K^{it}_{12}}{K^{it}_{21}} \tag{35}$$

(in such cases where these latter two sets of parameters can be measured).

Thus if the carrier or pore models hold, when one parameter is asymmetric across the membrane, all of these parameters are asymmetric. We note that the equilibrium exchange parameters are necessarily symmetric.

The symmetrical pore or carrier is a very special case of the general model in which R_{21} and R_{12} happen to be equal. There are no theoretical difficulties in handling the general case and there seems no reason why the assumption of symmetry need ever be made, until it can be experimentally verified.

One should note that the limiting permeability (defined as the value of the unidirectional flux divided by the substrate concentration, extrapolated to zero substrate concentration) has the value Q for the pore or of KR_{00} for the carrier. From Tables 1 and 3 it will be seen that the values of these terms are unchanged when the subscripts 1 and 2 are interchanged. They are invariant as to side, representing average values across the membrane for particular membrane properties. Hence even for a highly asymmetric system, transport properties measured at substrate concentrations well below those of the corresponding half-saturation concentrations will be apparently symmetric. Asymmetric systems behave symmetrically at limitingly low substrate concentrations.

8. Exchange diffusion and countertransport

Although the experimental procedures above are more than sufficient to test for and characterise simple pores and carriers, some other procedures of a venerable history (although somewhat difficult to interpret) are often used in transport studies. We proceed to discuss these.

(a) Exchange diffusion

The flow of substrate across a membrane at a particular substrate concentration is often compared in circumstances where there is substrate on one side of the membrane only and where there is substrate on both sides of the membrane. One can compare zero *trans* and equilibrium exchange conditions or zero *trans* and infinite *trans* conditions. Since substrate may flow in the opposite direction to that from which the flow is being measured, substrate molecules are said to "exchange" across the membrane, and the process is termed "exchange diffusion" [12].

We compare first equilibrium exchange and zero *trans* conditions at a substrate concentration S_1 at face I. For the simple pore:

$$\frac{\text{equil. exch. unidirect. flux}}{\text{zero } trans \text{ flow}} = \frac{Q + R_{12}S_1}{Q + (R_{12} + R_{21})S_1} \tag{36}$$

The ratio of the flows will vary from unity at low substrate concentrations to $R_{12}/(R_{12} + R_{21})$ at limitingly high substance concentrations. The ratio is never therefore, greater than unity and the finding of a ratio significantly greater than one, at any substrate concentration, is sufficient to reject the simple pore.

For the simple carrier:

$$\frac{\text{equil. exch. unidirect. flux}}{\text{zero } trans \text{ flow}} = \frac{KR_{00} + R_{12}S_1}{KR_{00} + R_{ee}S_1} \tag{37}$$

The "exchange ratio" varies between unity at low substrate concentration and R_{12}/R_{ee} at limitingly high substrate concentrations. Depending on the respective values of the resistances R_{12} and R_{ee}, the exchange ratio can be greater or less than unity at high substrate concentrations. A value for this ratio greater than one [13] is perfectly consistent with a simple carrier.

Concepts of mediated transport

A comparison of the infinite *trans* and zero *trans* fluxes cannot be made for the simple pore but for the carrier the ratio of the flow is:

$$\frac{\text{infinite } trans \text{ unidirectional flux}}{\text{zero } trans \text{ flow}} = \frac{KR_{00} + R_{12}S_1}{KR_{21} + R_{ee}S_1} \qquad (38)$$

This ratio lies between R_{00}/R_{21} and R_{12}/R_{ee} and can be greater than, equal to, or less than unity at any substrate concentration.

If the ratio of the unidirectional (exchange) flux to the zero *trans* flow is greater than unity, an interesting conclusion can be derived. From both Eqns. 38 and 39, at high substrate concentrations, we have $R_{12} > R_{ee}$ or, from Table 3 $1/b_2 + 1/k_2 > 1/b_1 + 1/b_2$, or $1/k_2 > 1/b_1$, or $b_1 > k_2$.

If, for the comparison involving the infinite *trans* condition, the relevant ratio is greater than unity at low substrate concentrations: $R_{00} > R_{21}$ or, from Table 3, $1/k_1 + 1/k_2 > 1/b_1 + 1/k_1$, $1/k_2 > 1/b_1$, and $b_1 > k_2$. Hence, if the unidirectional flux is greater than the zero *trans* flow, at any substrate concentration [14], the rate constant for the release of substrate from the substrate-carrier complex at side 1 of the membrane is greater than the rate constant for the return of free carrier from side 2 to side 1.

If one wishes to adopt the conventional carrier model, then the corresponding result is (from Table 3): $1/k_2 > 1/b_1 + 1/g_2 + g_1/b_1g_2$. Thus the resistance experienced by the free carrier during the transition across the membrane is greater than *both* of the following: the resistance for the transition of the carrier-substrate complex across the membrane, *or* its breakdown at the *cis* face to liberate free carrier.

(b) Countertransport

An experimental procedure which is often used as a criterion to reject a pore model (or a "fixed-site" model in the terms often used) is that known [15] as "countertransport" or "counterflow" (Fig. 8). Here one arranges an experiment in which a test (or "driven") substrate, whose flow is to be measured, is present at equal concentrations across the cell membrane, while a second (or "driving") substrate is present at a substantial gradient of concentration across the membrane. The second substrate will flow down its concentration gradient. If a carrier-type system is present and used by both substrates, this flow may drive the test substrate in the opposite direction, thus moving it against its own concentration gradient. In the course of time, as the driving substrate flows down its concentration gradient, equilibrium is reached across the membrane and the condition for demonstrating countertransport disappears. The phenomenon is thus a transitory one (unless an experiment can be contrived in which the driving substrate is continually fed into the membrane at one face and removed at the other).

We note that the two substrates need not have different *chemical* properties. One may be the isotopically labelled equivalent of the other. The (net) movement of

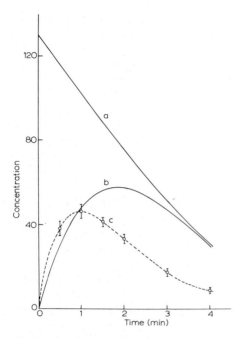

Fig. 8. Countertransport of glucose into human red blood cells at 20°C. Cells were loaded with 180 mM of unlabelled glucose and then transferred to a solution containing labelled glucose at 4.3 mM. The ordinate is the intracellular concentration of glucose in mmol/litre cell water. The open circles and bars represent the measured intracellular concentration of labelled glucose. This rises at first to well above the external level of 4.3 mM, and then falls as glucose leaves the cell. The solid curves are theoretically computed values of (a) the total concentration of glucose labelled and unlabelled and (b) the concentration of labelled glucose only within the cell, as functions of time. (Data from [34].) Figure taken, with permission, from Lieb and Stein [35].

isotopic label across the membrane will be in a direction opposite to the chemical gradient of substrate if countertransport is occurring.

The phenomenon of countertransport is a striking one (see Fig. 8) and provides a useful, since experimentally very accessible, rejection criterion for the pore; however, this criterion is in no way stronger than the criteria that we have discussed above, nor does the phenomenon provide any information that cannot be obtained by *conceptually* simpler experimental procedures.

To obtain the kinetic equations describing countertransport we will have to solve the carrier model for the situation in which two substrates S and P are present across the membrane, both of them capable of interacting with the carrier. The relevant formal model is given in Fig. 9, where we have used the diagram representing the simple carrier or in Fig. 10 where the representation is that of the conventional carrier. To solve the mathematics, we write a steady-state equation for each of the forms involving carrier and the two substrates and a conservation equation ex-

Concepts of mediated transport

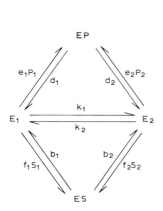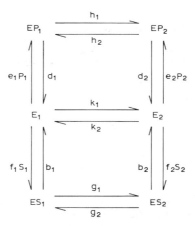

Fig. 9. Two substrates using the simple carrier. Formal representation of the kinetic scheme. Symbols as in Fig. 7, with the addition of rate constants e_1 and e_2 for the formation of the carrier-second substrate complex EP, and d_1 and d_2 for the breakdown of this complex. P is the second substrate.

Fig. 10. Two substrates using the conventional carrier. Formal representation of the kinetic scheme. Symbols as in Figs. 8 and 10, with the addition of rate constants h_1 and h_2 for the interconversion of the two forms of the substrate P-carrier complex EP_1 and EP_2.

pressing the fact that the sum of the carrier molecules in all their forms is unchanged during the interconversions required for transport. The resulting equation describing the unidirectional flux from side I to side II of the substrate S, in the presence of concentrations S_1, P_1 at side I, and S_2, P_2 at side II is:

$$f_{12}^S = \left\{ K^P S_1 + \frac{K^P}{K^S} S_1 S_2 + S_1 P_2 \right\}$$

$$\times \left\{ R_{00} K^S K^P + K^P(S_1 R_{12}^S + S_2 R_{21}^S) + K^S(P_1 R_{12}^P + P_2 R_{21}^P) \right.$$

$$\left. + R_{ee}^{SP} S_1 P_2 + R_{ee}^{PS} S_2 P_1 + \frac{K^P}{K^S} R_{ee}^S S_1 S_2 + \frac{K^S}{K^P} R_{ee}^P P_1 P_2 \right\}^{-1} \tag{39}$$

In Eqn. 39 we have had to introduce new symbols over those in Eqn. 30 to account for the presence of two substrates S and P. There is now an intrinsic dissociation constant K for each substrate. Resistances terms R_{12} and R_{21}, describing the zero *trans* situation, are defined for each substrate and also for the equilibrium exchange experiment R_{ee} involving just a single substrate S or P. There are also defined cross-resistances R_{ee}^{SP} and R_{ee}^{PS} for the situation where S is present only at one side of the membrane and P only at the other. These cross-resistances can be derived from the resistances obtainable by single substrate experiments (see Table 6 where these

TABLE 6
Cross-resistance terms for two substrates sharing a single carrier. Rate constants are defined in Fig. 9.

Terms in R^S are as defined in Table 3. Terms in R^P will be obtained by substituting d for b, and e for f, in the appropriate definitions in Table 3. In addition:

$$R_{ee}^{SP} = \frac{1}{b_2} + \frac{1}{d_1}$$

$$R_{ee}^{PS} = \frac{1}{b_1} + \frac{1}{d_2}$$

Also, $R_{ee}^{SP} = R_{12}^S + R_{21}^P - R_{00}$
and $R_{ee}^{PS} = R_{21}^S + R_{12}^P - R_{00}$

---

terms are defined). In all, seven parameters are sufficient (if the simple carrier model holds) to describe all experiments involving two substrates. These are the intrinsic dissociation constants K^P and K^S and the various resistance terms R_{ee}^S and R_{ee}^P, R_{12}^S and R_{12}^P, R_{21}^S anf R_{21}^P and R_{00}. That a single value of R_{00} characterises a simple carrier for any substrate or combination of substrates can be used as a rejection criterion for this model.

How does the carrier model and Eqn. 39 account for countertransport? We recall that we need to measure net flow (of S) in a particular direction in a situation where $S_1 = S_2 = S$. The net flow will be given by the difference between the two unidirectional fluxes obtained directly from Eqn. 39 as f_{12}^S and (by interchanging the subscripts 1 and 2) as f_{21}^S. Now of the terms in the numerator of Eqn. 39 the first two, $K^P S_1$ and $(K^P/K^S)S_1 S_2$, will disappear on making the substitution $S_1 = S_2$ and subtracting f_{21}^S from f_{12}^S. The third term will remain, however, and one finds that

$$\text{net flow of } S_{1\to 2} = \frac{S(P_2 - P_1)}{D} \tag{40}$$

where D is the denominator of the right-hand side of Eqn. 39, on putting $S_1 = S_2 = S$.

There will be a definite flow of substrate S in the direction opposite to that of the concentration gradient of P and countertransport will be observed. The actual value of the net flow will be given by Eqn. 40 and can be calculated if the seven parameters describing the transport of the substrates S and P are known from independent experiments.

An intuitive feeling for the phenomenon of countertransport can be appreciated as follows: the two substrates S and P compete with each other for combination with the free carrier E. At the face at which P is at the high concentration, it will compete more effectively with S than it will at the other face. Hence less S will flow from the face at which most P flows. There will be thus a net flow of S in a direction opposite to the flow of P. All this occurs because, on the carrier model, the interaction of substrates and carrier at one face of the membrane are separated from the interactions at the other face by the measurable activation energy barrier for the interconversion of the two forms of the free carrier.

Concepts of mediated transport

It is important to note that formally the phenomenon of countertransport is demonstrable only because of the existence of the cross terms involving S and P (S_1P_2 and S_2P_1) in Eqn. 39. If the rate constants k_1 and k_2 describing the rates of interconversion of the forms E_1 and E_2 become very high, terms involving k_1 and k_2 will drop out of the equation. It turns out that it is the cross terms which drop out of the equation, with the result that the phenomenon of countertransport becomes no longer observable. Thus the existence of countertransport is another example of a phenomenon which points to the existence of a detectable energy barrier for the interconversion of the two forms E_1 and E_2 of the free site. Countertransport rejects the pore model, but does so for the same reason as does the existence of accessible infinite *trans* and infinite *cis* half-saturation constants.

9. The kinetics of competition

The phenomenon of countertransport is a particular example of an experimental finding dependent on the presence of competition between two substrates for a single transport system. We can devise various other situations in which competition can be demonstrated. Since the flux Eqn. 39 is sufficiently general so as to describe all possible experimental situations for two substrates and since it contains only parameters measurable by experiments involving only individual substrates, there is in principle no new information yielded by studies of competition. Yet precisely because of this redundancy of information, the study of competition when combined with a study of the transport parameters involving single substrates yields criteria which can test the applicability of the carrier model. If experiments involving pairs of substrates are not consistent with the predictions from experiments involving single substrates, the appropriateness of the carrier model should be questioned.

To simplify the mathematical derivations, we will derive results only for the situation where one substrate S, whose flux is being measured, is present in very low concentrations—so that S_1 and S_2 can be neglected in comparison with the appropriate half-saturation parameters. With these assumptions, Eqn. 39 becomes:

$$f_{12}^S = \frac{(K^P/K^S)(K^P + P_2)S_1}{R_{00}(K^P)^2 + K^P(R_{12}^P P_1 + R_{21}^P P_2) + R_{ee}^P P_1 P_2} \tag{41}$$

The flux in the absence of the competitor P is given by putting $P_1 = P_2 = 0$ in Eqn. 41. The fractional inhibition is given by the ratio of the flux in the presence of inhibitor to that in the absence of inhibitor and is:

$$\text{fractional inhibition} = \frac{R_{00} K^P (K^P + P_2)}{R_{00}(K^P)^2 + K^P(R_{12}^P P_1 + R_{21}^P P_2) + R_{ee}^P P_1 P_2} \tag{42}$$

an equation which is formally closely related to Eqn. 30.

One can define three useful experimental procedures for studying competition.

(a) In the zero *trans* case, the concentration of the competitor at face 2 of the membrane is zero. Putting $P_2 = 0$ in Eqn. 42 we obtain after simplifying:

$$\text{fractional inhibition (zero } trans) = \frac{R_{00} K^P}{R_{00} K^P + R_{12}^P P_1} \tag{43}$$

An inhibition of 50% will be found when the concentration of $P_1 = R_{00} K^P / R_{12}^P$. This is exactly the half-saturation concentration for the zero *trans* experiment using pure P.

(b) In the equilibrium exchange situation, the concentration of the competitor at the two sides of the membrane is equal. We put $P_1 = P_2 = P$ in Eqn. 42 and obtain after simplifying:

$$\text{fractional inhibition (equilibrium exchange)} = \frac{R_{00} K^P}{R_{00} K^P + R_{ee} P} \tag{44}$$

An inhibition of 50% will be found when the concentration of $P = K^P (R_{00}/R_{ee}^P)$. This is exactly the half-saturation concentration for the equilibrium exchange procedure using pure P.

(c) Similarly an infinite *trans* ($P_2 = \infty$) procedure can be defined and Eqn. 42 solved for this case. The flux of S in the direction 1 to 2, when P_2 is limitingly high, will be reduced on increasing P_1 and is halved when the concentration of P_1 is equal to $K^P R_{21}^P / R_{ee}^P$. This is precisely the half-saturation concentration for the infinite *trans* procedure using pure P.

10. Secondary active transport

The discussion of countertransport provides an appropriate introduction to the phenomenon of "active transport", a phenomenon which has intrigued investigators of cell biology for many years. Active transport is the net movement of a substrate against its concentration gradient (or, if the substrate is charged, against the prevailing electrochemical gradient). It turns out that the phenomenon results from the coupling of two flows (of substrate or of chemical reaction) when two substrates share the same carrier. The driven substrate flows up its (electro-)chemical gradient while the driving substrate flows down its gradient or takes part in a chemical reaction which is not at equilibrium. How this coupling occurs will be the subject of the present section.

Consider Fig. 11. On the left-hand side of the figure is the simplest scheme for countertransport. A and B are two substrates which share the same carrier E. The carrier exists in the two forms E_1 and E_2, which, on this simplest of all possible models, can interconvert only through the transport reactions which move A and B across the membrane. Then, as we saw in a previous section and as is given also in the figure, a net flux of A will occur even when the concentration of A is equal on the two sides of the membrane, provided that the concentrations of B are unequal. This active transport arises because A and B share the same carrier and because the interaction between substrate and carrier at the two faces of the membrane are shielded from one another by the presence of a finite energy barrier. With these

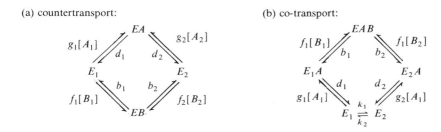

(a) counter:

$$J = \frac{1 - \dfrac{[A_1][B_2]}{[A_2][B_1]K^{eq}}}{\dfrac{1}{b_1} + \dfrac{1}{d_2} + \left(\dfrac{1}{b_2} + \dfrac{1}{d_1}\right)\dfrac{[A_1][B_2]}{[A_2][B_1]K^{eq}} + \dfrac{\left(1 + \bar{B}_1 + \dfrac{\bar{B}_1}{\bar{B}_2}\right)}{g_1[A_1]} + \dfrac{\left(1 + \bar{A}_2 + \dfrac{\bar{A}_2}{\bar{A}_1}\right)}{f_1[B_1]}}$$

(b) co-transport:

$$J = \left\{1 - \frac{[A_2][B_2]K}{[A_1][B_1]K^{eq}}\right\} \left\{ \frac{1}{b_2} + \frac{1}{k_2} + \frac{1}{d_2} + \left(\frac{1}{b_1} + \frac{1}{k_1}\right) \cdot \frac{[A_2][B_2]K}{[A_1][B_1]K^{eq}} + \frac{1 + K + \bar{A}_2 K}{g_1[A_1]} + \frac{\left(1 + \dfrac{b_1}{b_2}\right)\left(1 + \dfrac{1+K}{\bar{A}_1}\right)}{f_1[B_1]} \right.$$

$$\left. + \left(\frac{1}{b_1} + \frac{1}{b_2}\right) \cdot \frac{\bar{A}_2 K}{\bar{A}_1} + \frac{\bar{B}_2}{d_2 \bar{A}_1 \bar{B}_1} + \frac{\bar{A}_2 K}{k_1} + \left[\frac{1}{d_2} + \left(\frac{1}{k_1} + \frac{1}{g_1[A_1]}\right)\bar{A}_2 K\right] \cdot \left(\bar{B}_2 + \frac{\bar{B}_2}{\bar{B}_1}\right) \right\}^{-1}$$

Fig. 11. Active transport using the simple carrier. Formal kinetic schemes for (a) countertransport and (b) co-transport. A and B are the two substrates of the carrier E either of which (in countertransport) or both which (in co-transport) combine with E to form EA, EB or EAB. Subscripts 1 or 2 refers to substrate at side 1 or 2 of the membrane. The rate constants b, d, f, g, k are defined in the figure. K^{eq} is the equilibrium constant of the chemical reaction $A_1 \rightleftarrows A_2$ in primary active transport (and is equal to unity in the case of secondary active transport). $K = k_1/k_2$ in co-transport. The square brackets denote concentrations and terms such as $\bar{A}_1 = [A_1]/K_1^A$, where K_1^A is the relevant dissociation constant, here $= d_1/g_1$. J is the net flux in the $1 \to 2$ direction. (Figure taken, with permission, from [30].)

conditions satisfied active transport of substrate will occur. Many, many cases of such active transport by countertransport are known in biology and will be dealt with in later sections of this treatise. Common driven substrates are metabolites such as sugars [16], amino acids [17], intermediates of the tricarboxylic acid cycle [18], cations [19] and anions [20]. Common driving substrates are the cations proton, potassium, sodium and calcium. It must be stressed that it is quite arbitrary, in fact, which of the two substrates is considered to be the driving and which the driven one. The formulation of Fig. 12 is quite symmetric between these two possibilities and the system itself allows of no distinction. It is essential to accept that unless there is a flow of driving substrate there can be no active transport of the driven substrate. The driving substrate drives the carrier across the membrane and is itself brought across during this transformation. Coupling of the two flows of substrate must take place if active transport is to occur. The result of mathematical analysis of this model appears in Fig. 11(a).

But there is another mode of active transport other than countertransport. This is depicted on the right-hand side of Fig. 11. Here both substrates, the driving and the driven, can combine simultaneously with the carrier and accompany one another across the membrane. On this, the simplest model for what can be called co-transport, the uncombined carrier must be able to cross the membrane if any transport at all is to occur. The model depicts the case where the carrier combined with only one species of substrate cannot cross the membrane. The mathematical analysis of this model yields the result listed in the Fig. 11(b). Once again there will be a net flow of substrate A even when it is at the same concentration on either side of the membrane, provided that the substrate B is not at zero electrochemical gradient. There exist very many examples of active transport systems operating by the mechanism of co-transport. Once again, the driven substrate is a metabolite such as sugar [16] or amino acid [21] or an ion [22], while the driving substrate is very often sodium or perhaps less frequently proton.

It will be seen from Fig. 11 that in the case of countertransport, the steady-state distribution of the driving and driven substrates are such that both are highest on the same side of the membrane: $A_1/A_2 = B_1/B_2$, whereas in co-transport the "high" of the driven substrate is at the opposite face of the membrane to the "high" of the driving substrate: $A_1/A_2 = B_2/B_1$. The two mechanisms have also been called antiport and symport [23] for counter- and co-transport, respectively.

11. Primary active transport

Not all active transport is brought about by the coupling of two transport flows. It can also be the case that a transport flow is coupled to the progress of a chemical reaction, as we shall now discuss. We saw in Section 3 that there was a very close analogy between the formal description of an enzymatically catalysed chemical reaction and the formal description of transport. We can approach Fig. 11 in the same spirit. Consider countertransport, the left-hand figure. Here, B_1 and B_2 can, indeed, be substrate molecules present on side 1 and side 2 of the membrane, respectively. But we can take A_1 and A_2 to be two components of a chemical reaction $A_1 \rightleftarrows A_2$, where A_1 might be ATP while A_2 is $ADP + P_i$ (inorganic phosphate). Then, if the enzyme catalyses the chemical reaction of ATP to form $ADP + P_i$, while the two different forms of the enzyme E_1 and E_2 bind to ATP and to (ADP and P_i), respectively, flow of B can be coupled to the progress of the chemical reaction of ATP hydrolysis. It is immaterial whether one translates the equations of Fig. 11 in terms of transport substrates or as components of a chemical reaction, the mathematical analysis is the same. Indeed, as Mitchell was the first to point out in his chemi-osmotic theory [24], the description is not merely formal, it is mechanistic—and active transport of a transportable substrate can by such co- or countertransport systems be coupled to chemical reactions. Just as it is immaterial for the transport-transport systems which component was termed driving or driven, so in chemical-transport systems flow of substrate can drive a chemical reaction (such as synthesis of ATP) or the reverse can take place.

Such primary active transport systems are perfectly equivalent formally to the secondary active transport systems discussed in the previous section. Both systems require that true flows of substrate or progress of a chemical reaction must take place for coupling to occur, for it is these processes that drive the carrier from one conformation, state or side of the membrane to the other. Both systems require absolutely that the reactions between carrier and substrate at one side of the membrane are shielded from those at the other, so that a true displacement of the carrier-binding sites from "accessible at one side of the membrane" to "accessible at the other" must take place. In the case of the primary active transports, it is the chemical reaction which brings about this vectorial movement of the carrier, since the two components of the chemical reaction, product and reactant, must combine with different forms of the carrier. In the case of secondary active transport, it is transport of the driving substrate which brings about reorientation of the carrier.

Primary active transport systems that have been studied include the sodium + potassium pump [25,26] (which is the sodium + potassium-activated ATPase) and the calcium ATPase [27] and include also the ATP synthesising proton pumps of mitochondria [18,28].

12. Design principles for active transport systems [29,30]

An active transport system operating according to schemes such as those of Fig. 11 will pump actively whatever the fine details of its design and structure, provided only that the flows of substrates or of the progress of the chemical reaction are coupled, so that there is a conformational change of the system between the membrane faces [29]. No special design of storing energy in the conformation change need be postulated, nor any necessary change of affinity of the system for the transported substrate as the membrane is traversed. But such design specialisations will affect the *rate* at which net transport occurs and hence affect the efficiency of pumpong [30]. An efficient pump will be one which rapidly takes up substrate at the side *from* which net transport is occurring while rapidly releasing it at the side *to* which pumping is taking place. Hence the relative affinity of pump for substrate at the two faces of the membrane is a matter of some importance for efficient pump design, while having no bearing on the fundamental issue of whether pumping will or will not occur.

There seem to be three different ways in which the system can be designed so that substrate affinities are set differently at the two membrane faces. The first way is to ensure that the two forms of the free carrier are of different standard-state free energies. Thus E_1, say, may be of lower energy than E_2. This results in a lower affinity of the substrates A and B in the countertransport scheme and only A in the co-transport scheme at face 1 of the membrane, as compared with the affinities at face 2. Pumping will be more effective from face 2 than from face 1. One can think of the energy of the driving reaction as being trapped in the conformation energy of the carrier [31], since the driving substrate or chemical reaction moves the carrier from a form of lower to one of higher energy. (But, of course, this "trapping" of energy brings about only the ancillary changes in affinity and is not the cause of the fact of active transport.) The second and third design possibilities, available only for co-transport, are to ensure that the carrier after combination with driving substrate, i.e. the form E_1A in Fig. 11(b), has a higher intrinsic affinity for the driven substrate B than has the form E_2A. Then pumping will be more effective from side 1 than from side 2, and will be more efficient in overall. This differential affinity can, as one possibility, be achieved by having the two forms of E to be of different chemical nature. Thus, one can be a phosphorylated form, if the chemical reaction is ATP splitting. Part of the free energy of the chemical reaction is used to modify the affinity of the carrier-enzyme for the transported substrate. Finally, as another possibility here, the two forms of the carrier-enzyme can bind in one case the product or in the other the reactant of the chemical reaction. The resultant complexes can demonstrate differential affinity for the driven substrate. In this possibility, all of the free energy in the chemical reaction can be used to set the differential affinities of the carrier for the driven substrate. It appears that a systematic analysis of which of these design possibilities is realised in the many known biological examples is not yet available.

13. Conclusion

The carrier model has been enormously successful in leading to an understanding of membrane physiology and biochemistry. That movement of substrate by the carrier brings about the movement of the carrier itself and that this movement can be coupled to the flow of other substrates and to the progress of chemical reactions are concepts which have been immensely fertile. All of these can be seen now in a unified framework, giving a dynamic picture of the commerce of membrane activity. The working out of the molecular details of these flows and couplings will be the principal task of many researchers over the years to come.

Acknowledgements

I am very grateful to Chana Stein for preparing the original figures and for helpful criticism of the draft manuscript.

References

1. Lieb, W.R. and Stein, W.D. (1974) Biochim. Biophys. Acta 373, 165–177.
2. Lieb, W.R. and Stein, W.D. (1974) Biochim. Biophys. Acta 373, 178–196.
3. Dixon, M. and Webb, E.C. (1964) in Enzymes, 2nd ed., Longmans, Green, New York.
4. Cleland, W.W. (1963) Biochim. Biophys. Acta 67, 104–187.
5. Christensen, H.N. (1975) In Biological Transport, 2nd ed., Benjamin, Reading, MA.
6. Stein, W.D. and Eilam, Y. (1974) Meth. Memb. Biol. 2, 283–354.
7. Stein, W.D. and Lieb, W.R. (1973) Israel J. Chem. 11, 325–339.
8. Cabantchik, Z.I. and Ginsburg, H. (1977) J. Gen. Physiol. 69, 75–96.
9. Hankin, B.L., Lieb, W.R. and Stein, W.D. (1972) Biochim. Biophys. Acta 288, 114–126.
10. Dawson, A.C. and Widdas, W.F. (1964) J. Physiol. (Lond.) 172, 107–122.
11. Hoare, D.G. (1972) J. Physiol. (Lond.) 221, 311–331.
12. Ussing, H.H. (1949) Acta Physiol. Scand. 19, 43–56.
13. Hankin, B.L. and Stein, W.D. (1972) Biochim. Biophys. Acta 288, 127–135.
14. Lacko, L., Wittke, B. and Kromphardt, H. (1972) Eur. J. Biochem. 25, 447–454.
15. Rosenberg, T. and Wilbrandt, W. (1957) J. Gen. Physiol. 41, 289–296.
16. Geck, P. and Heinz, E. (1981) This volume, Chapter 11.
17. Eddy, A.A. (1978) in F. Bronner and A. Kleinzeller (Eds.), Current Topics in Membrane Transport, Vol. 10, Academic Press, New York, pp. 279–360.
18. Meyer, A.J. and van Dam, K. (1981) This volume, Chapter 9.
19. Baker, P.F. (1976) Fed. Proc. 35, 2589–2595.
20. Knauf, P. (1979) in F. Bronner and A. Kleinzeller (Eds.), Current Topics in Membrane Transport, Vol. 12, Academic Press, New York, pp. 251–365.
21. Vidaver, G.A. and Shepherd, S.L. (1968) J. Biol. Chem. 243, 6140–6152.
22. McManus, T.J. and Schmidt, W.F. (1977) in J.F. Hoffman (Ed.), Membrane Transport Processes, Vol. 1, Raven Press, New York, pp. 79–106.
23. Mitchell, P. (1967) Adv. Enzymol. 29, 33–88.
24. Mitchell, P. (1977) Annu. Rev. Biochem. 46, 996–1005.
25. Glynn, I.M. and Karlish, S.J.D. (1975) Annu. Rev. Physiol. 37, 13–55.
26. Schuurmans Stekhoven, F.M.A.H. and Bonting, S.L. (1981) This volume, Chapter 6.
27. Hasselbach, W. (1981) This volume, Chapter 7.
28. Ernster, L. (1977) Annu. Rev. Biochem. 46, 981–995.
29. Stein, W.D. and Honig, B. (1977) Mol. Cell. Biochem. 15, 27–45.
30. Honig, B. and Stein, W.D. (1978) J. Theor. Biol. 75, 299–305.
31. Boyer, P.D. (1975) FEBS Lett. 50, 91–94.
32. Weber, G. (1974) Ann. N.Y. Acad. Sci. 242, 286–296.
33. Ginsburg, H. and Stein, W.D. (1975) Biochim. Biophys. Acta 382, 353–368.
34. Miller, D.M. (1968) Biophys. J. 8, 1329–1338.
35. Lieb, W.R. and Stein, W.D. (1970) Biophys. J. 10, 585–609.
36. Lieb, W.R. and Stein, W.D. (1976) Biochim. Biophys. Acta 455, 913–927.

CHAPTER 6

Sodium-potassium-activated adenosinetriphosphatase

F.M.A.H. SCHUURMANS STEKHOVEN and S.L. BONTING

Department of Biochemistry, University of Nijmegen, Nijmegen, The Netherlands

1. Introduction

(a) Cation transport in cells

Most living animal cells are characterized by a high intracellular K^+ concentration and a low Na^+ concentration, whereas the surrounding plasma and tissue fluid have a high Na^+ and a low K^+ concentration. These cation gradients across the cell membrane are gradually abolished upon cell death or upon inhibition of the cellular energy metabolism by cooling, withholding of glucose or application of suitable inhibitors. This suggests that the maintenance of the cation gradients requires metabolic energy. Radioisotope experiments have indicated that during steady-state conditions there is a passive influx of Na^+ ions and a passive efflux of K^+ ions. These findings have led to the so-called "pump-leak concept": there is an active extrusion of Na^+ ions and influx of K^+ ions, balanced by a passive leak of these cations in the opposite directions.

Since all cells have a trans-membrane potential, which is negative inside, the Na^+ extrusion must take place against a concentration gradient as well as a potential gradient, and thus must always be an active, energy-requiring process. The K^+ influx takes place against a concentration gradient, but is assisted by the potential gradient, and thus need not always be an active process. From a knowledge of the cation concentrations and the membrane potential, it can be shown that in erythrocytes there is a coupled, active transport of Na^+ and K^+ ions, whereas in nerve and muscle Na^+ is actively transported and K^+ is passively distributed approximately according to the Nernst equation.

(b) Relation to energy metabolism

Two questions arise from these findings: (1) how is this cation pump system related to the cellular energy metabolism, and (2) what is the nature of the pump system. The question about the relation to energy metabolism has been answered first, and has led the way to an answer to the second question. In most cells, e.g. nerve,

Bonting/de Pont (eds.) Membrane transport
© *Elsevier/North-Holland Biomedical Press, 1981*

muscle, nucleated erythrocytes of birds and reptiles, oxidative metabolism is the source of energy for active cation transport. This is shown by the fact that anaerobiosis (absence of oxygen) or application of inhibitors of oxidative metabolism (cyanide, 2,4-dinitrophenol) inhibits active cation transport. On the other hand, in mammalian erythrocytes and lens, oxidative metabolism is absent or virtually absent. In these cases anaerobiosis does not inhibit cation transport, but removal of glucose or addition of glycolytic inhibitors (fluoride, iodoacetic acid) does. This suggests that the cation pump is not directly linked to some step in the glycolytic or oxidative phosphorylation chains, but rather that it is the adenosinetriphosphate (ATP) produced by these systems which provides the energy for cation transport. Direct evidence for this conclusion has come from experiments in which ATP is placed inside cells or cell membrane vesicles, which lack any other energy source. Incorporation of ATP into erythrocyte ghosts (lysed erythrocytes) that are in the process of resealing permits cation transport [1]. Injection of ATP into a squid giant axon poisoned with cyanide restores sodium efflux [2].

(c) Nature of the cation transport system

A first answer to the question of the nature of the cation pump has been supplied by Skou [3]. He reasoned that making the energy of ATP available to cation transport would require an ATPase, an enzyme which splits ATP into ADP and inorganic phosphate. Since such an ATPase would have to make this energy available to the translocation of Na^+ and K^+ ions, he assumed that it would require these cations for activity. In a study of a crab nerve particulate fraction he was able to demonstrate a Mg^{2+}-activated ATPase activity, which was increased considerably upon simultaneous addition of Na^+ and K^+ to the assay medium. In addition he showed that the digitalis glycoside ouabain, which had some years earlier been found to be a powerful and specific inhibitor of active cation transport in erythrocytes [4], completely inhibited the additional ATPase activity [5]. From these two observations Skou concluded that this ouabain-sensitive, $(Na^+ + K^+)$-stimulated ATPase activity (henceforth called Na-K ATPase) might be identical with or be part of the cation transport system.

In the ensuing ten years the occurrence of the enzyme and its function in active cation transport have been demonstrated in a great variety of animal cells and tissues (for a review, see [6]). The properties of the enzyme in different species and tissues are remarkably similar. In all cases it requires Mg^{2+}, Na^+ and K^+ for its activity, its optimal pH is 7.0–7.5, the K_m values for Na^+ and K^+ are 5–10 mM and 0.4–1.8 mM, respectively, and it is always inhibited by ouabain and other cardiac glycosides. By comparing the effects of ouabain on the enzyme and on the physiological process, a primary and rate-limiting role of Na-K ATPase in cation transport could be established in single-cell systems like erythrocyte and eye lens, in excitatory systems like nerve, muscle, cochlea and electric eel electroplax, as well as in secretory processes like aqueous humour formation, cerebrospinal fluid formation, salt secretion by marine bird salt gland and elasmobranch rectal gland, renal and toad bladder salt reabsorption and pancreatic fluid secretion. In all three types of systems the ratio of Na$^+$ transported per ATP hydrolysed is approx. 3 [7]. Remarkable is the versatility of the system: by generating active Na$^+$ extrusion it maintains osmotic equilibrium in cells, it drives coupled transport processes of metabolites like glucose and amino acids (see Chapter 11), it permits excitation of nerve, muscle etc. and it drives various secretory processes.

With the universal significance of the Na-K ATPase cation transport system rapidly becoming established, the interest in its mechanism of action was kindled. From the earlier work it was clear that the enzyme system is an intrinsic membrane protein, which is activated by internal Na$^+$ ions and external K$^+$ ions, reacts with intracellular ATP and is inhibited by extracellular ouabain. Most intriguing, however, is the fact that in reacting with a molecule of ATP the enzyme moves 3 Na$^+$ ions across the membrane to the outside and 2 K$^+$ ions in the opposite direction. What has been learned in the past 10–15 years about this intriguing process will be summarized in the subsequent sections of this chapter.

2. Reaction mechanism

(a) Substrate binding

The overall Na-K ATPase reaction leads to the hydrolysis of ATP to ADP and inorganic phosphate and transport of 3 Na$^+$ ions to the extracellular compartment coupled to transport of 2 K$^+$ ions to the intracellular compartment per molecule of ATP split. This reaction is under control of the ligands, which bind to the ATPase molecule and are involved in its hydrolytic action and transport function: ATP, Mg^{2+}, Na$^+$ and K$^+$. Although binding of the various ligands at their respective intra- and extracellular sites would seem to occur simultaneously [8], we shall for the sake of clarity treat their binding sequentially with the corresponding step of the reaction mechanism.

The first step in the reaction mechanism involves the random binding of Na$^+$ and ATP at their intracellular sites of the enzyme molecule. The affinity for Na$^+$ to the native enzyme is relatively high: $K_D = 0.19-0.26$ mM [9,10]. For ATP it is $K_D = 0.1$ -0.2 μM [11,12]. The binding of one molecule of ATP is accompanied by the cooperative binding of 3 Na$^+$ ions [13]. However, the binding of Na$^+$ to the native enzyme does not appear to be dependent on ATP, nor the binding of ATP on Na$^+$. Intracellular K$^+$ competitively antagonizes the binding of ATP [14] as well as of Na$^+$. Intracellular Na$^+$, in turn, antagonizes the decrease in ATP binding caused by the intracellular K$^+$ [15]. K_i-values for the displacement of Na$^+$ and ATP by intracellular K$^+$ range from 0.1 to over 8 mM, depending on whether equilibria [11–13] or rates [16] are studied. The higher K_i-value determined from rate studies matches the K_D for intracellular K-sites (9 mM) determined from Na$^+$ efflux rates in the presence of increasing intracellular K$^+$ concentration [9]. The discrepancy between the K_i-values determined from equilibrium and rate studies is caused by the two-step mechanism exerted by K^+:

$$E_1 + K^+ \underset{}{\overset{K_1}{\rightleftharpoons}} E_1.K \underset{}{\overset{K_2}{\rightleftharpoons}} E_2.K,$$

by shifting the equilibrium from the native conformation E_1 with high affinity for Na^+ and ATP to a conformation $E_2.K$ with low affinity for Na^+ and ATP; $K_1 = [E_1][K^+]/[E_1.K]$ and $K_2 = [E_2.K]/[E_1.K]$ [16]. In rate measurements the low affinity of K^+ for E_1 in the first step is determined, whereas in equilibrium studies the K^+ concentration giving the half-maximal shift is determined by the ratio $K_1/(K_2 - 1)$. Since the equilibrium of the second step lies far to the right ($K_2 \gg 1$ and $K_2 > K_1$), it is clear that this will give higher values for K^+ affinity than evident from rate measurements of the transition.

Ouabain, which is a highly specific and potent inhibitor of Na-K ATPase, inhibits ATP binding [17] by reacting at the opposite, i.e. outer membrane side.

(b) Phosphorylation of the enzyme

The second step in the reaction mechanism is the phosphorylation of the enzyme by the γ-linked phosphate from ATP. The reaction is reversible, which gives rise to a slow ADP-ATP exchange:

$$ATP + E_1 \overset{Na^+, Mg^{2+}}{\rightleftharpoons} E_1 \sim P + ADP.$$

This reaction is dependent on tightly bound Mg^{2+} ($K_D = 5$ μM [18,19]). A model for the active centre, based on measurements of proton relaxation rates [20,21] has been presented (Fig. 1). In this model the γ-phosphate of ATP displaces one of the hydroxyl anions from the Mg-coordination sphere and binds as monoanion to Mg, Na^+ and the β-carboxyl anion of an aspartyl residue in the active centre.

In addition, the β-phosphate group, the 6-NH_2 group of the purine ring and the 2-OH group of the ribose moiety appear to be involved in the binding of ATP to the enzyme. This is inferred from increased dissociation constants of ATP analogues with these groups missing or substituted, such as AMP, ITP and deoxy-ATP [12,22].

Functional groups on the enzyme have been probed by means of group-specific inhibitors for their involvement in the binding of ATP (ATP protecting against the inhibition). Four such groups have been found. The first is the guanidino group of arginine [23], which could interact with the β- or α-phosphate group. The second is a tyrosyl–OH group [24,25], which may be involved in hydrogen bonding. The third is the imido group in the imidazole ring of histidine, as inferred from the pH dependence of ATP binding [26] and by photooxidation with protection by ATP [27], which may interact with the β- or α-phosphate group. The fourth is a cysteinyl–SH group, which probably interacts with the purine moiety of ATP [28]. It appears that ATP in these cases protects by steric hindrance rather than by

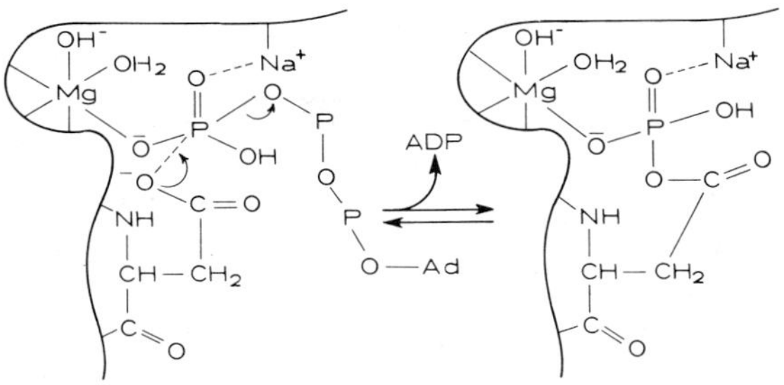

Fig. 1. Active centre of ATP binding and phosphorylation [20].

conformational change of the enzyme, since Na^+, which stabilizes the ATP binding E_1-conformation, enhances inhibition by the arginine-directed butanedione [23] and by the tyrosine-directed 7-Cl-4-NO_2-benzo-2-oxa-1,3-diazole (NBD-Cl) [25], whereas the E_2-inducing K^+ antagonizes inhibition as well as the binding of ATP. The adenine-binding –SH group has been probed by 6-SH derivatives of ATP, which were hydrolysed, whereas the derivatives of AMP were unreactive [28].

The actual phosphorylation step is the transfer of the ATP γ-phosphate to an aspartyl-β-carboxyl group [29,30] with release of ADP (Fig. 1). Some authors [31] assume an intermediate, in which ADP is still bound to the phosphoenzyme, which is feasible in view of the reversibility of the reaction. Under normal conditions, however, the phosphorylated intermediate does not react with ADP, unless the bound Mg^{2+} is replaced by Ca^{2+} [32,33], the enzyme has been pretreated with the sulfhydryl reagent N-ethylmaleimide [34] or has undergone limited proteolysis in the presence of Na^+ [35]. In all of these conditions the overall Na-K ATPase activity is inhibited, apparently through interference with a conformational transition from an ADP-sensitive to an ADP-insensitive phosphointermediate, the third step in the reaction mechanism.

(c) Transformation of the phosphoenzyme

The third step in the reaction mechanism then would be the transition from an ADP-sensitive to an ADP-insensitive phosphointermediate: $E_1 \sim P \rightleftharpoons E_2 - P$.

The latter intermediate undergoes spontaneous hydrolysis stimulated by K^+. The sensitivity to K^+ is decreased by the treatments mentioned above: replacement of Mg^{2+} by Ca^{2+}, treatment with N-ethylmaleimide and limited tryptic proteolysis in the presence of Na^+. From these data we can infer that the ADP-sensitive $E_1 \sim P$ is K^+-insensitive and that the reverse is true for $E_2 - P$.

Oligomycin also appears to block the transition, as it inhibits the K^+-dependent dephosphorylation of the phosphoenzyme ($E_1 \sim P$) [36], but not the formation of $E_1 \sim P$ by reaction with ATP [37,38]. This inhibitor stabilizes the E_1 conformation, since inhibition by oligomycin is uncompetitive to Na^+ and ATP [39]. Another mitochondrial ATPase inhibitor, quercetin, has a similar action [40]. In addition, lipid removal from the enzyme preparation also appears to interfere preferentially with this conformational change (Section 4). The transition is conformational, but apparently does not involve a phosphate shift from the aspartyl-β-carboxyl to another group. Phosphorylation of N-ethylmaleimide treated Na-K ATPase, followed by proteolysis of the phosphoenzyme with pronase, leads to a phosphotripeptide with identical electrophoretic behaviour as that of the control [34].

The transition does not only involve a reduced affinity for ADP, but the affinities for Na^+ and K^+ are changed drastically as would be expected for a cation transport step. Three Na^+ ions are released per phosphorylation site and two K^+ ions bound [13]. This means that in the $E_2 - P$ conformation the affinity for Na^+ is decreased and that for K^+ is increased as compared to those for the $E_1 \sim P$ conformation. The K_D for K^+ of $E_2 - P$ is approx. 0.1 mM, whereas that for Na^+ competing at that site is 14 mM [10]. The release of Na^+ and the binding of K^+ by $E_2 - P$ occur at the outside of the cell membrane.

The affinity for K^+ is in good agreement with the $K_{0.5}$ of 0.14 mM (extracellular K^+) for the K^+ influx by erythrocytes [41]. The ratio of Na^+ release and K^+ binding by the enzyme corresponds with the ratio of 3 Na^+ ions expelled for 2 K^+ ions taken up per ATP hydrolysed by the erythrocyte [42,43] and squid giant axon [44,45]. These findings support the assumption that this transition is the actual transport step.

The affinity of $E_2 - P$ for extracellular Na^+ appears to depend on the presence or absence of extracellular K^+ [46]. In the absence of extracellular K^+ a site of high affinity for Na^+ ($K_D = 0.5$ mM) is exposed through which Na^+ inhibits the spontaneous hydrolysis of $E_2 - P$ [46,47] or its reformation following hydrolysis [13]. Increasing the external Na^+ concentration abolishes this inhibition and increases the K_D for Na^+ to about 30 mM [9]. The $E_1 \sim P \rightleftharpoons E_2 - P$ transition, like the primary phosphorylation, is reversible. Increasing the Na^+ concentration raises the fraction of ADP-sensitive phosphointermediate [40] and decreases the intrinsic tryptophan fluorescence evoked by phosphorylation with ATP to $E_2 - P$ [48].

(d) Hydrolysis of the phosphoenzyme

The fourth step on the reaction mechanism is the hydrolysis of the $E_2 - P.K$ complex to $E_2.K$ and P_i or the spontaneous but much slower hydrolysis of $E_2 - P$ to E_2 and P_i. Phosphate is released to the intracellular space, like ADP in the second step [42]. This hydrolysis step is, like ATP binding, inhibited by the cardiac glycoside ouabain. Ouabain stabilizes the $E_2 - P$ intermediate [49] by binding at the outer membrane side. If the inhibition by low Na^+ ($K_D = 0.5$ mM) would act on the reformation of E_2-P rather than on its spontaneous hydrolysis [13], then its high affinity for the extracellular side may be for the E_2 rather than for the E_2-P state and the K_D of 30 mM [9] would be the true dissociation constant of the E_2-P state.

As mentioned before, the $E_2.K$ conformation is inhibitory to ATP binding, but when $E_2.K$ is formed by hydrolysis of E_2-P its inhibition is not antagonized either by intracellular or by extracellular Na^+ [15], but it is antagonized by high intracellular ATP ($K_D = 0.45$ mM [47]). The non-sensitivity to Na$^+$ indicates that E_2.K contains K$^+$ in an occluded form [50].

(e) Return to the native enzyme form

The fifth and final step is the transformation of the enzyme from its occluded E_2.K form to a non-occluded E_1.K form, which readily exchanges its K$^+$ for intracellular Na$^+$ and releases it to the intracellular compartment. In the absence of ATP, but with saturating Na$^+$, this transition is extremely slow with a rate constant of 0.26/s for enzyme from pig and rabbit kidney outer medulla [47] or 6/s for enzyme from electrical eel electroplax [51]. It is substantially enhanced by binding of ATP to E_2.K at a site of low affinity ($K_D = 0.45$ mM). At saturating ATP concentrations the rate constant of the transition is increased to 54/s for the kidney enzyme [47]. In the absence of ATP the equilibrium is towards E_2.K, as the rate constant for the $E_1 \xrightarrow{K^+} E_2$.K transition is in the order of 60–100/s [16,51]. The effect of ATP probably depends on its binding only, since studies with non-phosphorylating ATP analogues revealed that they can support K$^+$ uptake in exchange for an equivalent release of K$^+$ (the so-called K$^+$–K$^+$ exchange, Section 5) just as well [52]. This does not necessarily mean that ATP binds to E_2.K at a site different from the high-affinity phosphorylating site. It may simply mean that the affinity for E_2.K is much lower than that for E_1 and that the enzyme acquires the proper high-affinity conformation for phosphorylation after binding of Na$^+$ (Fig. 1).

Besides its regulatory role in modulating the K$^+$ affinity via the E_2.K → E_1.K transition, binding of MgATP at the low-affinity sites also inhibits the ATP–ADP exchange, which effect is mimicked by Mg^{2+} binding to these sites ($K_D = 0.8$ mM [53,54]). K$^+$ ions antagonize the effect of MgATP via their sites on the inside of the membrane, but they do not antagonize the Mg^{2+} effect [55]. This leads us to the role of Mg^{2+} in millimolar concentrations (about 1 mM in excess of the optimal ATP concentration [56]), required for the overall ATPase activity and the K$^+$-stimulated phosphatase activity (see Section 2f). Originally, it was believed that millimolar concentrations of Mg^{2+} shift the equilibrium $E_1 \sim P \rightleftharpoons E_2 - P$ to the right, since they inhibit the ADP–ATP exchange involving $E_1 \sim P$ [57]. This conclusion has been opposed by Klodos and Skou [18,58] who found no difference in reactivity of the phosphointermediate to K$^+$ or ADP at millimolar and micromolar Mg^{2+} levels, suggesting that high Mg^{2+} would be required at a later stage. Robinson [59] assumes that MgATP is the true substrate at the low-affinity substrate sites, which are involved in the E_2.K → E_1.K transition. The K$_m$ for MgATP for the overall ATPase reaction is 0.48 mM, whereas free ATP and Mg^{2+} competitively inhibit with K$_i$-values of 4.8 and 40 mM, respectively [53]. The high concentrations of Mg^{2+} would then be required for complexation of ATP at the low-affinity substrate site. However, the K_D value of 0.45 mM for the E_2.K.ATP complex mentioned above was determined in the absence of Mg^{2+} [47]. Hence, the role of Mg^{2+} in millimolar concentration and the step at which it acts are still open to question. However, there is evidence that Mg^{2+} and $Mg^{2+} + ATP$ in millimolar concentrations induce a conformational effect in the enzyme [48] and that Mg^{2+} increases cooperative interactions between the Na^+ sites in the overall ATPase reaction [60]. The latter observation could mean that high Mg^{2+} causes subunit interactions, which are necessary for the operation of the dimeric Na-K ATPase (Section 3). On the other hand, the monomerically operating p-nitrophenyl phosphatase (Section 3) also requires high Mg^{2+} concentrations.

Following the exchange of K^+ for Na^+ at the internal membrane sites, the enzyme can start a new cycle of phosphorylation and dephosphorylation as outlined in a scheme given by Karlish and Yates [47], in which Mg^{2+} is omitted:

$$\begin{array}{c}
Na_i \cdot E_1 \cdot ATP \xrightarrow{k_1 = 11\,000/\text{min}} Na_i \cdot E_1 P + ADP \\
\\
k_4 = 3240/\text{min} \uparrow \quad \begin{array}{c} -K_i^+ \\ +Na_i^+ \end{array} \quad \quad \begin{array}{c} -Na_o^+ \\ +K_o^+ \end{array} \quad \downarrow k_2 = 4500/\text{min} \\
\\
E_2 \cdot ATP\,(K) \underset{K_D}{\overset{ATP}{\rightleftharpoons}} E_2 \cdot (K) \xleftarrow[k_3]{-P_i} K_o \cdot E_2 P \\
\\
= 0.45\,\text{mM} \quad\quad\quad\quad = 14\,000/\text{min}
\end{array}$$

It is clear from the indicated rate constants (all at 20°C) that the $E_2.K \to E_1.K$ and $E_1 P \to E_2 P$ transitions are rate-limiting as compared to the phosphorylation and dephosphorylation steps. From the turnover number of the overall ATPase reaction and the indicated rate constants it can be calculated that in the steady state at a saturating ATP level 53% of the enzyme molecules is present in the form of $E_2.K$, 11% is in the E_1-state, 27% is phosphorylated to $E_1 P$ and only 9% to $E_2 P$ [47]. The relatively high K_D value for ATP at its low-affinity sites ($K_D = 0.45$ mM) implies that at low ATP concentrations extracellular K^+ is inhibitory and not stimulatory [46], since ATP is then unable to prevent further accumulation of $E_2.K$, which inhibits binding of the nucleotide.

Other nucleotides are poor substitutes for ATP. CTP is the best substrate after ATP, giving 15% of the activity for ATP [56,61], mainly because of its low affinity for $E_2.K$ ($K_D = 2$ mM at the low-affinity site [62]). CTP is the only other natural nucleotide triphosphate supporting $K^+ - K^+$ exchange in red-blood cells with a rate constant one quarter of that with ATP [63].

The specificity of Na^+ for phosphorylation is absolute [64], but the effect of K^+ on dephosphorylation is shared by $Tl^+ > Rb^+ = K^+ > Cs^+ = NH_4^+ > Li^+$ in order of decreasing efficiency [65]. Mg^{2+} can be replaced by Mn^{2+} and Co^{2+}, which have 10% of its efficiency. Fe^{2+}, Ca^{2+}, Zn^{2+}, Cu^{2+}, Ba^{2+}, Sr^{2+} and Be^{2+} inhibit the Mg^{2+}-catalysed hydrolysis of ATP [66].

(f) K^+-stimulated phosphatase activity

The enzyme also displays a K^+ stimulated phosphatase activity, which is an inherent part of the Na-K ATPase system and which hydrolyses substrates containing phosphate bonds of an energy intermediate between that of an acid anhydride and of an ester bond. The hydrolysis of p-nitrophenylphosphate has been investigated most intensively. Its hydrolysis is normally stimulated by intracellular K^+ via the transition $E_1 + K^+ \rightleftharpoons E_1.K \rightleftharpoons E_2.K$ [15]. The K_m for K^+ for the phosphatase reaction (1 mM [10]) lies in the same range as that for this transition. The K_m values for p-nitrophenylphosphate and Mg^{2+} are 3.3 mM and 0.6–0.9 mM, respectively [53,54,62], both substances acting at intracellular sites [67,68]. Na^+ and ATP each inhibit the activation by K^+ [69], conceivably by antagonizing the $E_1 \xrightarrow{K^+} E_2.K$ transition. They also inhibit the activation of the activity by p-nitrophenylphosphate [70]. ATP inhibits activation by p-nitrophenylphosphate through competition at the low affinity nucleotide binding sites ($K_i ATP = 0.1$ mM), and not at the high-affinity nucleotide binding sites [69]. Hence, it has been inferred that hydrolysis of p-nitrophenylphosphate occurs via the low-affinity ATP binding sites of $E_2.K$.

On the other hand, ATP plus Na^+ at suboptimal K^+ concentration stimulate the phosphatase activity via phosphorylation [69] and thus via the high-affinity ATP binding sites. Hydrolysis of the intermediate $E_2 - P$ leads, like the $E_1 \xrightarrow{K^+} E_2.K$ transition, to $E_2.K$, but is stimulated by extracellular K^+ [15]. Since $E_2 - P$ has a higher affinity for K^+ than E_1, the K^+ affinity of the phosphatase activity is increased via the phosphorylation pathway [71]. However, the maximal velocity is decreased as compared to that occurring via the $E_1 \xrightarrow{K^+} E_2.K$ transition (rate constant 6000/min) because of the slower $E_1 \sim P \rightarrow E_2 - P$ transition (rate constant 4500/min). The $E_2.K$ intermediate by virtue of its relative stability can undergo at least ten phosphorylation-dephosphorylation cycles from p-nitrophenylphosphate before decomposing into E_2 and K^+ [50]. Despite the relatively fast $E_1 \xrightarrow{K^+} E_2.K$ transition p-nitrophenylphosphatase activity is only 15% as high as the overall ATPase activity, presumably because phosphorylation of $E_2.K$ by p-nitrophenylphosphate, the rate limiting step in the phosphatase reaction [72], is much slower than phosphorylation of E_1 to $E_2 - P$ by ATP. Other phosphatase substrates, like 2,4-dinitrophenylphosphate and β-(2-furyl)acryloyl phosphate, are better substrates for $E_2.K$ than p-nitrophenylphosphate; they are hydrolyzed 1.9–4.1 × as fast as ATP and their hydrolysis is not stimulated by $Na^+ + ATP$ [73]. Ouabain inhibits the dephosphorylation rather than the phosphorylation step [74].

3. Structural aspects

(a) Subunit structure and composition

Na-K ATPase is composed of two different subunits, one of about 100 000 M_r (α-subunit) and one of about 50 000 M_r (β-subunit), as shown by sodium dodecyl-sulfate (SDS) gel electrophoresis [75,76]. Since both subunits are glycoproteins and therefore bind different amounts of SDS as compared to normal proteins, their SDS gel electrophoretic mobility, and hence their apparent molecular weights can deviate considerably. Recently the molecular weights of the separated subunits of Na-K ATPase from rabbit kidney outer medulla have been determined more accurately by sedimentation equilibrium analysis in the absence of detergents [77]. The value for the α-subunit thus determined is 131 000 (120 600 for its protein part) and 61 800 for the β-subunit (42 800 for its protein part).

The number of each subunit in the total enzyme molecule has been uncertain until recently, values ranging from $\alpha_2\beta_1$ to $\alpha_2\beta_4$ and $\alpha_8\beta_x$ having been proposed. Peters et al. [77] have determined the α/β ratio in three independent ways, by determining the α/β protein mass ratio through amino acid analysis, by determining the molar absorption coefficients and the α/β absorbance ratio at 280 nm and by calculation from greatly differing amino acid contents in the two subunits. All three approaches yield an α/β molar ratio of 1:1. In combination with published molecular weight data for the complete enzyme molecule, ranging from 250 000–380 000 [78–80], the most likely subunit composition is an $\alpha_2\beta_2$ tetramer (calculated molecular weight 386 000, protein molecular weight 327 000). High magnification electron microscopy of freeze-fractured rat kidney membranes support a tetrameric structure of the enzyme complex [81].

Cross-linking studies have underlined the juxtaposition of the α- and β-subunits. The α-chain can be cross-linked to the β-chain by means of the bifunctional amino-reagent dimethylsuberimidate and dimethyl-3,3'-dithiobispropionimidate [82,83]. However, a sulfhydryl reagent like cupric phenanthroline, which catalyses oxidation of $-SH$ groups to a disulfide bridge, leads only to cross-linking of two α-subunits [84–86], while the β-subunits are oxidised to high molecular weight aggregates in a much slower process [85]. This is understandable, since only 2 of the 36 sulfhydryl groups of the Na-K ATPase are located on the β-subunit [87].

The α-subunit carries the ATP binding site [88], the phosphorylation site [89–91] and the specific sites for inhibition by N-ethylmaleimide [92] and ouabain [93–95]. Since ouabain and ATP bind at opposite sides of the membrane, this implies that the α-subunit must traverse the membrane. Antibodies have been prepared against the α- and β-subunits [96]. Those against the α-subunit inhibited phosphorylation but not ouabain binding, whereas those against the β-subunit inhibited ouabain binding but not the phosphorylation. This means that, since ouabain binds to the outer membrane side, the β-subunits must, at least partially, be oriented to that side.

It has been claimed that, despite the presence of two α-subunits per Na-K ATPase molecule, only one molecule of ATP binds and reacts and only one molecule of ouabain binds per enzyme molecule [75,97]. This is the more striking, since both α-subunits display the same gel electrophoretic behaviour and sensitivity to tryptic proteolysis and the amino acid sequence from the N-terminus shows no difference up to the twelfth residue [98]. From the linear relationship between fractional loss of ATP binding capacity and tryptic cleavage of the α-chains in KCl medium (Section 3b) Jørgensen and Petersen [97] concluded that only one of these proteins in the dimer contains a functioning ATP binding centre. This conclusion can be extended to the sites of phosphorylation and ouabain binding, which are restricted to one and the same α-subunit in the dimer [99]. The 1:1 molar exclusion of nucleotide binding by phosphorylation (Schuurmans Stekhoven et al., unpublished data) also could be most easily explained by assuming that their respective sites would be restricted to one and the same α-subunit, and reside in the same active centre (Fig. 1, Section 2). The other α-subunit in the dimer would then only have a regulatory function. There is, however, considerable doubt about this conclusion, as former stoichiometries for ATP binding or phosphorylation have been expressed per 250000 M_r and have been based on protein determination by the Lowry method with reference to bovine serum albumin. The protein molecular weight of Na-K ATPase now has become more probably 327000 (i.e. 30% higher than previously accepted), whereas the Lowry protein determination gives 35% higher values than amino acid analysis. This leads to a phosphorylation capacity of 1.9 mol/mol enzyme (2α-subunits) [100].

Although the function of the α-subunit as centre of phosphorylation and dephosphorylation has been well established, the function of the β-subunit is still uncertain. It is a sialoglycoprotein, containing 13–31% carbohydrates, viz. glucosamine and galactosamine as amino sugars and fucose, galactose, mannose and glucose as neutral sugars [76,77,82]. The α-subunit also contains carbohydrates (amino sugars, neutral sugars and sialic acid), but only 5–20% of the amount found in the β-subunit [77,101]. Isoelectric focussing gel electrophoresis shows that the β-subunit is heterogeneous as far as the sugar moiety is concerned, due in part to variation in the sialic acid content [102]. Removal of the sialic acid residues with neuraminidase reduced the number of protein bands from 9 to 4. Inhibition of the overall Na-K ATPase reaction by an antibody against the β-subunit suggests that this subunit is involved in that reaction [103]. Determination of lipid bilayer conductance has shown that the β-subunit displays ionophoric properties with specificity for Na^+ imposed upon it through interaction with the α-subunit [104]. Optimal Na^+ ionophoric properties are obtained at a molar ratio of one α-subunit to two β-subunits. K^+ and Cs^+, which do not increase the conductance of the 1:2 complex, inhibit its Na^+ ionophoric activity [104]. On the other hand, K^+ provides better protection against tryptic degradation of the β-subunits than Na^+, indicating that K^+ may have a more profound influence on the conformation of this protein than Na^+ [105]. Thus the β-subunit may act as a Na^+ ionophore, but this action is influenced by the conformational state of the entire Na-K ATPase complex.

The phosphatase activity is inhibited by radiation in agreement with a molecular weight of 140000, which is half of that reported for the overall Na-K ATPase activity [78]. This means that the rate-limiting step in this reaction would occur via an ($\alpha\beta$)-monomer, while that in the Na-K ATPase reaction occurs via an ($\alpha\beta$)-dimer. However, molecular weight determinations from radiation inhibition have not been reported on other partial reactions of the Na-K ATPase enzyme. Yet, the ATP-driven $E_2.K \to E_1.K$ transition is one of the rate limiting steps in the Na-K ATPase reaction scheme (Section 2) and thus could be a likely candidate for a dimeric enzyme action. This is supported by the finding that the sulfhydryl reagent thimerosal (ethylmercurithiosalicylate), which appears to inhibit by interference with subunit interactions [106], inhibits a step in the overall Na-K ATPase reaction with low affinity for ATP [107].

(b) Conformational states

In Section 2 we have discussed two major conformational enzymatic transitions, i.e. the $E_1 \sim P \rightleftharpoons E_2 - P$ and $E_2.K \rightleftharpoons E_1K$ transition as probed by inhibitors, fluorescent probes, intrinsic tryptophan fluorescence and sensitivity to tryptic proteolysis. We describe the proteolysis method in more detail, since it gives a clear insight in the conformational nature of the transitions, even though it cannot trace the rate of these transitions and may not be sensitive enough to detect other minor conformational changes. A change in the proteolysis pattern indicates that previously hidden peptide bonds become exposed (or vice versa) as a consequence of a conformational change. Bearing this in mind we shall consider the findings of Jørgensen et al. [35,48,97,108]. They found that in the presence of NaCl, the enzyme displays different kinetics of inactivation by trypsinolysis than in the presence of KCl (Fig. 2).

In the presence of NaCl the inactivation is biphasic. First, there is a relatively fast

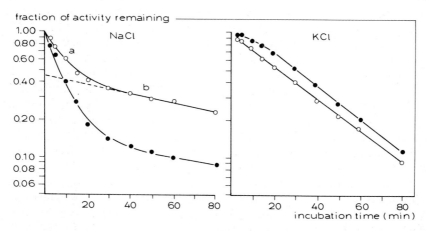

Fig. 2. Semilogarithmic plots of tryptic inactivation of Na-K ATPase (O) and K$^+$-p-nitrophenylphosphatase (●) activities in the presence of 150 mM NaCl (left) or 150 mM KCl (right). The fractional loss of activity (55%) in the rapid phase (rate constant a) is determined by extrapolation of the slower phase (rate constant b) to zero time. (From [97,108].)

reaction down to 60% inactivation with a rate constant a, which is linearly proportional to the amount of trypsin used. This phase is followed by a slow phase of inactivation with a rate constant b, which on the average is only 5% of a [108]. The first phase removes a small part of the α-subunit (no change in gel electrophoretic behaviour), but the second phase removes a larger part, leaving a 78 000 M_r product. During the first phase ATP binding and phosphorylation levels remain at 90% of those of the control enzyme, the ADP-ATP exchange activity increases to 150% of the control [97], the K^+-dependent dephosphorylation is halved and the K^+-dependent phosphatase activity is reduced to 15–20% (Fig. 2), but the ADP-sensitive $E_1 \sim P$ is increased 2–3-fold [35]. This leads them to the conclusion that the K^+-activation sites are remote from the relatively unharmed ATP binding and phosphorylation centre. During the second phase there is parallel loss of the partial reactions and the overall Na-K ATPase activity.

In the presence of KCl the α-subunit is split in fragments of about 58 000 and 48 000 M_r [108], accompanied by parallel inactivation of nearly all partial activities and the overall Na-K ATPase activity, with transient deviations for the ADP–ATP exchange and phosphatase reaction (Fig. 2). This occurs at a rate constant intermediate between a and b (18% of a) [108]. Thus the Na(E_1) and K(E_2) conformations are clearly distinguishable by trypsin digestion with regard to pattern of inactivation of Na-K ATPase activity and formation of proteolytic fragments from the α-subunit. Addition of ATP to the potassium enzyme yields the Na^+-pattern of inactivation via binding of ATP to the low-affinity site (half max. concentration 0.2 mM), which also promotes the $E_2.K \rightarrow E_1.K$ transition. On the other hand, addition of ATP plus Mg^{2+} to the sodium enzyme yields the K^+-inactivation pattern, representative of $E_2.K$ and $E_2.P$.

Other conformational changes have been investigated, which give the Na^+-inactivation pattern with changes in the rate constants a and/or b but not in the extent of inactivation in the first phase. These are: ATP (0.15–1 mM) in the presence or absence of Na^+ (a unaffected, b decreased), Mg^{2+} (2 mM) with or without Na^+ (a unaffected, b increased) and Mg^{2+} (2 mM) plus ATP (1–2 mM) (a decreased, b increased more than by Mg^{2+} alone). The latter effect is typical for ATP since ADP (2 mM) antagonizes the Mg^{2+} effect [48]. Thus Mg^{2+} and MgATP in millimolar concentrations cause conformational changes, but these effects cannot yet be ascribed to a particular step in the reaction mechanism (Section 2).

4. Phospholipid involvement

(a) Phospholipid headgroups

Na-K ATPase is an intrinsic membrane protein, which requires phospholipids for its activity. Lipid removal from crude microsomal preparations by detergents, organic solvents or phospholipase treatment leads to partial or complete inactivation of the enzyme (for refs., see [109]). Reactivation is obtained upon reconstitution with phospholipids.

Since optimal reactivation was obtained with the acidic phospholipids phosphatidyl serine (PS) and phosphatidyl inositol (PI), it was concluded that negatively charged phospholipids [110–113] and in particular PS [114], are essential for activity. This would be in accord with the partition of spin-labelled stearyl compounds between the protein-associated boundary layer and the fluid bilayer in a purified enzyme preparation from electric eel electroplax [115]. The negatively charged methyl phosphate derivative was most strongly protein-bound, the neutral dimethyl phosphate and alcohol had intermediate affinities, whereas the positively charged trimethyl ammonium compound was least strongly held. From the extent of release of spin label from the protein-associated boundary layer upon salt addition, the authors calculate that 51 mol of lipid/mol Na-K ATPase belong to the boundary layer at 130 mol total phospholipids present/mol enzyme.

The conclusion that PS is essential for enzyme activity has been challenged by De Pont et al. [116]. They have quantitatively removed PS and PI without delipidating the enzyme preparation from rabbit kidney outer medulla. A purified enzyme preparation was treated with PS-decarboxylase converting all PS to phosphatidyl ethanolamine or with PI-specific phospholipase C converting all PI to the diglyceride

TABLE 1

EFFECT OF SPECIFIC PHOSPHOLIPID MODIFICATION ON PURIFIED Na-K ATPase FROM RABBIT KIDNEY OUTER MEDULLA

	No. treatment 1	PS-decarboxylase 1	PI-Plase C (S. aureus) 2	Plase C (Cl. welchii) 3	1 + 3	2 + 3	1 + 2
	Number of phospholipid molecules per molecule of enzyme						
Total	267	266	249	99	99	90	254
Phosphatidylcholine	95	104	100	5	8	7	104
Phosphatidyl ethanolamine	74	100	74	25	53	16	96
Phosphatidylserine	35	0	29	34	0.5	42	2
Phosphatidylinositol	15	14	0	18	19	0.5	0.5
Sphingomyelin	48	48	46	17	18	24	51
Enzyme activity:	% of control activity						
Na-K ATPase	100	87 ± 7	101 ± 3	80 ± 6	90 ± 15	82 ± 6	56 ± 2
K-phosphatase	100	n.d.	99 ± 4	90 ± 8	75 ± 16	85 ± 6	64 ± 8
Number of detns.	15	7	11	7	7	3	4

(From [116].)

without affecting the activity (Table 1). Only combined action of PS-decarboxylase and PI-phospholipase C led to partial inactivation. Hence, it was concluded that the acidic phospholipids are not essential for activity. The results in this table also show that the enzyme can function maximally down to 90–100 mol of phospholipid/mol of the enzyme. This figure compares well with the value of 72–102 derived from reactivation titration data of a lipid-depleted Na-K ATPase preparation from dog fish rectal glands [117].

(b) Fatty acid groups

Another important parameter for optimal functioning of the enzyme is lipid fluidity, determined by the degree of unsaturation of the fatty acid residues in the phospholipids. Delipidated enzyme preparations could be reactivated by a number of diacyl phosphatidyl glycerols, ranking in order of increasing reactivation: distearoyl (18:0; 18:0) < dipalmitoyl (16:0; 16:0) < dimyristoyl (14:0; 14:0) < dioleoyl (18:1; 18:1) [118]. Di-unsaturated derivatives are more effective than monounsaturated ones [119]. Thus the shorter and the more unsaturated the acyl-chains, the more effective the phosphatidylglycerol derivative is in reactivating Na-K ATPase activity. This requirement of lipid fluidity is conceivable in view of the conformational changes, which the enzyme has to undergo during its reaction cycle (Section 3b).

(c) Role of phospholipids

Lipids may well be involved in the subunit interactions in the Na-K ATPase complex, since this would explain why the monomeric K^+-activated phosphatase is less inhibited upon delipidation and is reactivated at lower lipid concentrations than the overall Na-K ATPase activity [117,120]. This requirement of subunit interaction is supported by the finding that in the presence of detergents covalent cross-linking of α-subunits by Cu^{2+} or Cu^{2+}-phenanthroline is inhibited and is replaced by cross-linking of an α- to a β-subunit [121], despite the very low SH-group content of the β-subunit (Section 3a). In a lipid-depleted enzyme preparation the ATP-dependent phosphorylation level is reduced less than the overall Na-K ATPase activity [122]. Addition of K^+ to the phosphorylated lipid-depleted enzyme did not stimulate P_i production, whereas addition of K^+ to the lipid-reactivated preparation increased hydrolysis. This implies that the $E_1 \sim P \rightleftharpoons E_2 - P$ conformational transition is blocked in the lipid-depleted preparations, and that this is one of the steps in which subunit interaction is involved.

Drastic delipidation, which inhibits K^+-stimulated phosphatase activity nearly completely, apparently also interferes with the $E_1.K \rightleftharpoons E_2.K$ transition. Although the capacity for binding of ADP was not impaired and the affinity for ADP was only reduced by 50%, the antagonism between binding of K^+ and ADP was lost and was restored upon lipid reactivation [123]. This may indicate subunit interaction in the ATP-driven transition of $E_2.K$ to $E_1.K$. The latter hypothesis is supported by the finding that all modes of ouabain- and oligomycin-sensitive Na^+-fluxes (Section 5), and hence the coupled uptake of Rb^+ (a K^+-analogue) by thimerosal treated red blood cells are inhibited to the same extent as the Na-K ATPase activity, whereas the K^+-stimulated phosphatase, occurring via the $E_1.K \rightleftharpoons E_2.K$ transition is not inhibited at all [106].

Another consequence of partial lipid depletion is that the residual K^+-stimulated phosphatase is not inhibited by Na^+ nor undergoes stimulation by $Na^+ + ATP$ at low K^+ concentrations [117,124]. This has led to the suggestion that the K^+-stimulated phosphatase activity in preparations containing a full lipid complement is a monomeric enzyme function by virtue of a K^+-induced dissociation of the dimer into monomers. Na^+ would inhibit the phosphatase by inducing dimerization. In lipid-depleted preparations, having only one of the monomers saturated with lipid as a prerequisite for phosphatase activity to occur, Na^+ would be unable to cause dimerization. Hence, it does neither inhibit the phosphatase activity nor stimulate Na-K ATPase activity for which dimerization is essential. As a consequence, the Na^+-dependent reactions (ATP-phosphorylation and the ADP-ATP exchange) would be expressions of the dimer and the K^+-dependent reactions (K^+-stimulated dephosphorylation and phosphatase) expressions of the monomer. However, thimerosal, which would inhibit monomer interaction, does not inhibit any of these partial reactions, but increases the affinities for K^+ [125,126] and lipid depletion reduces the Na^+-dependent phosphorylation less than the overall Na-K ATPase activity [122]. It is clear from these studies that an unequivocal answer as to the monomeric and dimeric nature of the partial reactions in the overall Na-K ATPase reaction and their lipid dependence must await determination of their functional molecular weights.

5. Transport mechanism

(a) Normal and reversed Na^+-K^+ exchange transport

The relationship between Na-K ATPase activity and active trans-membrane transport of Na^+ and K^+, discussed in detail in earlier reviews [6,127,128] rests on the following arguments. Both Na-K ATPase and Na^+-K^+ transport are activated by the simultaneous presence of internal ATP, Mg^{2+} and Na^+ and external K^+ and both are inhibited by externally present cardiac glycosides like ouabain. The half-maximal activating concentrations of Na^+ and K^+, the K_m values for ATP and the half-inhibitory concentrations of ouabain are nearly equal for the two activities in the systems where they have been determined. For a large variety of tissues there is a remarkably constant ratio of 3 Na^+ transported per ATP hydrolyzed ([6] pp. 271–272; [128] p. 158). The 3 $Na^+/2 K^+$ stoichiometry for the transport agrees with the ratio of Na^+ released to K^+ bound upon phosphorylation of the enzyme (Section 2). Definitive proof for the involvement of the enzyme in transmembrane transport of Na^+ and K^+ has come from reconstitution studies in which a purified

Na-K ATPase preparation was incorporated into phosphatidyl choline vesicles containing radioactive Na^+ and K^+ [129]. Upon incubation of the vesicles, half of which were inside out, and thus could respond to external ATP, in a medium containing ATP, Mg^{2+}, Na^+ and K^+, uptake of 3 mol Na^+ and extrusion of 2 mol K^+ per mol ATP hydrolyzed was observed. Transport and ATPase activity were more than 90% inhibited by intravesicular ouabain, thus at the side opposite to the ATP-activation side.

In addition to the normal transport mode of 3 mol Na^+ extruded and 2 mol K^+ accumulated/mol ATP hydrolyzed the Na-K ATPase system can catalyze at least 4 other transport modes, depending on the cation concentrations on either side of the membrane and the concentrations of ATP, ADP and P_i on the internal membrane side. These transport modes will be illustrated with the aid of the scheme, depicted in Fig. 3. In cycle 1 of the scheme the relation between Na^+ transport and the relevant steps of the reaction mechanism are shown, in cycle 2 this is done for K^+ transport.

Three comments are in order before discussing the various transport modes by means of this scheme.

(1) In cycle 1 binding of Na^+ and ATP at the high-affinity sites in E_1 are indicated as two consecutive steps 1 and 2, although their sequence is random (Section 2a). This is done to illustrate the inhibition by Na^+ of phosphorylation by P_i in the presence of Mg^{2+} (reversal of steps $7' + 6'$). Half-maximal inhibition of P_i-phosphorylation occurs at the same Na^+ concentration as half-maximal stimulation of ATP-phosphorylation [130]. Conversely, inhibition of Na^+ and ATP binding

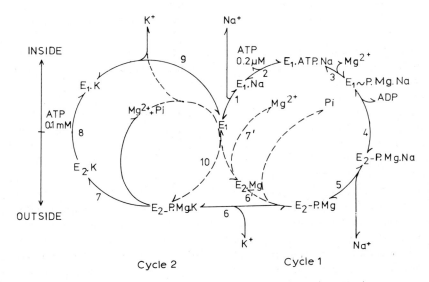

Fig. 3. Reaction scheme for Na-K ATPase in relation to its Na^+ and K^+ transport modes. "Inside" stands for the cytoplasmic side and "outside" for the extracellular side.

(reaction 1 + 2) as well as inhibition of the ADP-ATP exchange (reaction 2 + 3 in forward and reverse direction) by Mg^{2+} ($I_{50} = 0.5-0.8$ mM [59]) may all occur via reversal of step 7'.

(2) In cycle 2 we have depicted steps 7 and 8 as irreversible, but made provision for reversibility via step 10 in order to obtain a symmetrical bicyclic system.

(3) We have omitted the stoichiometry of cation binding and release for the sake of simplicity.

The normal transport mode is given by steps 1 through 9 (solid lines). Extrusion of Na^+ via steps 4 + 5 is provided for by high-affinity ATP-phosphorylation, and uptake of K^+ via steps 8 + 9 by low-affinity ATP binding without phosphorylation. When the extracellular K^+ concentration is decreased, step 6 is inhibited because external Na^+ inhibits both K^+ influx [41] and ATPase activity [131,132]. A high intracellular P_i concentration inhibits Na-K ATPase activity [133] at step 10, and consequently both the K^+ influx and the Na^+ efflux. The same effect is obtained by inhibition of step 8 through reduction of the intracellular ATP concentration [134]. This explains why at decreased extracellular K^+ and intracellular ATP, but at increased intracellular P_i the cycle starts running backward through step 10 (broken line) and the reversal of steps 1–6. In the absence of external K^+, but at high extracellular Na^+ the reversal becomes complete, as demonstrated by Na^+ influx and K^+ efflux as well as formation of ATP [135,136]. This is called *the reversed transport mode*.

(b) Na^+-Na^+ exchange transport

Cycle 1 can illustrate two other Na^+ transport modes: 1:1 Na^+-Na^+ exchange and uncoupled Na^+ efflux. *The 1:1 Na^+-Na^+ exchange* occurs in the presence of internal Na^+ and K^+ and at high external Na^+ in the absence of external K^+ [9,137,138]. As can be deduced by comparing the conditions required for the reversed transport mode and the Na^+-Na^+ exchange mode, both transport modes are superimposed on one another, but the Na^+ efflux accompanying the Na^+-Na^+ exchange is two- to four-fold the K^+ efflux accompanying the reversed mode [135]. The Na^+-Na^+ exchange has a rather high requirement for ADP. The accompanying Na^+ efflux increases linearly with the ADP concentration up to 0.3 mM [139]. The requirement for ATP is low, but present [138]. The K_D-values for intracellular Na^+ and K^+ are the same as for the normal transport mode, 0.19 and 9 mM respectively, and the K_D-value for extracellular Na^+ is 30 mM [9]. Coupled with its inhibition by oligomycin [140] at step 4, we may infer that the Na^+-Na^+ exchange occurs via steps 1–5 in forward and reverse direction. The requirement of intracellular K^+, which has been reported also for the Na^+-K^+ exchange mode [9] may be explained by substrate inhibition at step 2 in the presence of high intracellular Na^+ [133]. This effect may be overcome by intracellular K^+ via a reduction of the affinity for Na^+ [9]. Extracellular K^+ inhibits the Na^+-Na^+ exchange and the reversed mode via steps 6 and 7, leading to E_2.K which inhibits the binding of ATP at step 2. The reversibility of step 6 explains the mutual exclusion of the Na^+-Na^+ exchange and the Na^+-K^+ exchange mode [138].

(c) Uncoupled Na^+ efflux

The uncoupled Na^+ efflux occurs when both external Na^+ and K^+ are absent [46]. It lacks the back pressure of ADP and extracellular Na^+ at steps 4 and 5, respectively. It can thus be seen as a clockwise running of cycle 1 since it is accompanied by hydrolysis of ATP. The efficiency of Na^+ extrusion is about 3 mol/mol of ATP hydrolyzed [46]. The K_m values for ATP (<1 μM [46]) and for Na^+ (<1 mM [141]) correspond with those for high-affinity binding (Section 2a; steps 1 and 2 in cycle 1) and with those for the subsequent phosphorylation (step 3 in cycle 1), which has a K_m for ATP $= 0.5$ μM [64] and a K_D for $Na^+ = 0.2$ mM [142], and with those for Na-ATPase activity (cycle 1 in full), which has a K_m for ATP $= 0.3$ μM [62] and a K_D for $Na^+ = 0.4$ mM [142]. The equality of the K_m and K_D values indicates that a common step, probably the $E_1 \sim P \rightleftharpoons E_2 - P$ transition (step 4), is rate-limiting in this sequence of steps. This is supported by the finding that oligomycin inhibits the uncoupled Na^+ efflux and the Na-ATPase activity [141]. Oligomycin also inhibits Na^+ efflux in the $Na^+ - Na^+$ and normal $Na^+ - K^+$ exchange modes [140], indicating that the $E_1 \sim P \rightleftharpoons E_2 - P$ transition is the actual Na^+ transport step, depending on high-affinity ATP binding and phosphorylation. External K^+ and Na^+ in low concentrations inhibit the uncoupled Na^+ efflux [141] as well as the Na-ATPase activity [46]. The effect of K^+ is apparently caused by dissipation of $E_2 - P$ via steps 6 and 7, but is overcome in the normal transport mode by the presence of high intracellular ATP via step 8. Inhibition of Na-ATPase activity by low extracellular Na^+ is overcome at high extracellular Na^+ [46], which may be an effect on step 7' in cycle 1 if it is dealing with reformation of E_2-P rather than with its hydrolysis (Section 2). The effect of high extracellular Na^+ on the uncoupled Na^+ efflux, i.e. at low intracellular ADP, has not been reported, however.

(d) $K^+ - K^+$ exchange transport

Another altered transport mode involving intracellular and extracellular K^+ and requiring internal ATP and P_i, but absence of internal Na^+ [63] is *the 1:1 $K^+ - K^+$ exchange mode*. It can be visualized by cycle 2 turning clockwise, coupled to step 6 in forward and reverse direction. The relevant K_m values are: for ATP 0.1 mM, for P_i 1.7 mM, for intracellular K^+ 10 mM [63] and for extracellular K^+ (in the presence of 5 mM Na^+) 0.3 mM [135]. External Na^+ has little effect on K^+ efflux via $K^+ - K^+$ exchange transport, except that it increases the K_m value for external K^+ (1 mM, in the presence of 150 mM Na^+ [135]). Intracellular Na^+, however, inhibits K^+ efflux half-maximally at 1.5 mM in the presence of 9 mM intracellular K^+ [63], presumably through dissipation of E_1 via cycle 1. This explains the mutual exclusion of the $K^+ - K^+$ exchange and the normal $Na^+ - K^+$ exchange modes at increasing or decreasing intracellular Na^+ [143]. ATP can be replaced in the $K^+ - K^+$ exchange, but not in the $Na^+ - K^+$ exchange, by non-phosphorylating analogues like adenylyl imidodiphosphate [52] indicating that step 8 occurs by mere binding of ATP. The K_m value for ATP binding of 0.1 mM matches that for the overcoming by intracellular ATP of Na-K ATPase inhibition by extracellular K^+ [46], and is close to the K_m value (0.38 mM) for ATP in the $Na^+ - K^+$ exchange [141]. This suggests that the low-affinity, non-phosphorylating binding of ATP provides for the transport of K^+ in the normal transport mode. The requirement of P_i in step 10 of cycle 2 for the $K^+ - K^+$ exchange can be understood, if one assumes that E_2-P.Mg.K, in contrast to E_2.K, is a non-occluded K^+ containing conformation, which can exchange its K^+ at the outer surface of the membrane.

The separate existence of the $Na^+ - Na^+$ exchange and uncoupled Na^+ extrusion on the one hand and the $K^+ - K^+$ exchange on the other might indicate that the $Na^+ - K^+$ exchange is a sequential rather than a simultaneous process. This would require that the conditions for $Na^+ - K^+$ exchange inhibit the other transport modes via external K^+ and internal Na^+, respectively. Strong evidence that Na^+ efflux and K^+ influx must involve sequential steps, is provided by the finding that ATP-phosphorylation leads to release of Na^+ and binding, but apparently not to transport of K^+ [13]. Phosphorylation excludes nucleotide binding (Schuurmans Stekhoven et al., unpublished data), so that P_i must be released in step 7 of cycle 2 before ATP can bind in step 8 and give rise to K^+-influx.

(e) Phosphatase reaction as non-transporting system

The phosphatase activity can be visualized by step 10, followed by step 7, when P_i is replaced by an organic phosphate like *p*-nitrophenylphosphate (*p*-NPP). Actually, P_i competitively inhibits the phosphatase activity ($K_i = 1.5$ mM [144]) and exchanges with enzyme-bound phosphate derived from *p*-NPP [74]. Likewise, ATP through binding to the low-affinity binding site inhibits step 10 (comprising the reversal of steps 7-9) and competitively inhibits *p*-nitrophenylphosphatase activity ($K_i = 0.1$ mM [62]). The competitive inhibition by Na^+ with respect to K^+ and *p*-NPP [67] will conceivably take place at the inner membrane side via step 1, if the reaction occurs via the $E_1 \xrightarrow{K^+} E_2.K$ transition [15]. Mg^{2+}, like *p*-NPP activates intracellularly [68] and the K_m (0.6-0.9 mM) matches that for phosphorylation by P_i (0.8 mM) via the reversal of steps $6' + 7'$ [130]. The latter reaction also is inhibited by Na^+, conceivably via step 1. The relatively stable occluded $E_2.K$ conformation can undergo a number of phosphorylation and dephosphorylation steps in agreement with the finding that K^+ is required for dephosphorylation and phosphorylation as well [72,145,146].

The *p*-nitrophenylphosphatase activity (step 10, followed by step 7) does not lead to transport of Na^+ and K^+ ions. Rather, *p*-NPP inhibits $Na^+ - K^+$ exchange and $Na^+ - Na^+$ exchange [147] by withdrawal of E_1 via step 10. Obviously, *p*-NPP hydrolysis will not lead to ouabain-sensitive Na^+ transport, as it does not involve the partial reactions of cycle 1 involved in Na^+ transport. Neither will it lead to K^+ influx since this is dependent on nucleotides via step 8. It would, however, be conceivable that *p*-NPP hydrolysis by formation of E_2 – P.Mg.K via step 10, followed by reversal of step 6 would lead to uncoupled K^+ efflux, just as formation of $E_2 - $P.Mg.Na in cycle 1 can give rise to uncoupled Na^+ efflux. The difference with the uncoupled Na^+ efflux would be the slower phosphorylation of E_2.K by p-NPP than phosphorylation of E_1.Na by ATP to $E_2 - $P (Section 2), so that the enzyme would be most of the time in the occluded E_2.K conformation. The second difference is that $E_2 - $P.Mg has a lower affinity for Na^+ than for K^+, so the uncoupled K^+ efflux would soon come to a stop. The third difference is that p-NPP leads to a highly increased passive permeability for K^+, more than for Na^+ [147], which could make it indistinguishable from a slow active efflux. Fourthly, and perhaps most importantly, p-nitrophenylphosphatase activity is the expression of a monomeric form of the enzyme (Section 3), whereas cation transport may depend on interaction within the dimeric form [106].

6. Concluding remarks

Coming to the end of this chapter on Na-K ATPase, the reader will probably agree with the authors that this cation transport system is not only an extremely important component of animal cells but also a very complex system. Even after more than 20 years of studies by a great number of investigators, who have produced many hundreds of publications, our understanding of the actual mechanism of this system is still very incomplete. This is primarily due to its being a large intrinsic membrane protein, on which the currently known methods of protein structure determination, like X-ray diffraction, cannot be used. Barring an immediate breakthrough allowing the use of such techniques, it is to be expected that many more studies along the lines indicated in this chapter will be needed in order to advance our knowledge and understanding of the system bit by bit.

References

1 Gardos, G. (1954) Acta Physiol. Acad. Sci. Hung. 6, 191–199.
2 Caldwell, P.C., Hodgkin, A.L., Keynes, R.D. and Shaw, T.I. (1960) J. Physiol. 152, 561–590.
3 Skou, J.C. (1957) Biochim. Biophys. Acta 23, 394–401.
4 Schatzmann, H.J. (1953) Helv. Physiol. Acta 11, 346–354.
5 Skou, J.C. (1960) Biochim. Biophys. Acta 42, 6–23.
6 Bonting, S.L. (1970) in E.E. Bittar (Ed.), Membranes and Ion Transport, Vol. 1, pp. 257–363, Wiley Interscience, London.
7 Bonting, S.L. and Caravaggio, L.L. (1963) Arch. Biochem. Biophys. 101, 37–46.
8 Garrahan, P.J. and Garay, R.P. (1976) in F. Bronner and A. Kleinzeller (Eds.), Current Topics in Membranes and Transport, Vol. 8, pp. 29–97, Academic Press, New York.
9 Garay, R.P. and Garrahan, P.J. (1973) J. Physiol. 231, 297–325.
10 Robinson, J.D. (1977) Biochim. Biophys. Acta 482, 427–437.
11 Nørby, J.G. and Jensen, J. (1971) Biochim. Biophys. Acta 233, 104–116.
12 Hegyvary, C. and Post, R.L. (1971) J. Biol. Chem. 246, 5234–5240.
13 Yamaguchi, M. and Tonomura, Y. (1979) J. Biochem. 86, 509–523.
14 Nørby, J.G. and Jensen, J. (1974) Ann. N.Y. Acad. Sci. 242, 158–167.

15 Blostein, R., Pershadsingh, H.A., Drapeau, P. and Chu, L. (1979) in J.C. Skou and J.G. Nørby (Eds.), Na,K-ATPase, Structure and Kinetics, pp. 233-245, Academic Press, New York.
16 Karlish, S.J.D., Yates, D.W. and Glynn, I.M. (1978) Biochim. Biophys. Acta 525, 252-264.
17 Hansen, O., Jensen, J. and Nørby, J.G. (1971) Nature New Biol. 234, 122-124.
18 Klodos, I. and Skou, J.C. (1977) Biochim. Biophys. Acta 481, 667-679.
19 Swann, A.C. and Albers, R.W. (1978) Biochim. Biophys. Acta 523, 215-227.
20 Grisham, C.M. and Mildvan, A.S. (1974) J. Biol. Chem. 249, 3187-3197.
21 Grisham, C.M., Gupta, R.K., Barnett, R.E. and Mildvan, A.S. (1974) J. Biol. Chem. 249, 6738-6744.
22 Jensen, J. and Nørby, J.G. (1971) Biochim. Biophys. Acta 233, 395-403.
23 De Pont, J.J.H.H.M., Schoot, B.M., Van Prooyen-Van Eeden, A. and Bonting, S.L. (1977) Biochim. Biophys. Acta 482, 213-227.
24 Masiak, S.J. and d'Angelo (1975) Biochim. Biophys. Acta 382, 83-91.
25 Cantley, Jr., L.C., Gelles, J. and Josephson, L. (1978) Biochemistry 17, 418-425.
26 Brodsky, W.A. and Shamoo, A.E. (1973) Biochim. Biophys. Acta 291, 208-228.
27 Robinson, J.D. (1971) Nature 233, 419-421.
28 Patzelt-Wenczler, R., Pauls, H., Erdmann, E. and Schoner, W. (1975) Eur. J. Biochem. 53, 301-311.
29 Degani, C., Dahms, A.S. and Boyer, P.D. (1974) Ann. N.Y. Acad. Sci. 242, 77-79.
30 Nishigaki, I., Chen, F.T. and Hokin, L.E. (1974) J. Biol. Chem. 249, 4911-4916.
31 Fukushima, Y. and Tonomura, Y. (1973) J. Biochem. 74, 135-142.
32 Tobin, T., Akera, T., Baskin, I. and Brody, T.M. (1973) Mol. Pharmacol. 9, 336-349.
33 Fukushima, Y. and Post, R.L. (1978) J. Biol. Chem. 253, 6853-6862.
34 Post, R.L., Kume, S., Tobin, T., Orcutt, B. and Sen, A.K. (1969) J. Gen. Physiol. 54, 306s-326s.
35 Jørgensen, P.L., Klodos, I. and Petersen, J. (1978) Biochim. Biophys. Acta 507, 8-16.
36 Fahn, S., Koval, G.J. and Albers, R.W. (1968) J. Biol. Chem. 243, 1993-2002.
37 Whittam, R., Wheeler, K.P. and Blake, A. (1964) Nature 203, 720-724.
38 Israel, Y. and Titus, E. (1967) Biochim. Biophys. Acta 139, 450-459.
39 Inturrisi, C.E. and Titus, E. (1968) Mol. Pharmacol. 4, 591-599.
40 Kuriki, Y. and Racker, E. (1976) Biochemistry 15, 4951-4956.
41 Garrahan, P.J. and Glynn, I.M. (1967) J. Physiol. 192, 175-188.
42 Sen, A.K. and Post, R.L. (1964) J. Biol. Chem. 239, 345-352.
43 Whittam, R. and Ager, M.E. (1965) Biochem. J. 97, 214-227.
44 Baker, P.F. (1965) J. Physiol. 180, 383-423.
45 Baker, P.F., Blaustein, M.P., Keynes, R.D., Manil, J., Shaw, T.I. and Steinhardt, R.A. (1969) J. Physiol. 200, 459-496.
46 Glynn, I.M. and Karlish, S.J.D. (1976) J. Physiol. 256, 465-496.
47 Karlish, S.J.D. and Yates, D.W. (1978) Biochim. Biophys. Acta 527, 115-130.
48 Jørgensen, P.L. and Petersen, J. (1979) in J.C. Skou and J.G. Nørby (Eds.), Na,K-ATPase, Structure and Kinetics, pp. 143-155, Academic Press, New York.
49 Sen, A.K., Tobin, T. and Post, R.L. (1969) J. Biol. Chem. 244, 6596-6604.
50 Post, R.L., Hegyvary, C. and Kume, S. (1972) J. Biol. Chem. 247, 6530-6540.
51 Froehlich, J.P., Albers, R.W. and Hobbs, A.S. (1979) in J.C. Skou and J.G. Nørby (Eds.), Na,K-ATPase, Structure and Kinetics, pp. 129-141, Academic Press, New York.
52 Simons, T.J.B. (1975) J. Physiol. 244, 731-739.
53 Robinson, J.D. (1974) Biochim. Biophys. Acta 341, 232-247.
54 Robinson, J.D. (1976) Biochim. Biophys. Acta 440, 711-722.
55 Robinson, J.D. (1977) Biochim. Biophys. Acta 484, 161-168.
56 Skou, J.C. (1974) Biochim. Biophys. Acta 339, 246-257.
57 Fahn, S., Koval, G.J. and Albers, R.W. (1966) J. Biol. Chem. 241, 1882-1889.
58 Klodos, I. and Skou, J.C. (1975) Biochim. Biophys. Acta 391, 474-485.
59 Robinson, J.D. and Flashner, M.S. (1979) in J.C. Skou and J.G. Nørby (Eds.), Na,K-ATPase, Structure and Kinetics, pp. 275-285, Academic Press, New York.
60 Robinson, J.D. (1972) Biochim. Biophys. Acta 266, 97-102.

61 Matsui, H. and Schwartz, A. (1966) Biochim. Biophys. Acta 128, 380–390.
62 Robinson, J.D. (1976) Biochim. Biophys. Acta 429, 1006–1019.
63 Simons, T.J.B. (1974) J. Physiol. 237, 123–155.
64 Post, R.L., Sen, A.K. and Rosenthal, A.S. (1965) J. Biol. Chem. 240, 1437–1445.
65 Robinson, J.D. (1975) Biochim. Biophys. Acta 384, 250–264.
66 Dahl, J.L. and Hokin, L.E. (1974) Annu. Rev. Biochem. 43, 327–356.
67 Garrahan, P.J., Pouchan, M.I. and Rega, A.F. (1969) J. Physiol. 202, 305–327.
68 Rega, A.F., Garrahan, P.J. and Pouchan, M.I. (1970) J. Membrane Biol. 3, 14–25.
69 Swann, A.C. and Albers, W. (1975) Biochim. Biophys. Acta 382, 437–456.
70 Skou, J.C. (1974) Biochim. Biophys. Acta 339, 258–273.
71 Robinson, J.D. (1969) Biochemistry 8, 3348–3355.
72 Robinson, J.D. (1971) Biochem. Biophys. Res. Commun. 42, 880–885.
73 Gache, C., Rossi, B. and Lazdunski, M. (1977) Biochemistry 16, 2957–2965.
74 Inturrisi, C.E. and Titus, E. (1970) Mol. Pharmacol. 6, 99–107.
75 Jørgensen, P.L. (1974) Biochim. Biophys. Acta 356, 53–67.
76 Perrone, J.R., Hackney, J.F., Dixon, J.F. and Hokin, L.E. (1975) J. Biol. Chem. 250, 4178–4184.
77 Peters, W.H.M., De Pont, J.J.H.H.M., Koppers, A. and Bonting, S.L. (1981) Biochim. Biophys. Acta 641, 55–70.
78 Kepner, G.R. and Macey, R.I. (1968) Biochim. Biophys. Acta 163, 188–203.
79 Atkinson, A., Gatenby, A.D. and Lowe, A.G. (1971) Nature New Biol. 233, 145–146.
80 Hastings, D.F. and Reynolds, J.A. (1979) Biochemistry 18, 817–821.
81 Haase, W. and Koepsell, H. (1979) Pflügers Arch. 381, 127–135.
82 Kyte, J. (1972) J. Biol. Chem. 247, 7642–7649.
83 De Pont, J.J.H.H.M. (1979) Biochim. Biophys. Acta 567, 247–256.
84 Kyte, J. (1975) J. Biol. Chem. 250, 7443–7449.
85 Giotta, G.J. (1976) J. Biol. Chem. 251, 1247–1252.
86 Giotta, G.J. (1977) Arch. Biochem. Biophys. 180, 504–508.
87 Schoot, B.M., De Pont, J.J.H.H.M. and Bonting, S.L. (1979) in J.C. Skou and J.G. Nørby (Eds.), Na,K-ATPase, Structure and Kinetics, pp. 193–204, Academic Press, New York.
88 Haley, B.E. and Hoffman, J.F. (1974) Proc. Natl. Acad. Sci. USA 71, 3367–3371.
89 Kyte, J. (1971) Biochim. Biophys. Res. Commun. 43, 1259–1265.
90 Uesugi, S., Dulak, N.C., Dixon, J.F., Hexum, T.D., Dahl, J.L., Perdue, J.F. and Hokin, L.E. (1971) J. Biol. Chem. 246, 531–543.
91 Schuurmans Stekhoven, F.M.A.H., Van Heeswijk, M.P.E., De Pont, J.J.H.H.M. and Bonting, S.L. (1976) Biochim. Biophys. Acta 422, 210–224.
92 Hart, W.M. and Titus, E.O. (1973) J. Biol. Chem. 248, 4674–4681.
93 Alexander, A. (1974) FEBS Lett. 45, 150–154.
94 Ruoho, A. and Kyte, J. (1974) Proc. Natl. Acad. Sci. USA 71, 2352–2356.
95 Kott, M., Spitzer, E., Beer, J., Malur, J. and Repke, K.R.H. (1975) Acta Biol. Med. Germ. 34, K19–K27.
96 Jean, D.H. and Albers, R.W. (1977) J. Biol. Chem. 252, 2450–2451.
97 Jørgensen, P.L. and Petersen, J. (1977) Biochim. Biophys. Acta 466, 97–108.
98 Hopkins, B.E., Wagner, Jr., H. and Smith, T.W. (1976) J. Biol. Chem. 251, 4365–4371.
99 Forbush, B. and Hoffman, J.F. (1979) Biochemistry 18, 2308–2315.
100 Peters, W.H.M., Swarts, H.G.P., De Pont, J.J.H.H.M., Schuurmans Stekhoven, F.M.A.H. and Bonting, S.L. (1981) Nature 290, 338–339.
101 Churchill, L., Peterson, G.L. and Hokin, L.E. (1979) Biochem. Biophys. Res. Commun. 90, 488–490.
102 Marshall, P.J. and Hokin, L.E. (1979) Biochem. Biophys. Res. Commun. 87, 476–482.
103 Rhee, H.M. and Hokin, L.E. (1975) Biochem. Biophys. Res. Commun. 63, 1139–1145.
104 Shamoo, A.E. and Myers, M. (1974) J. Membrane Biol. 19, 163–178.
105 Churchill, L. and Hokin, L.E. (1976) Biochim. Biophys. Acta 434, 258–264.
106 Askari, A., Huang, W. and Henderson, G.R. (1979) in J.C. Skou and J.G. Nørby (Eds.), Na,K-ATPase, Structure and Kinetics, pp. 205–215, Academic Press, New York.

107 Jensen, J., Nørby, J.G. and Ottolenghi, P. (1979) in J.C. Skou and J.G. Nørby (Eds.), Na,K-ATPase, Structure and Kinetics, pp. 227–230, Academic Press, New York.
108 Jørgensen, P.L. and Petersen, J. (1975) Biochim. Biophys. Acta 401, 399–415.
109 De Pont, J.J.H.H.M., Van Prooyen-Van Eeden, A. and Bonting, S.L. (1973) Biochim. Biophys. Acta 323, 487–494.
110 Taniguchi, K. and Tonomura, Y. (1971) J. Biochem. 69, 543–557.
111 Hokin, L.E. and Hexum, T.D. (1972) Arch. Biochem. Biophys. 151, 453–463.
112 Kimelberg, H.K. and Papahadjopoulos, D. (1972) Biochim. Biophys. Acta 282, 277–292.
113 Wheeler, K.P., Walker, J.A. and Barker, D.M. (1975) Biochem. J. 146, 713–722.
114 Wheeler, K.P. and Whittam, R. (1970) J. Physiol. 207, 303–328.
115 Brotherus, J.R., Jost, P.C., Griffith, O.H., Keana, J.F.W. and Hokin, L.E. (1980) Proc. Natl. Acad. Sci. USA 77, 272–276.
116 De Pont, J.J.H.H.M., Van Prooyen-Van Eeden, A. and Bonting, S.L. (1978) Biochim. Biophys. Acta 508, 464–477.
117 Ottolenghi, P. (1979) Eur. J. Biochem. 99, 113–131.
118 Kimelberg, H.K. and Papahadjopoulos, D. (1974) J. Biol. Chem. 249, 1071–1080.
119 Walker, J.A. and Wheeler, K.P. (1975) Biochim. Biophys. Acta 394, 135–144.
120 Wheeler, K.P. and Walker, J.A. (1975) Biochem. J. 146, 723–727.
121 Huang, W.H. and Askari, A. (1979) Biochim. Biophys. Acta 578, 547–552.
122 Wheeler, K.P. (1975) Biochem. J. 146, 729–738.
123 Jensen, J. and Ottolenghi, P. (1976) Biochem. J. 159, 815–817.
124 Hansen, O., Jensen, J. and Ottolenghi, P. (1979) in J.C. Skou and J.G. Nørby (Eds.), Na,K-ATPase, Structure and Kinetics, pp. 217–226, Academic Press, New York.
125 Henderson, G.R. and Askari, A. (1976) Biochem. Biophys. Res. Commun. 69, 499–505.
126 Henderson, G.R. and Askari, A. (1977) Arch. Biochem. Biophys. 182, 221–226.
127 Glynn, I.M. and Karlish, S.J.D. (1975) Annu. Rev. Physiol. 37, 13–55.
128 Bonting, S.L. and De Pont, J.J.H.H.M. (1977) in G.A. Jamieson and D.M. Robinson (Eds.), Mammalian Cell Membranes, Vol. 4, pp. 145–183, Butterworth, London.
129 Goldin, S.M. (1977) J. Biol. Chem. 252, 5630–5642.
130 Schuurmans Stekhoven, F.M.A.H., Swarts, H.G.P., De Pont, J.J.H.H.M. and Bonting, S.L. (1980) Biochim. Biophys. Acta 597, 100–111.
131 Glynn, I.M. (1961) J. Physiol. 160, 18–19P.
132 Whittam, R. and Ager, M.E. (1964) Biochem. J. 93, 337–348.
133 Robinson, J.D., Flashner, M.S. and Marin, G.K. (1978) Biochim. Biophys. Acta 509, 419–428.
134 Garay, R.P. and Garrahan, P.J. (1975) J. Physiol. 249, 51–67.
135 Glynn, I.M., Lew, V.L. and Lüthi, U. (1970) J. Physiol. 207, 371–391.
136 Glynn, I.M. and Lew, V.L. (1970) J. Physiol. 207, 393–402.
137 Garrahan, P.J. and Glynn, I.M. (1967) J. Physiol. 192, 159–174.
138 Garrahan, P.J. and Glynn, I.M. (1967) J. Physiol. 192, 189–216.
139 Glynn, I.M. and Hoffman, J.F. (1971) J. Physiol. 218, 239–256.
140 Garrahan, P.J. and Glynn, I.M. (1967) J. Physiol. 192, 217–235.
141 Karlish, S.J.D. and Glynn, I.M. (1974) Ann. N.Y. Acad. Sci. 242, 461–470.
142 Flashner, M.S. and Robinson, J.D. (1979) Arch. Biochem. Biophys. 192, 584–591.
143 Sachs, J.R. (1977) J. Physiol. 273, 489–514.
144 Robinson, J.D. (1970) Biochim. Biophys. Acta 212, 509–511.
145 Bond, G.H., Bader, H. and Post, R.L. (1971) Biochim. Biophys. Acta 241, 57–67.
146 Schoot, B.M. (1978) Ph.D. Thesis, University of Nijmegen, The Netherlands.
147 Garrahan, P.J. and Rega, A.F. (1972) J. Physiol. 223, 595–617.

CHAPTER 7

Calcium-activated ATPase of the sarcoplasmic reticulum membranes

WILHELM HASSELBACH

Max-Planck-Institut für medizinische Forschung Abteilung Physiologie, Heidelberg, F.R.G.

1. Introduction

In contrast to sodium and potassium ions, ionized calcium specifically affects a great variety of cellular functions at extremely low concentrations. In many cells numerous activities are regulated by changing the cytoplasmic concentration of ionized calcium. A low intracellular calcium level is an essential prerequisite for the ions' regulatory effect. In fact, in erythrocytes and giant nerve and muscle fibers, where the activity of ionized calcium can be measured, it was found to be lower than 1 μM. This concentration is more than 1000-fold smaller than that existing in the extracellular fluid. There is increasing evidence which indicates that hardly any cell can tolerate ionized calcium in its cytoplasm. Consequently, all cells should be equipped with calcium transporting systems in their plasma membranes which are able to maintain a low internal calcium level. The transport activity of these systems is most often quite low, so that in many cases their existence has been difficult to prove. A low calcium transport activity in the plasma membrane is sufficient in most cases, because the rates at which calcium ions invade the cell are usually low due to the relative calcium impermeability of the cell membrane. One of the transport systems in plasma membranes acts by exchanging internal calcium ions for external sodium ions. The energy for calcium extrusion is thus furnished by the sodium potential. For the removal of the exchanged sodium ions the sodium-potassium pump becomes active. Thus this calcium transport is indirectly driven by the sodium-potassium pump. In addition to this system other cells like the erythrocytes are equipped with a calcium transport mechanism sui generis which is indirectly fueled by ATP.

The contractile proteins of muscle were the first biological structure in which the role of ionized calcium as an activity-regulating agent was proved. Direct and indirect evidence has been furnished showing that the rapid activation and inactivation cycle of muscle depends on the sudden release of calcium ions and their subsequent complete removal. The quantities of calcium ions which have to be set free to cause the sudden start of chemical and mechanical activity are quite large, because the contractile proteins need for saturation the considerable quantity of approx. 0.2 μmol calcium/ml. Calcium depots with fairly high capacity on the one hand, and a powerful calcium removing system, on the other hand, must also be present in the intimate vicinity of the protein filaments of the contractile apparatus. Only in very thin and/or slowly contracting muscle fibers can the gates for calcium and the calcium-removing system be located in the plasma membrane alone. In thick and/or fast contracting fibers the structures involved in calcium release and removal are the membranous network of the sarcoplasmic reticulum. While the mechanism of calcium release is virtually unknown, considerable evidence concerning the mechanism of calcium withdrawal has been obtained which shows that the rapid removal of calcium is brought about by a very active ATP-driven calcium transport system in the sarcoplasmic reticulum membranes.

These membranes possess nearly all the properties with which membrane systems, ideally suited for studies on the mechanism of active ion translocation, should be endowed:

(1) The vesicular fragments of the sarcoplasmic reticulum allow the simultaneous study of ion translocation and the cleavage of ATP, which is the energy-yielding reaction.

(2) The preparation allows the quantification of transport energetics because the concentrations of the involved ions can be estimated not only in the external fluid but also in the internal space of the membrane vesicles.

(3) The formation and disappearance of transport intermediates resulting from the interaction of the energy-yielding substrates with the transport protein can be followed during ion translocation at every moment.

(4) The transport protein is the main constituent of the sarcoplasmic reticulum membranes and thus can be isolated in large quantities.

The present review deals with calcium transport and the accompanying enzymatic reactions displayed by the sarcoplasmic reticulum membranes. Subjects such as the molecular organization of the membranes, transport energetics, activity modulation, the development of the calcium transport system and its role in the regulation of muscular activity could not be considered. These problems have been partially reviewed previously (cf. [1–6]).

2. *The sarcoplasmic reticulum membranes, a structural component of the muscle cell—organization, isolation and identification*

The sarcoplasmic reticulum membranes are the main constituents of muscle's intracellular membranes, consisting of the transverse tubular membrane system and the longitudinally oriented sarcoplasmic reticulum proper [7,8]. The transverse tubules are narrow invaginations of the plasma membrane along which the electrical excitation spreads inwards [9]. They must be considered as differentiated plasmalemma [10–12]. In contrast, the sarcoplasmic reticulum membranes constitute a closed intracellular membrane system not connected with the external fluid [13,14]. It consists of different structural elements, (1) the large cisternae which are opposed to the transverse tubules or the plasmalemma, and (2) the so-called free sarcoplasmic reticulum which consists of narrow tubules coalescing in some areas to network-like structures. These structural elements differ in their development and organization in different muscles [15].

The relevance of the sarcoplasmic reticulum to muscle function became apparent with series of discoveries starting with Marsh's finding of a physiological relaxing factor present in the myoplasma of striated muscles [16–19]. This factor was later identified as consisting of vesicular membranes, endowed with a calcium storing mechanism which was responsible for the relaxing activity [19–22]. The identification of the relaxing particles as elements of the sarcoplasmic reticulum relies on their ability to store calcium ions. This was visualized in the electron microscope as electron-dense calcium precipitation inside the isolated vesicular membranes and in the interior of the membrane structure in situ [23,24]. The demonstration of a causal linkage between calcium uptake and ATP splitting by isolated sarcoplasmic reticulum vesicles provided evidence that calcium storage results from the activity of a membrane-bound calcium transport system, fueled by the hydrolysis of ATP [20,21]. The abundance of these highly specialized membranes in striated muscles together with an almost total absence of interfering activities made these membranes an ideal system for studying transport phenomena of biologically active membranes. In addition these studies yielded considerable information concerning the problem of excitation contraction coupling in muscle [25].

Most studies were performed with fragments of the sarcoplasmic reticulum membranes isolated from rabbit skeletal muscle. Membranes from a great variety of other muscles, i.e. from chicken [26], rat [27], frog [28], and lobster [29,30] were also investigated. An intracellular membrane system, analogous to muscle sarcoplasmic reticulum is present even in slime-mold plasmodia [31,32]. The isolation of pure and functioning membranes from some muscles, especially from cardiac [33], red [34] and smooth muscles [35,36], poses difficulties due to the presence of contaminants for example mitochondria, a highly active calcium-independent, magnesium-activated ATPase or a rapidly decaying transport activity.

The methods used for the isolation and purification of the sarcoplasmic reticulum membranes are modifications of the procedure originally described by Makinose and Hasselbach [1,21]. This method yielded three fractions, of which the lightest exhibited the highest activity. Contractile proteins, which occasionally contaminate the preparations, can be removed by treating the membranes with concentrated solutions of potassium-chloride [37,38]. The discarded heavier fraction is enriched in cisternal elements and accessorial proteins, some of which originate from the transverse tubules and the sarcolemma [39–41]. The separation of the crude microsomal muscle extract very much depends on the ionic strength of the media used for the isolation and preparation. At low ionic strength the heavy fraction, consisting of terminal cisternae partially in contact with T-tubules, predominates. At higher ionic strength, the intermediate and the light fractions increase. A relative selective separation of the T-tubules from the terminal cisternae has been achieved with high shear forces in a French press. The T-tubules were monitored by the binding of ouabain to their sodium-potassium ATPase [11,12].

3. Phenomenology of calcium movement

(a) Energy-dependent calcium accumulation

The sarcoplasmic reticulum vesicles rapidly take up considerable quantities of calcium from solutions containing in addition to calcium (0.1 μM–100 μM), ionized magnesium (approx. 0.1 mM) and ATP (approx. 1 mM). A maximum calcium uptake capacity of 150 nmol/mg protein is reached when calcium ions are offered in excess [21,44]. When less calcium is present, the solution is depleted of calcium and a minimal level of ionized calcium in the external medium in the range of 1–5 nM remains [45,46]. Since the ATP-dependent calcium uptake only occurs as long as the sarcoplasmic reticulum membranes exist as tightly sealed vesicular structures, it can be excluded that calcium uptake is brought about by an ATP-mediated binding to the vesicular membranes [22,44,47]. The more recent finding that the stored calcium is released from the sarcoplasmic reticulum vesicles when calcium ionophores such as X537A (Lasalocid) (Hoffmann La Roche), A23187 (Lilly) and Ionomycin (Hoffmann La Roche) [48,49] are added to the vesicular suspension, additionally supports the concept of an ATP-dependent calcium accumulation. The concentration of soluble calcium inside the vesicles can roughly be estimated to reach 1–5 mM based on the vesicular volume [50] and calcium binding to internal structures [42,43]. This internal concentration is at least 100 to 1000 times higher than the concentration of ionized calcium remaining in the external solution after cessation of net calcium uptake. When concentrations of ionized calcium are offered which optimally activate the uptake mechanism, the storage process ceases after a couple of seconds, either due to the limited calcium-storing capacity of the vesicles or with limited calcium supply due to the high activity of the uptake mechanisms. Time resolution of the process therefore requires the application of optical methods using calcium indicator dyes [48,51] or procedures which allow rapid interruption of the uptake process, avoiding liberation of stored calcium [52].

When, however, the same experiments are performed in media containing calcium precipitating and membrane permeable anions such as oxalate, phosphate or pyrophosphate, the calcium-storing capacity of the vesicles is increased by a factor of 100 [21,53,54]. Under these conditions calcium and the anions are taken up in nearly stoichiometrical amounts and calcium precipitates are formed inside the vesicles, thus maintaining the internal calcium concentration at low levels. When these agents are used, adjustment of the concentration of ionized calcium precipitation is necessary. This can easily be achieved by the addition of calcium complexing agents like ethyleneglycol bis-(2-aminoethyl)-N,N,N',N'-tetraacetic acid (EGTA) [21]. Due to the increase of the calcium-storing capacity of the vesicles in the presence of calcium-precipitating anions, the time course of calcium uptake is prolonged and can be measured over a considerable period of time. This enables the separation of calcium-loaded vesicles by precipitation [21], millipore filtration [27] or by flow dialysis [55]. The optimal rate of calcium uptake in its steady phase, observed in the presence of 5 mM oxalate, reaches 1–2 μmol/mg/min at 20°C which is approx.

50% of the initial rate measured in oxalate-free media. This difference is presumably due to the fact that under these conditions, oxalate permeation limits the rate of calcium uptake [56].

The specificity of the calcium transport system with respect to its ionic transport substrate, ionic cofactors and its energy-yielding substrates must be considered to be quite low. The sarcoplasmic reticulum membrane handles the calcium analogue, strontium, similarly to its natural transport substrate, with respect to rate and storing capacity. Manganese can substitute for magnesium as an ionic cofactor [57] and instead of ATP a great variety of phosphate compounds can be used as substrates [58–63]; these have been denoted as pseudosubstrates. While with all substrates the same storing capacity is reached, the rates of calcium uptake and of calcium-dependent phosphate liberation are dependent on the nature of the substrate declining in the following order:

$$ATP > GTP \approx ITP > CTP > UTP \approx DNPP > ACP \approx PNPP$$

(b) Coupling between calcium accumulation and ATP splitting

ATP is cleaved by most sarcoplasmic reticulum preparations at low rates in the absence of calcium ions. This activity has been denoted as basal activity [20]. When calcium accumulation is initiated, ATP is rapidly hydrolyzed in a calcium-dependent activity, reaching its optimum at a calcium concentration of 10 μM, and which is severely suppressed by the rising calcium concentrations in the interior of the vesicles [45,64,65]. The calcium-dependent activity was early characterized as the activity of an enzyme distinctly different from the calcium-independent enzyme. In contrast to the calcium-dependent ATPase, the calcium-independent enzyme is quite insensitive to thiol or amino group reagents. Conversely, the calcium-independent activity can be abolished by low concentrations of detergents which do not reduce the activity of the calcium-dependent enzyme [66]. The two enzymatic activities further differ in their nucleotide specificity and affinity, as well as in their magnesium and temperature dependences. The basal activity most likely originates from plasmalemma and T-tubules membranes [41].

The uptake of calcium in the absence, as well as in the presence of calcium precipitating ions, is strictly correlated in time with the calcium-dependent hydrolytic cleavage of ATP or other pseudosubstrates. The qualitative proof for a causal connection between calcium uptake and calcium-dependent ATP splitting emerged from the observation that all reagents which activate or inhibit the calcium-dependent ATPase activity, in parallel, stimulate or reduce calcium uptake.

In order to establish a quantitative correlation between calcium transport and ATP splitting, i.e. the degree of coupling of the transport system, calcium uptake and calcium-dependent ATP splitting must be measured simultaneously.

As mentioned before, in the absence of calcium-precipitating anions the uptake of calcium and the related calcium-dependent ATP cleavage last only a short period of time. The monitoring of both events requires methods which allow to initiate and to interrupt the reactions in less than a second. This is possible with the use of an apparatus consisting of mixing chambers in series [67,68]. The ATPase reaction is usually stopped by acid quenching. The time course of calcium uptake has been measured either by photometric recording of the spectral changes of metallochromic indicators, or by interrupting calcium uptake with high concentrations of EGTA or lanthanum [52]. These methods, however, do not permit the simultaneous monitoring of both events. Parallel measurements, during the initial phase of calcium uptake, demonstrate the occurrence of an early burst of calcium uptake presumably related to the transfer of the terminal phosphate residue of ATP to the enzyme [69,70]. In the subsequent steady-state phase, calcium uptake and phosphate liberation were found to be coupled with a ratio of 2, the same as that obtained under conditions where calcium uptake and ATP splitting were measured synchronously during the steady-state period of calcium accumulation in the presence of oxalate (cf. [2]).

Larger and smaller, and sometimes variable coupling ratios have repeatedly been described [63,71]. Coupling ratios greater than 2 were found when, in the presence of calcium-precipitating agents, calcium precipitation occurred spontaneously [71].

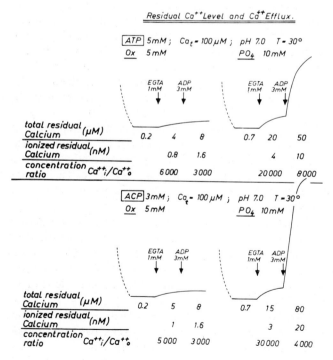

Fig. 1. Minimum levels of ionized calcium approached in the presence of different calcium-precipitating agents and energy-yielding substrates. Note that ADP induces a calcium release only in the presence of phosphate.

Larger coupling ratios can also erroneously be obtained when calcium uptake ceases before the splitting of ATP is interrupted. In this case, the calcium-dependent ATPase activity is underestimated. Similar deviations can be expected if the preparation contains an enzyme which transports calcium without being activated by calcium ions, as it is the case for sarcoplasmic reticulum preparations isolated from red skeletal muscles [34]. On the other hand, smaller ratios were found when the preparation contained a sizeable fraction of open vesicles; this fraction only contributes to the calcium-dependent ATPase activity and not to calcium transport. A reduced coupling ratio can also result from an increase in calcium permeability occurring for instance, at elevated temperatures or at alkaline pH [72].

It is sometimes difficult to distinguish between a permeability increase or an increase of the non-storing fraction. However, the fact that under numerous conditions a constant coupling ratio exists indicates that calcium translocation across the sarcoplasmic membranes and phosphate liberation are obligatory coupled. The rigid coupling ratio remains invariant even if the pump is driven by energy-yielding substrates which are only slowly consumed. Conflicting findings most likely result from the presence of variable amounts of open vesicles in the preparation. According to the rules of irreversible thermodynamics, a transport system, in which transport and the energy-yielding reaction are stoichiometrically coupled with a coefficient of one, should transform energy with a high efficiency. The reported maximum concentration ratios Ca_i/Ca_o support this reasoning. The mass law relation derived from the calcium/ATP stoichiometry of 2, yields with $K = 10^5$, Ca_i/Ca_o ratios which are close to the experimentally observed values (Fig. 1)

$$2\,Ca_o + ATP \rightleftarrows 2\,Ca_i + ADP + P_i; \qquad \frac{ADP \cdot P_i}{ATP} = K$$

$$\frac{Ca_i}{Ca_o} \leq \sqrt{\frac{ATP \cdot K}{ADP \cdot P_i}}$$

In agreement with this relation, the concentration ratio declines when the concentration of the splitting product, ADP and inorganic phosphate, are elevated.

(c) Passive calcium efflux

The same close coupling as observed for ATP-driven calcium uptake was found under conditions where the calcium pump operates as an ATP synthetase during calcium efflux (cf. [2]). The essential prerequisite for such a tight coupling between ion movement and ATP consumption or synthesis is a low passive calcium permeability of the membranes.

The sarcoplasmic reticulum vesicles start to release calcium passively when the activity of the calcium pump is blocked, either by inhibition of the transport enzyme or by depletion of energy-yielding substrates [45]. The rate of passive calcium efflux from vesicles loaded in the absence of calcium-precipitating agents, is approx. 20

nmol/mg/min if EGTA is present in the external fluid at neutral pH and 25°C. The membrane-permeability constant for calcium can be estimated from the observed release rate, if assumptions are made concerning the internal calcium concentration and the size of the surface of the vesicles. That the release rate is dependent on the internal calcium concentration is shown when the release experiments are performed in the presence of calcium-precipitating agents. The release rate increases as the concentrations of oxalate or phosphate are reduced. The calcium permeability of the membranes has a pronounced minimum at pH 6.3 and increases considerably when the pH or the temperature is raised [72].

The observed passive calcium flux across the sarcoplasmic membranes corresponds well to the calcium flux across the plasma membranes of resting muscle. The calcium permeability of the native membranes must be considered to be relatively high when compared with the calcium fluxes across liposome membranes, prepared from lipids isolated from sarcoplasmic membranes. As to the mechanism of passive calcium permeation, one might suggest that calcium passes through the transport protein per se or defects of the lipid bilayer. That charged membrane constituents seem to have little importance is shown by the fact that calcium permeability is scarcely affected by the ionic strength of the medium. In contrast, however, low concentrations of chaotropic anions which interact with the transport protein, specifically accelerate calcium release [73]. The membranes display an even lower calcium permeability when magnesium or magnesium together with ATP are present in the calcium-free external medium [52,74]. This slow passive calcium release makes it possible to analyze the effect of the occupation of internal calcium-binding sites on the interaction of the transport ATPase and its various substrates [75–77].

(d) Calcium efflux coupled to ATP synthesis

In the course of permeability studies performed with calcium-loaded vesicles, it was discovered that the slow calcium release could be accelerated 10–50-fold when the release medium was supplemented with phosphate and ADP [78]. The fact that the accelerated calcium efflux only occurred when both splitting products of ATP were present in the medium, strongly suggested that the reaction might be the reversal of the ATP-driven calcium uptake. In fact, a net synthesis of ATP can easily be demonstrated during calcium release [79–82]. The stoichiometry of the calcium-efflux-dependent ATP synthesis is identical with that of active calcium accumulation. The ratio of 2 for calcium/ATP in the synthesis reaction is valid, when the total amount of ATP synthesized is compared with the amount of calcium released by ADP from vesicles loaded with 0.2–1 μmol calcium/mg protein. This coupling is observed when conditions are chosen such that the reaction runs unidirectionally to completion. This can be achieved when an increase in the concentration of ionized calcium in the external medium is prevented and the level of ATP is kept low, by transferring its terminal phosphate group to glucose. Under these conditions, phosphate incorporation from ATP into glucose proceeds until the calcium stores are nearly completely depleted.

4. Reaction sequence: substrate binding

(a) Calcium binding

Calcium transport depends on the occupance of high-affinity calcium-binding sites on the external surface of the membranes (cf. [2,21,44,83–85]). There is general agreement that two high-affinity calcium-binding sites are involved and that the saturation of these sites and the activation of transport and ATP hydrolysis occur in the same concentration range.

It has repeatedly been stated that calcium affinity depends on the nature of the substrate. However, when experiments are performed under conditions where the effect of internal calcium can be excluded, i.e. by measuring the calcium-dependent ATPase of open vesicles, the calcium affinity proved to be the same for ATP and dinitrophenylphosphate and even for the very weak substrate nitrophenylphosphate [86]. Calcium ions can completely and easily be removed by EGTA from these sites [83,87]. Further, the binding sites are located in a section of the molecule which is only very little or not at all affected when cleaved by trypsin. Trypsin digestion performed in a Collowick cell did not accelerate calcium release as would be expected if trypsin cleavage destroys the calcium-binding sites. The occupation of the high-affinity sites presumably causes a number of changes in the molecule leading to an enhanced reactivity of 4–5 thiol groups towards 5,5'-dithiobis(2-nitrobenzoic acid) (DTNB) [88] and dansylcysteine mercurinitrate [89] and to a modulation of the mobility of thiol groups labelled with spin probes [90]. That the small changes in tryptophan fluorescence of native membranes [91,92] produced by addition and removal of calcium ions are not directly related to the state of occupance of calcium-binding sites in the transport protein, was shown when no fluorescence change could be observed in solubilized ATPase preparations [93]. The kinetics of calcium binding and dissociation has been approached by monitoring phosphoprotein formation [94,95]. When phosphoprotein formation is initiated by the addition of a saturating concentration of [^{32}P]ATP to the calcium-protein complex, the half time of the reaction is 10 ms (20°C) (Fig. 2a). If, however, phosphoprotein formation is started by adding calcium ions to the ATP-containing assay, phosphorylation proceeds much more slowly. An "on-rate" of 50/s for the formation of the phosphate-accepting calcium-protein complex has been estimated from these measurements. The overshoot of phosphoprotein formation, often observed when the reaction is started by the addition of ATP to the calcium-protein complex, can be explained by the low "on-rate" of calcium. As a consequence of the low "on-rate", the level of the phosphate-accepting complex is lower during the steady state than the initial level which prevails when the reaction is started by the addition of ATP to the calcium-saturated enzyme [68,94,96]. The dissociation rate of the calcium-protein complex has been obtained from the amounts of phosphoprotein formed, when started after different time intervals following the addition of EGTA to the calcium-protein complex (Fig. 2b). The first-order rate constant for calcium dissociation depends very much on pH and temperature. This observation indicates

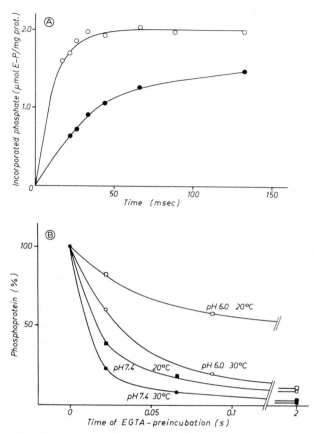

Fig. 2. Time course of formation and decline of phosphoprotein. (a) Phosphoprotein formation was started by the addition of 0.1 mM ATP to the calcium-containing assay (O— — —O) or by addition of calcium to the sarcoplasmic vesicles suspended in a solution containing EGTA and ATP (●— — —●). (b) To measure the decline of phosphoprotein following calcium removal, calcium was removed from the preparation by the addition of EGTA in a flow-quench apparatus, and at the time indicated [^{32}P]ATP was added. Phosphoprotein formation was terminated by quenching with perchloric acid after 150 ms [95].

that the slow dissociation rate is not only attributable to the relatively high stability of the calcium-protein complex.

In contrast to the well established existence of external high-affinity calcium-binding sites the permanent presence of low-affinity sites in the molecule seems ambiguous. This uncertainty is inherent in the scheme used to describe the reaction sequence of phosphate transfer and calcium translocation (cf. [3,4]). In this scheme, low-affinity sites are assumed to result from the phosphorylation of the transport protein. The low-affinity calcium binding sites in the unphosphorylated protein are generally considered as non-existent. The inhibition of calcium transport by increasing concentrations of internal calcium cannot be taken as a proof for the existence of permanent internal low-affinity sites, because under these conditions calcium-binding sites might be formed as a result of phosphorylation [64,65,97].

Yet, a number of observations strongly indicate that low-affinity calcium-binding sites are present in the non-phosphorylated enzyme. Ikemoto et al., have shown that the reactivity of a number of essential thiol groups of the protein is reduced under non-phosphorylating conditions in the presence of millimolar concentration of calcium [89]. However, these groups might belong to the great number of low-affinity calcium-binding sites located at the external surface of the vesicles [94]. The number of these groups can be obtained by measuring calcium binding to closed vesicles. During a short incubation period only the external calcium-binding sites are occupied because the calcium ions only slowly permeate the intact membranes of the vesicles. On the other hand, the external as well as the suspected internal calcium-binding sites are occupied when the experiments are performed with purified ATPase preparations. The difference of the binding values of intact vesicles and of calcium-permeable preparations should furnish the number of internal binding sites. Since lipid deprivation does not measurably reduce the total number of calcium-binding sites, the difference found can be assigned to the internal calcium-binding sites of the transport ATPase yielding 2–4 sites per molecule. The direct titration of the number of low-affinity sites facing the interior of the vesicles is inherently uncertain because the value determined is a small difference of large binding values. The obtained figures can also be ascertained by calorimetric titration of the low-affinity binding sites. The heat developed when calcium is bound by closed vesicles can be assigned to the occupation of 2 external binding sites per molecule characterized by a specific binding heat of 4000 J/mol. In ATPase preparations, approx. 4 binding sites per molecule of the same heat were found. The difference must be due to the presence of low-affinity calcium binding sites which are not accessible in closed vesicles and therefore face the internal space of the vesicles (Koenig and Hasselbach, unpubl. results).

(b) ATP binding

The evaluation of the enzyme's affinity for ATP is based either on binding or kinetic studies. Meissner [85] found that in the absence of ionized calcium the purified ATPase only binds ATP as its magnesium chelate, and that only one binding site per molecule with an affinity of $3 \cdot 10^6 \, M^{-1}$ is involved. More recent binding studies indicate that the enzyme likely possesses two classes of ATP-binding sites. Half of the ATP is assumed to be bound with an affinity of $10^5 \, M^{-1}$ and the other half with an affinity of only $10^3 \, M^{-1}$ [98]. The finding of Froehlich and Taylor [96] that ATP after being bound is rapidly hydrolyzed also in the absence of calcium, is difficult to reconcile with the result of binding experiments in the course of which the substrate molecules are decomposed. However, the existence of two binding sites with different affinities is supported by experiments performed with ATP analogues which are not hydrolyzed by the enzyme (Linder and Hasselbach, unpubl. results).

The presence of two binding sites for ATP seems to correlate well with the rather complex enzymatic properties of the calcium-dependent ATPase. The enzyme's activity increases over more than four orders of substrate concentration. The increase occurring between 0.1 μM and 10 μM ATP is followed by a less pronounced increment between 0.1 mM and 10 mM. This type of activity profile is at variance with the observation of Taylor et al. [99], who found that the main activity rise occurs between 0.1 mM and 10 mM magnesium ATP. A similar activity profile of the enzyme was observed by Rossi et al. [63], but only in an assay medium of alkaline pH. Possible causes for the complex kinetic behavior of the enzyme have been discussed. The alternatives are (1) two families of active sites; (2) one active site with two or more reaction pathways; (3) negative cooperativity between two sites; and (4) coupling between the enzymatic site and a regulatory site. However, the occurrence of two binding sites for ATP seems to depend on the state of the enzyme. When the turnover of the enzyme is accelerated either by cleaving the membrane phospholipids with phospholipase A_2 or by replacing the natural lipids by Triton X-100, other single-chain lipids or detergents, or when sufficient quantities of a calcium ionophore are added to native vesicles, the activity of the enzyme reaches its maximum at a magnesium-ATP concentration of 10 μM indicating the disappearance of the low-affinity binding site [100,101]. A single active site also accounts for the interaction of the native ATPase with other energy-yielding substrates such as ITP, GTP [102,103], and dinitrophenyl phosphate and nitrophenyl phosphate (Fig. 3).

The value of the apparent affinity for the high-affinity sites, monitored by steady-state ATP splitting, considerably deviates from the results of studies measuring initial phosphorylation velocities. These studies showed a 10-fold lower affinity for magnesium-ATP, indicating that the reaction steps following ATP binding lead to a considerable increase of the apparent affinity of ATP [68]. The same is true for

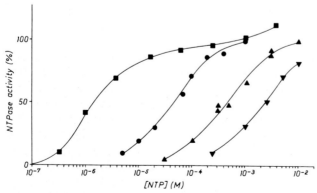

Fig. 3. The dependence of phosphate liberation on the concentration of various substrates: ■, ATP; ●, GTP; ▲, ITP; ▼, nitrophenylphosphate.

GTP as it can be seen by comparing Fig. 3 and Fig. 5. In contrast the apparent affinities for ITP derived from steady-state splitting and the initial rate of phosphorylation coincide.

(c) ADP binding

In contrast to the nucleoside triphosphates which enter the reaction cycle of the calcium transport system only as their magnesium complexes, the nucleoside diphosphates interact without ligands. Meissner [85] first proposed that free ADP might be the acceptor substrate for the phosphoryl group of the enzyme. An analysis of the magnesium requirement of the ATP–ADP exchange reaction performed by Makinose furnished additional evidence for the role of unliganded ADP as phosphate acceptor [103]. The fact that the synthesis of ATP from ADP and inorganic phosphate during calcium release is only suppressed when the concentration of magnesium is reduced below 0.1 mM, is also in support of free ADP as phosphate acceptor. An apparent affinity constant of $5 \cdot 10^5$ M^{-1} was found in measurements of the rate of ATP formation and of calcium release dependent on the ADP concentration in magnesium-containing media. The estimate is based on the regular Michaelis–Menten kinetics of the reaction. When magnesium chelation is taken into account an affinity of the free ADP for the enzyme of 10^6 M^{-1} results. In contrast experiments performed with the unphosphorylated enzyme yielded much lower affinities for unliganded ADP 10^4 M^{-1} (Linder, unpubl. results). Evidently the affinity of the enzyme for ADP depends on its functional state; the high affinity for ADP is a property of the phosphorylated enzyme, where phosphorylation occurred either by ATP or by inorganic phosphate. Among the different NDPs which can effectively function as phosphate acceptors, ADP has the highest affinity for the phosphorylated enzyme (cf. [2]).

(d) Magnesium binding

The indispensibility of magnesium ions as a cofactor for calcium transport and calcium-dependent ATPase activity is undisputed. Yet different mechanisms for its interaction with the enzyme had been proposed. It has been shown that the activation of calcium transport and calcium-dependent ATPase activity produced by added magnesium follow the increase of the concentration of the magnesium-ATP complex [46,105]. On the other hand, in more recent experiments it was observed that the terminal phosphate group can be transferred to the enzyme and appreciable amounts of phosphoprotein can accumulate even if no magnesium has been added to the assay [106,107]. This observation indicates that obviously very small quantities of magnesium ions which may be present as contaminations in the solvent or in the protein are sufficient for phosphoprotein formation. In fact, when the preparations are treated with ionic exchange resins phosphoprotein formation is drastically reduced. In the presence of very small quantities of magnesium ions phosphoprotein can accumulate even if its rate of formation is low, because the decay of phosphoprotein is much lower than the rate of formation. The absolute requirement of magnesium ions for phosphoprotein formation is most convincingly demonstrated under conditions allowing the transfer of the phosphoryl group of ATP back to ADP. When increasing concentrations of magnesium were added to assays containing ATP and ADP or GTP and GDP, phosphoprotein formation increases [103,104] (Fig. 4). This is possible only when magnesium-ATP reacts as the phosphate donor and unliganded ADP as the phosphate acceptor. Analysis of this reaction further reveals that magnesium ions are not only required for the formation of the substrate complex but also as cofactor reacting directly with the protein. Presumably these protein-bound magnesium ions participate in the exchange reaction as well as in the further processing of the intermediates leading to phosphate liberation. Corroboration for this idea comes from the observation that for calcium transport supported by nitrophenylphosphate and dinitrophenylphosphate, it can be shown that the enzyme is directly activated by ionized magnesium at the same concentrations effective for the reaction supported by magnesium-ATP. A further interesting feature of the dinitrophenylphosphate and nitrophenylphosphate supported reaction, is the finding that the unliganded pseudosubstrates are most likely used as substrates. The enzyme can only use the magnesium complexes of the nucleoside triphosphates as substrates, while the monophosphate substrates react in their unliganded state; in this case magnesium might be required for the reduction of an electrostatic repulsion between the highly charged polyphosphate residue of the nucleoside triphosphates and the protein. The reaction of the calcium transport ATPase as an ATP synthetase, also requires magnesium ions as a cofactor. The magnesium concentrations which satisfy this requirement are quite low and are in a similar range as those found for the direct activating effect when the pump uses ATP. The enzyme exhibits the same high affinity for magnesium ions when the

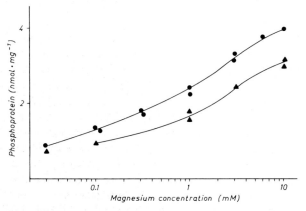

Fig. 4. Elevation of steady-state phosphoprotein level by magnesium. The sarcoplasmic reticulum vesicles were phosphorylated by 1 mM [^{32}P]GTP in the presence of 2 mM GDP at various magnesium concentrations [103].

incorporation of inorganic phosphate, thought to be the initial step in the reaction sequence, is studied provided that calcium ions are present at high concentration in the interior of the vesicles. If this is not the case, the incorporation of inorganic phosphate needs a tenfold higher magnesium concentration [76]. The proposal that magnesium ions might be transported across the membranes from inside to outside during active calcium influx has gained no experimental support [109]. Such a mechanism would require a very high permeability of the membranes for magnesium, to replenish the internal magnesium during massive calcium storage. The membranes, however, seem to be rather impermeable for ionized magnesium [56].

(e) Phosphate binding

The enzyme exhibits very different affinities for P_i depending on the experimental conditions. When, in the presence of calcium ions, the external calcium-binding sites of the membranes are occupied, the affinity for phosphate is very low and cannot be determined by binding experiments. The non-competitive inhibition of the calcium-stimulated phosphate liberation from nitrophenylphosphate by inorganic phosphate yields an inhibitory constant of approx. 50 mM. The inhibition of phosphate liberation from other substrates seems to require even higher phosphate concentrations [110,111]. In the absence of calcium ions on both surfaces of the membranes, phosphorylation with inorganic phosphate furnished a dissociation constant of the phosphoprotein complex of 10 mM [75,108]. This dissociation constant increases considerably at low temperatures corresponding to a heat of formation of 48 000 J [108]. The highest phosphate affinity is found when a calcium gradient exists across the membranes. Under these conditions the dissociation constant of the phosphoprotein complex is reduced to 0.1 mM [75,110]. All of these interactions of inorganic phosphate with the membranes require the presence of ionized magnesium; the phosphoprotein presumably exists as a magnesium-containing complex and is formed in a random sequence [76].

5. Reaction sequence: phosphoryl transfer reaction

(a) Phosphorylation of the transport protein in the forward and the reverse mode of the pump

In the presence of phosphate donating or accepting reactants, the translocation of calcium ions across the sarcoplasmic membranes is linked with phosphoryl transfer reactions leading to the phosphorylation of the transport protein. During calcium accumulation, the terminal phosphate group of ATP or of the other phosphate donors is rapidly transferred to the transport protein from which it is subsequently liberated by hydrolytic cleavage. The phosphoryl group in the protein is acid-stable and can therefore be stabilized in acidic quench media [112–114].

The phosphoryl transfer is activated by ionized calcium and magnesium in the external medium.

When the sarcoplasmic calcium transport system operates in the reverse mode and synthesizes ATP from ADP and inorganic phosphate during calcium release, inorganic phosphate reacts with the transport protein also leading to the formation of a phosphoprotein [115–117]. This reaction also requires ionized magnesium but is suppressed when the concentration of ionized calcium in the medium exceeds 10 μM. In the transport protein of the sodium-potassium system, analogous cation dependent phosphoryl transfer reactions take place. It is difficult, however, to directly correlate phosphorylation and ion movement in these membranes.

The linkage between calcium movement and the occurrence of intermediates in the calcium-transport protein has been investigated under conditions of continuously proceeding transport as well as during the initial phase of calcium accumulation and calcium release. At room temperature, the ATP-fueled transport system reaches a constant rate of calcium transport and ATP hydrolysis in a fraction of a second. This state can be maintained under suitable conditions for several minutes, i.e. in the presence of calcium-precipitating anions. As long as calcium transport continues the protein carries phosphoryl residues, the level of which increases in parallel with the rate of phosphate liberation and calcium transport (cf. [2,5]). This parallelism strongly supports a causal connection between phosphoprotein turnover and calcium transport. The finding that at high pH and high calcium concentrations, calcium transport and ATP splitting can be blocked without impairing the level of phosphoprotein does not contradict the central role of phosphoprotein as a transport intermediate. Only the demonstration of calcium transport without phosphoprotein formation would nullify this concept.

When calcium transport proceeds under optimal conditions a phosphoprotein level of 3–4 nmol/mg protein which corresponds to 0.5 equiv./mol ATPase has been observed in most studies [3,114]. This level of phosphoprotein is not significantly elevated when its hydrolytic decay is inhibited by the presence of millimolar concentrations of ionized calcium. The steady-state level of phosphoprotein which is found during optimally proceeding calcium-driven ATP synthesis is considerably lower reaching 0.3 nmol/mg [75]. Higher quantities of inorganic phosphate were found to be incorporated when the rate of net synthesis is retarded at low levels of ADP.

(b) ATP–P_i exchange

Inorganic phosphate is not only incorporated into the transport protein during synthesis of ATP but also when the transport system splits ATP or other energy-yielding substrates, provided that inorganic phosphate is present at sufficient concentrations [118–120]. The transfer of the protein-bound phosphate group to newly formed or to added NDP leads to the formation of labelled NTP. Since the formation of labelled NTP occurs during NTP splitting we are dealing with a reaction exchanging the terminal phosphate residue of NTP with inorganic phosphate. This reaction proceeds with relatively high rates as long as the medium calcium concentration is low and the intravesicular calcium concentration is high. As expected, the rate of the exchange reaction also depends on the concentration of the three phosphate-containing reactants NTP, NDP and inorganic phosphate in the system. Activation of the exchange by inorganic phosphate displays quite different characteristics depending on the nucleoside-phosphate reactants [103,119]. When the calcium concentration on both sides of the membranes becomes equal the exchange reaction subsides. At a pCa of 6 to 7 the rate of the exchange reaction falls to a few nmol/mg/min. Calcium concentrations in the millimolar range seem to enhance the exchange activity which, however, never reaches the high value observed when a calcium gradient exists across the membranes [103,120].

(c) Phosphate exchange between ATP and ADP

The participation of a phosphoprotein as a reaction intermediate in ATP-driven calcium transport, was first indicated by the finding of a nucleoside diphosphokinase activity in the sarcoplasmic reticulum vesicles which required ionized calcium [121,122]. It was assumed that a protein-bound phosphate residue might participate in the exchange reaction transferring the terminal phosphate group of ATP to ADP or between other nucleotide couples. The exchange reaction reflects the steady-state of the reaction sequence in which the transport protein is phosphorylated by ATP (initial step) and dephosphorylated by ADP (terminal step). The rate with which the phosphate group is exchanged between ATP and ADP is much faster than the rate of net calcium transport and phosphate liberation [122]. The activity of the exchange reaction rises at the same calcium concentration as the level of phosphoprotein, calcium transport and ATP splitting. As to the rate with which the phosphate residue is transferred, the velocity of the ATP–ADP exchange is exceptionally high and exchange between GTP and GDP, and ITP and IDP proceeds much more slowly [103,122]. The calcium dependence of the exchange reaction is not nucleotide-dependent. For the GTP–GDP exchange it has been shown that it is activated in the same range of medium calcium as the ATP–ADP exchange, calcium transport and GTP splitting. The NTP–NDP exchange reaction can most favorably be analyzed after net calcium uptake ceases, i.e. when the system approaches equilibrium [103]. Under these conditions the high internal calcium concentration (>10 mM) suppresses severely the calcium-dependent ATPase activity, thus reducing changes of the nucleotide concentrations to a minimum. The dependence of the exchange reaction on the concentration of its substrates, the magnesium chelate of NTP and unliganded NDP, indicates that the reaction exhibits ping-pong kinetics and that its equilibrium constant is not far from unity. When the constraint exerted by the internal calcium ions is removed either by the addition of a membrane-permeable calcium-precipitating anion or by making the vesicles calcium-permeable, net calcium transport and calcium-dependent splitting resume full activity. This activity transition affects the exchange activity only very little when the system performs ATP–ADP exchange, while in the case of GTP–GDP exchange the rate considerably declines but does not disappear. The decline of the exchange activity is thought to reflect the reduction of the concentration of the phosphorylated intermediate caused by the activation of the hydrolytic pathway (Table 1).

TABLE 1

Interrelation between NTPase activity, NTP, NDP and NTP-P$_i$ exchange reaction of sarcoplasmic reticulum vesicles

The reactions were measured at 22°C at pH 7.0 in the presence of 2 mM NDP and 5 mM inorganic phosphate.

Activities measured	ATP µmol/mg prot./min	GTP µmol/mg prot./min
Closed vesicles		
NTPase activity	0.053	0.053
NTP-NDP exchange	3.80	0.24
NTP-P$_i$ exchange	0.006	0.035
Opened vesicles		
NTPase activity	0.350	0.580
NTP-NDP exchange	3.10	0.08
NTP-P$_i$ exchange	0.002	0.004

(d) ADP-insensitive and ADP-sensitive phosphoprotein

Formation and decay of the phosphorylated transport protein has recently been resolved into a sequence of discrete reaction steps. At room temperature and under optimal conditions phosphoprotein formation from ATP proceeds so rapidly that the transfer of the phosphoryl group is difficult to follow. The kinetic data obtained by flow quench technique [68,94,102] assume a rapid formation of an acid-stable phosphoprotein which is subsequently transformed into an acid-labelled phosphocompound and from which inorganic phosphate is liberated.

$$E + T \underset{50/s}{\overset{k_2 \cdot T=160/s}{\rightleftarrows}} ET \underset{k_{-3} \cdot D=50/s}{\overset{160/s}{\rightleftarrows}} E \sim P \underset{24/s}{\overset{+D \quad 20/s}{\rightleftarrows}} E \cdot P \overset{10/s}{\rightleftarrows} E^+ \; {+P}$$

The phosphoryl transfer reaction occurs considerably more slowly when instead of ATP, ITP or GTP are used as phosphate donors [102,103] (Fig. 5). Yet the analysis of the reaction did not furnish new information concerning the existence of additional phosphoprotein species.

Experiments performed at low temperature and at very low concentration of ionized magnesium and ATP allow the demonstration of the sequential appearance of different acid-stable phosphoproteins. The phosphoprotein which appears first can transfer its phosphoryl group to ADP and is subsequently transformed to an ADP insensitive phosphoprotein. This transformation, which takes 2 to 3 s, has been followed, after phosphorylation was interrupted either by the removal of calcium with EGTA or by adding excess cold ATP [101,107,123–125]. The transition of the intermediate is retarded by the presence of calcium concentrations in the millimolar range, and to a lesser extent by monovalent cations [125]. The application of high

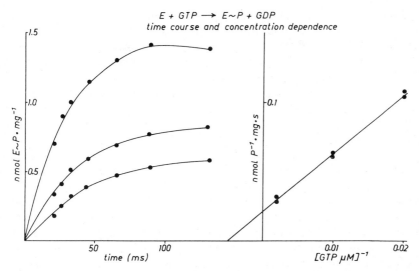

Fig. 5. Time course and concentration dependence of phosphoprotein formation supported by GTP. The reaction was started by the addition of GTP to the vesicular suspension containing 5 mM magnesium, and 0.1 mM calcium. The final concentrations of GTP were 5 µM, 10 µM and 50 µM. T=20°C, pH 7.0.

concentrations of ionized calcium to the phosphoprotein should have reversed the transition [107,125]. Monovalent cations on the other hand, considerably accelerate the hydrolysis of the ADP-insensitive phosphoprotein species. As to the nature of the two phosphoproteins, Takisawa and Tonomura [101] have assumed that the ADP-sensitive intermediate is a calcium-containing complex because its transition to the ADP-insensitive species can be induced by complexing calcium with EGTA. The finding that the transition also occurs when phosphoprotein formation is interrupted by the addition of cold ATP suggests that during this spontaneous transition calcium might be released from the protein. The reported parallelism between ATP–ADP exchange activity and the level of phosphoprotein needs evaluation in the light of the existence of two different phosphoproteins. Since the phosphoprotein comprises not only the phosphoprotein species directly involved in the exchange reaction, the ADP-sensitive fraction but also the ADP-insensitive intermediate—a parallelism between exchange activity and the level of total phosphoprotein—can occur only when the two species are present in constant proportions.

The incorporation of inorganic phosphate into the transport protein, which is the initial step in the reaction sequence of ATP formation, presumably leads also to different phosphoprotein species. When calcium is removed from the internal as well as from the external calcium binding sites of the membranes, i.e. when no calcium gradient exists, saturation of phosphate incorporation requires the presence of high concentrations of magnesium and phosphate and the absence of monovalent cation [126,127]. As mentioned before, the reaction proceeds endothermically with a heat of formation of 48 000 J/mol. The phosphoryl group of the resulting intermediate exchanges rapidly with medium P_i, and a corresponding exchange of the oxygen atoms between phosphate and water takes place [128,129]. The rate with which the incorporation of phosphate into the calcium-transport protein proceeds is quite high. At 30°C the half time of phosphoprotein formation is approx. 30 ms [127,129]. Rate constants for the hydrolytic cleavage of the phosphoenzyme were obtained in experiments monitoring its decay after a sudden reduction of the phosphate concentration in the medium or the incorporation of radioactive phosphate into the unlabelled phosphoenzyme at equilibrium by following the rate of "phosphate–water" exchange. The kinetic analysis of phosphoprotein formation is made difficult by the complex dependence of the reaction rate on the concentration of magnesium and phosphate ions [127]. Addition of ADP causes a rapid disappearance of the phosphoryl group. The residue is not transferred to ADP but released into the medium. Readdition of calcium ions to the assay, which elevates the calcium concentration, also induces phosphoprotein decay. Its time course can be resolved in a fast and a slow phase indicating the presence of two different intermediates. Because the decay rates of both fractions are much lower than the decay rates observed in the absence of calcium ions, one might assume that the addition of calcium produces a transient stabilization of the phosphoprotein species before it causes phosphate release [127]. Evidently a phosphoenzyme with quite different properties than those displayed in the absence of calcium ions seems to exist transiently. The phosphoryl group of a small fraction of this transiently appearing intermediate can possibly be transferred to ADP, giving rise to a small net synthesis of ATP [129,130]. The characterization of this ADP-sensitive intermediate formed in the presence of calcium from the ADP-insensitive species can most clearly be performed when calcium ions are present at high concentration in the interior of the sarcoplasmic vesicles and at low concentration in the external medium, i.e. when a calcium gradient exists. Under these conditions which can be maintained for a considerable period of time due to the low calcium permeability of the vesicles a phosphoprotein is formed which can immediately transfer its phosphoryl group to ADP [77,116,117]. This phosphoenzyme is further characterized as having a lower heat of formation, higher affinities for magnesium and phosphate ions, and a lesser sensitivity towards alkaline ions than the protein formed in the absence of a gradient [2,75]. As reported by Suko, the phosphoprotein level depends on the square of the concentration of calcium inside the vesicles with the concentration of calcium in the external medium kept constant [76]. Under suitable conditions the presence of internal calcium effects an increase of the phosphoprotein level from 0.5 to 3.5 nmol/mg. The existence of this calcium-gradient-dependent phosphoprotein species has recently been questioned on the basis of kinetic measurements yielding decay rates which were very similar for gradient-dependent and gradient-independent phosphoprotein. The authors assumed that it is the mere presence of calcium ions inside the vesicles which conveys ADP sensitivity [77]. The inferred kinetic similarity of the phosphoryl group, formed in the absence or in the presence of a calcium gradient, is at variance with older reports which indicated differences. These, however, were also not sufficient to explain the pronounced difference between the two intermediates with respect to other properties [75]. An immediate effect of calcium ions on the kinetic properties of the phosphoenzyme irrespective of the existence of a calcium gradient emerged from phosphate–water exchange experiments with calcium-permeable vesicles [132]. In the presence of 0.1 mM calcium the oxygen turnover is 10 times slower than in the absence of calcium ions. Under these conditions the calcium concentration at the external surface of the calcium-permeable vesicles determines the extent of phosphoprotein formation, being 6 times lower in the presence than in the absence of calcium. The large difference in the turnover number which was calculated taking into account the respective phosphoprotein level, reflects presumably the effect of calcium on the luminal calcium-binding site of the membrane.

(e) Phosphoryl transfer and calcium movement

The important problem of how the coupling between phosphoryl transfer and calcium movement might be brought about was first approached by Ikemoto [51]. Simultaneous measurements of phosphorylation by a small quantity of ATP and of the binding of calcium ions to non-accumulating ATPase preparations revealed that calcium ions are released when the ATPase is phosphorylated. To explain the finding that the maximum of enzyme phosphorylation was reached considerably earlier than that of calcium release, a formation of two phosphorylated intermediates was assumed. The first intermediate is formed when the external high-affinity calcium-binding sites are occupied. Calcium is released during the transition of the first to the second intermediate having a lower affinity for calcium. It remains to be demonstrated that the affinity decline leading to calcium liberation is comparable to the affinity difference which must be produced when calcium is concentrated during active accumulation by more than a factor of 1000. Similar experiments were performed measuring calcium uptake and phosphorylation of native vesicles when ATP was used as energy-yielding substrate, an initial burst of calcium uptake seems to occur and to be related to the formation of the phosphoenzyme with a stoichiometry of two to one [69,102]. In contrast to Ikemoto's finding calcium translocation is obviously directly coupled to phosphoprotein formation. A possible explanation for this discrepancy might be the use of different techniques for monitoring calcium movement. Ikemoto used calcium-sensitive dyes while Inesi quenched calcium movement by the addition of EGTA and subsequent filtration. The application of the EGTA method for monitoring the time course of calcium transport and phosphoprotein formation in heart sarcoplasmic reticulum seems to result in even larger discrepancy [94]. An appreciable quantity of calcium was found to be transported even before phosphate is transferred from ATP to the enzyme. The same surprising burst of calcium uptake was observed for GTP-supported calcium uptake when the reaction was initiated by the addition of GTP to the calcium-containing assay (Fig. 6a). In better agreement with the idea that calcium translocation is coupled to a transition in the state of the phosphoenzyme are the observations, in which the more slowly reacting ITP is used in place of ATP [102] (Fig. 6b) and uptake and phosphoprotein formation could be monitored. The repeatedly postulated coupling between formation of phosphoprotein and calcium translocation is

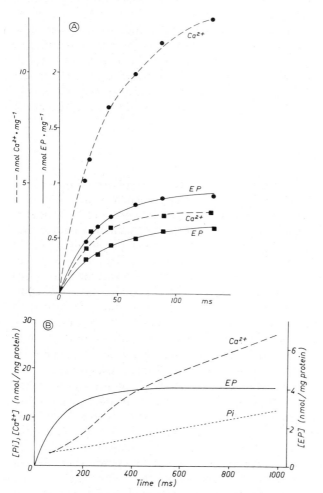

Fig. 6. (a) Calcium uptake and phosphoprotein formation supported by GTP. Calcium uptake (●------●) and phosphoprotein formation (●— — —●) were started by the addition of 0.3 mM GTP to the vesicular suspension containing 5 mM magnesium, 0.5 mM calcium, pH 7.0. Initial calcium uptake burst is nearly absent when the reaction is started by the addition of calcium to the assay containing GTP and EGTA (2 mM). Calcium uptake ■------■, phosphoprotein formation ■— — —■. (b) Calcium uptake and phosphoprotein formation supported by ITP (according to [102]). The reactions were started by the addition of 1 mM ITP. Calcium uptake in both experiments was interrupted by EGTA quenching. Note that the GTP supported calcium uptake occurs much faster than phosphoprotein formation. In contrast to the ATP supported calcium uptake, the calcium uptake burst can be resolved in time. Note further that the burst of the ITP induced calcium uptake is delayed with respect to enzyme phosphorylation.

difficult to reconcile with the behaviour of the calcium transport system near equilibrium, i.e. when after the cessation of net calcium uptake, a high calcium gradient exists. If the assumed coupling prevails, one should expect that the calcium exchange observed could be directly related to the phosphate exchange between NTP and NDP. Calcium entry should be accompanied by the phosphorylation of the enzyme and calcium exit should be paralleled by the transfer of the phosphoryl group to ADP. This, however, is not the case (Waas, unpubl. results). When the reaction is investigated in the presence of ATP and ADP the calcium exchange rate is 10 times slower than the rate of phosphate exchange. Furthermore, the rate of calcium exchange does not depend on the donor-acceptor couple although the phosphate exchange exhibits considerable nucleotide specificity.

Acknowledgements

The author is grateful to Mrs. Räker and Mrs. von der Schmitt for their patience during the preparation of the manuscript. He wishes to thank Dr. Nestruck-Goyke for many helpful corrections and Dr. Fassold, Dr. Rauch and Dr. Waas for providing partially unpublished results.

References

1. Hasselbach, W. (1964) Progr. Biophys. Mol. Biol. 14, 167–222.
2. Hasselbach, W. (1979) Topics in Current Chemistry 78, pp. 1–56, Springer, Heidelberg.
3. Inesi, G. (1979) Membrane Transport in Biology, pp. 357–393, Springer, New York.
4. de Meis, L. (1979) Annu. Rev. Biochem. 48, 275–292.
5. Tada, M., Yamamoto, T. and Tonomura, Y. (1978) Physiol. Rev. 58, 1–79.
6. MacLennan, D.H. and Holland, P.C. (1975) Annu. Rev. Biophys. Bioeng. 4, 377–404.
7. Peachey, L.D. (1965) J. Cell Biol. 25, 209–231.
8. Franzini-Armstrong, C. (1975) Fed. Proc. 34, 1382–1388.
9. Huxley, A.F. (1959) Ann. New York Acad. Sci. 81, 215–510.
10. Lau, Y.H., Caswell, A.H. and Brunschwig, J.P. (1977) J. Biol. Chem. 252, 5565–5574.
11. Lau, Y.H., Caswell, A.H., Brunschwig, J.P., Baerwald, R.J. and Garcia, M. (1979) J. Biol. Chem. 254, 540–546.
12. Caswell, A.H., Lau, Y.H., Garcia, M. and Brunschwig, J.P. (1979) J. Biol. Chem. 254, 202–208.
13. Huxley, H.E. (1964) Nature 202, 1067–1071.
14. Somlyo, A.V., Shuman, H. and Somlyo, A.P. (1977) J. Cell Biol. 74, 828–857.
15. Sommer, J.R. and Waugh, R.A. (1976) Am. J. Pathol. 82, 192–221.
16. Marsh, B.B. (1952) Biochim. Biophys. Acta 9, 247–260.
17. Portzehl, H. (1957) Biochim. Biophys. Acta 26, 373–377.
18. Ebashi, S. (1958) Arch. Biochem. Biophys. 76, 410–423.
19. Nagai, T., Makinose, M. and Hasselbach, W. (1960) Biochim. Biophys. Acta 43, 223–238.
20. Hasselbach, W. and Makinose, M. (1961) Biochem. Z. 333, 518–528.
21. Hasselbach, W. and Makinose, M. (1963) Biochem. Z. 339, 94–111.
22. Ebashi, S. and Lipmann, F. (1962) J. Cell Biol. 14, 389–400.
23. Hasselbach, W. (1964) Fed. Proc. 23, 909–912.
24. Constantin, L.L., Franzini-Armstrong, C. and Podolski, R.J. (1965) Science 147, 158–160.

25 Ebashi, S. (1976) Annu. Rev. Physiol. 38, 293–313.
26 Sabbadini, R., Scales, D. and Inesi, G. (1975) FEBS Lett. 54, 8–12.
27 Martonosi, A. and Feretos, R. (1964) J. Biol. Chem. 239, 648–658.
28 Muscatello, U., Andersson-Cedergren, E., Azzone, G.F. and von der Decken, A. (1961) J. Biophys. Cytol. Suppl. 10, 201–218.
29 van der Kloot, W.G. and Glovsky, J. (1965) Comp. Biochem. Physiol. 15, 547–565.
30 Baskin, R.J. (1971) J. Cell Biol. 48, 49–60.
31 Kato, T. and Tonomura, Y. (1977) J. Biochem. 81, 207–213.
32 Zubrzycka-Gaarn, E., Korczak, B. and Osinska, H.E. (1979) FEBS Lett. 107, 335–339.
33 Suko, J. and Hasselbach, W. (1976) Eur. J. Biochem. 64, 123–130.
34 Heilmann, C. and Pette, D. (1979) Eur. J. Biochem. 93, 437–446.
35 Wuytack, F. and Casteels, R. (1980) Biochim. Biophys. Acta 595, 257–263.
36 Raeymaekers, L., Agostini, B. and Hasselbach, W. (1980) Histochemistry 65, 121–129.
37 Martonosi, A. (1968) J. Biol. Chem. 243, 71–81.
38 de Meis, L. and Hasselbach, W. (1971) J. Biol. Chem. 246, 4759–4763.
39 Meissner, G., Conner, G.E. and Fleischer, S. (1973) Biochim. Biophys. Acta 298, 246–269.
40 Meissner, G. (1975) Biochim. Biophys. Acta 389, 51–68.
41 Malouf, N.N. and Meissner, G. (1979) Exp. Cell Res. 122, 233–250.
42 Ikemoto, N., Bhatnagar, G.M., Nagy, B. and Gergely, J. (1972) J. Biol. Chem. 246, 7835–7837.
43 Ostwald, T.J., MacLennan, D.H. and Dorrington, K.J. (1974) J. Biol. Chem. 249, 5867–5871.
44 Carvalho, A.P. (1966) J. Cell. Physiol. 67, 73–84.
45 Makinose, M. and Hasselbach, W. (1965) Biochem. Z. 343, 360–382.
46 Weber, A., Herz, R. and Reiss, J. (1966) Biochem. Z. 345, 329–369.
47 Carvalho, A.P. and Leo, B. (1967) J. Gen. Physiol. 50, 1327–1352.
48 Scarpa, A., Baldassare, J. and Inesi, G. (1972) J. Gen. Physiol. 60, 735–749.
49 Beeler, T.J., Jona, I. and Martonosi, A. (1979) J. Biol. Chem. 254, 6229–6231.
50 Duggan, P.F. and Martonosi, A. (1970) J. Gen. Physiol. 56, 147–167.
51 Ikemoto, N. (1976) J. Biol. Chem. 251, 7275–7277.
52 Chiesi, M. and Inesi, G. (1979) J. Biol. Chem. 254, 10370–10377.
53 Beil, F.U., von Chak, D., Hasselbach, W. and Weber, H.H. (1977) Z. Naturforsch. 32c, 281–287.
54 Newbold, R.P. and Tume, R.K. (1979) J. Membr. Biol. 48, 205–213.
55 Mermier, P. and Hasselbach, W. (1976) Eur. J. Biochem. 69, 79–86.
56 Kometani, T. and Kasai, M. (1978) J. Membr. Biol. 41, 295–308.
57 Kalbitzer, H.R., Stehlik, D. and Hasselbach, W. (1978) Eur. J. Biochem. 82, 245–255.
58 Makinose, M. and The, R. (1965) Biochem. Z. 343, 383–393.
59 de Meis, L. (1969) J. Biol. Chem. 244, 3733–3739.
60 Pucell, A. and Martonosi, A. (1977) J. Biol. Chem. 246, 3389–3397.
61 Inesi, G. (1971) Science 171, 901–903.
62 Nakamura, Y. and Tonomura, Y. (1978) J. Biochem. 83, 571–583.
63 Rossi, B., de Assis-Leone, F., Gache, C. and Lazdunski, M. (1979) J. Biol. Chem. 254, 2302–2307.
64 Hasselbach, W. (1966) Ann. N.Y. Acad. Sci. 137, 1041–1048.
65 Weber, A. (1971) J. Gen. Physiol. 57, 50–63.
66 Walter, H. and Hasselbach, W. (1973) Eur. J. Biochem. 36, 110–119.
67 Kanazawa, T., Saito, M. and Tonomura, Y. (1970) J. Biochem. 67, 693–711.
68 Froehlich, J.P. and Taylor, E.W. (1975) J. Biol. Chem. 250, 2013–2021.
69 Kurzmack, M., Verjovski-Almeida, S. and Inesi, G. (1977) Biochem. Biophys. Res. Commun. 78, 772–776.
70 Verjovski-Almeida, S. and Inesi, G. (1979) J. Biol. Chem. 254, 18–21.
71 Ebashi, F. and Yamanuchi, I. (1964) J. Biochem. 55, 504–509.
72 Hasselbach, W. (1969) Biochemistry of Intracellular Structures, pp. 145–152, PWN, Warsaw.
73 The, R. and Hasselbach, W. (1975) Eur. J. Biochem. 53, 105–113.
74 Sumida, M. and Tonomura, Y. (1974) J. Biochem. 75, 283–297.

75 Beil, F.-U., von Chak, D. and Hasselbach, W. (1977) Eur. J. Biochem. 81, 151–164.
76 Punzengruber, C., Prager, R., Kolossa, N., Winkler, F. and Suko, J. (1978) Eur. J. Biochem. 92, 349–359.
77 Chaloub, R.M., Guimaraes-Motta, H., Verjovski-Almeida, S., de Meis, L. and Inesi, G. (1979) J. Biol. Chem. 254, 9464–9468.
78 Barlogie, B., Hasselbach, W. and Makinose, M. (1971) FEBS Lett. 12, 267–268.
79 Makinose, M. and Hasselbach, W. (1971) FEBS Lett. 12, 271–272.
80 Panet, R. and Seliger, Z. (1972) Biochim. Biophys. Acta 255, 34–42.
81 Deamer, D.W. and Baskin, R.J. (1972) Arch. Biochem. Biophys. 153, 47–54.
82 Yamada, S., Sumida, M. and Tonomura, Y. (1972) J. Biochem. 72, 1537–1548.
83 Fiehn, W. and Migala, A. (1971) Eur. J. Biochem. 20, 245–248.
84 Chevallier, J. and Butow, R.A. (1971) Biochemistry 10, 2733–2737.
85 Meissner, G. (1973) Biochim. Biophys. Acta 298, 906–926.
86 The, R. and Hasselbach, W. (1977) Eur. J. Biochem. 74, 611–621.
87 Miyamoto, H. and Kasai, M. (1979) J. Biochem. 85, 765–773.
88 Murphy, A.J. (1976) Biochemistry 15, 4492–4496.
89 Ikemoto, N., Morgan, J.T. and Yamada, S. (1978) J. Biol. Chem. 253, 8027–8033.
90 Coan, C., Verjovski-Almeida, S. and Inesi, G. (1979) J. Biol. Chem. 254, 2968–2974.
91 Dupont, Y. (1976) Biochem. Biophys. Res. Commun. 71, 544–550.
92 Dupont, Y. and Barrington-Leigh, J. (1978) Nature 273, 396–398.
93 Nakamura, Y., Tonomura, Y. and Hagihara, B. (1979) J. Biochem. 86, 443–446.
94 Sumida, M., Wang, T., Mandel, F., Froehlich, J.P. and Schwartz, A. (1978) J. Biol. Chem. 253, 8772–8777.
95 Rauch, B., v. Chak, D. and Hasselbach, W. (1978) FEBS Lett. 93, 65–68.
96 Froehlich, J.P. and Taylor, E.W. (1976) J. Biol. Chem. 251, 2307–2315.
97 Ikemoto, N. (1975) J. Biol. Chem. 250, 7219–7224.
98 Dupont, Y. (1977) Eur. J. Biochem. 72, 185–190.
99 Taylor, J.S. and Hattan, D. (1979) J. Biol. Chem. 254, 4402–4407.
100 Dean, W.L. and Tanford, C. (1978) Biochemistry 17, 1683–1690.
101 Takisawa, H. and Tonomura, Y. (1979) J. Biochem. 86, 425–441.
102 Verjovski-Almeida, S., Kurzmack, M. and Inesi, G. (1978) Biochemistry 17, 5006–5013.
103 Ronzani, N., Migala, A. and Hasselbach, W. (1979) Eur. J. Biochem. 101, 593–606.
104 Makinose, M. and Boll, W. (1979) Cation Flux Across Biomembranes, pp. 89–100, Academic Press, New York.
105 Vianna, A.L. (1975) Biochim. Biophys. Acta 410, 338–406.
106 Garrahan, P.J., Rega, A.F. and Alonso, G.L. (1976) Biochim. Biophys. Acta 448, 121–132.
107 Takakuwa, Y. and Kanazawa, T. (1979) Biochem. Biophys. Res. Commun. 88, 1209–1216.
108 Kanazawa, T. (1975) J. Biol. Chem. 250, 113–119.
109 Kanazawa, T., Yamada, S., Yamamoto, T. and Tonomura, T. (1971) J. Biochem. 70, 95–123.
110 de Meis, L. and Carvalho, G.C. (1974) Biochemistry 13, 5032–5038.
111 de Meis, L. (1976) J. Biol. Chem. 251, 2055–2062.
112 Makinose, M. (1966) 2nd Int. Congr. Biophys., Wien, Abstr. No. 276.
113 Yamamoto, T. and Tonomura, Y. (1967) J. Biochem. 62, 558–575.
114 Makinose, M. (1969) Eur. J. Biochem. 10, 74–82.
115 Makinose, M. (1972) Cold Spring Harb. Symp. Quant. Biol. XXXVII, 681–684.
116 Makinose, M. (1972) FEBS Lett. 25, 113–115.
117 Yamada, S. and Tonomura, Y. (1972) J. Biochem. 71, 1101–1104.
118 Makinose, M. (1971) FEBS Lett. 12, 269–270.
119 de Meis, L. and Carvalho, M.G.C. (1974) Biochemistry 13, 5032–5037.
120 Carvalho, M.G.C., de Souza, D.G. and de Meis, L. (1976) J. Biol. Chem. 251, 3629–3636.
121 Hasselbach, W. and Makinose, M. (1962) Biochem. Biophys. Res. Commun. 7, 132–136.
122 Makinose, M. (1966) Biochem. Z. 345, 80–86.

123 Shigekawa, M. and Dougherty, J.P. (1978) J. Biol. Chem. 253, 1451–1457.
124 Shigekawa, M. and Dougherty, J.P. (1978) J. Biol. Chem. 253, 1458–1464.
125 Shigekawa, M. and Akowitz, A.A. (1979) J. Biol. Chem. 254, 4726–4730.
126 Masuda, H. and de Meis, L. (1973) Biochemistry 12, 4581–4585.
127 Rauch, B., von Chak, D. and Hasselbach, W. (1977) Z. Naturforsch. 32c, 828–834.
128 Kanazawa, T. and Boyer, P.D. (1973) J. Biol. Chem. 248, 3163–3172.
129 Boyer, P.D., de Meis, L., Carvalho, M.G.C. and Hackney, P.D. (1977) Biochemistry 16, 136–139.
130 de Meis, L. and Tume, R.K. (1977) Biochemistry 16, 4455–4463.
131 Vieyra, A., Scofano, H.M., Guimaraes-Motta, H., Tume, R.K. and de Meis, L. (1979) Biochim. Biophys. Acta 568, 437–445.
132 Boyer, P.D. and Ariki, M. (1980) Biochemistry 19, 2001–2004.

CHAPTER 8

Anion-sensitive ATPase and $(K^+ + H^+)$-ATPase

J.J.H.H.M. DE PONT and S.L. BONTING

Department of Biochemistry, University of Nijmegen,
6500 HB Nijmegen, The Netherlands

1. Introduction

The energy required for the active transport of ions can in principle be obtained from the hydrolysis of ATP. In the previous two chapters the properties of the transport enzymes $(Na^+ + K^+)$-ATPase [1] and $(Ca^{2+} - Mg^{2+})$-ATPase [2] have been described and evidence has been provided for a role of the former enzyme in the active transport of sodium and potassium and of the latter enzyme in the transport of calcium ions.

Other ATPases have been described, which are assumed to be involved in the active transport of specific ions. However, in order to make sure that an ATPase is involved in a specific active transport process two ways can be followed: (1) a specific inhibitor against the enzyme must be found or prepared (e.g. an antibody) and this inhibitor must be able to inhibit the specific transport process; (2) the enzyme must be isolated in a pure state and after its incorporation in liposomes must then support the particular transport process. In the case of a plasma membrane transport process, the transport ATPase must of course be located in the plasma membrane.

In this chapter two ATPases will be described for which a role in active transport has been assumed. The first enzyme is an anion-sensitive ATPase, which is present in many tissues and which is assumed to play a role in anion transport. The second enzyme is a K^+-sensitive ATPase, which has so far only been found in the gastric mucosa of several species and which seems to be involved in gastric acid secretion. The properties of these two enzymes and the evidence for their role in active transport will be described.

2. Anion-sensitive ATPase

(a) Definition and assay of enzyme activity

A role of an anion-sensitive ATPase in anion transport across the plasma membrane was first suggested by Durbin and Kasbekar [3] after they found such an activity in a microsomal fraction of frog gastric mucosa [4]. In analogy to the $(Na^+ + K^+)$-ATPase system, it would be activated by two anions, bicarbonate and chloride, resulting in exchange transport of these anions across the plasma membrane of the acid-secreting cells of the gastric mucosa. Carbonic acid would be formed inside the plasma membrane from CO_2 and H_2O by carbonic anhydrase, which is also present in the microsomal fraction. The bicarbonate ion would be transported into the cell, while a chloride ion would be extruded together with the proton from the carbonic acid, leading to HCl secretion into the gastric lumen. An important supporting argument for their hypothesis was the finding that the enzyme activity is inhibited by thiocyanate, which also inhibits gastric acid secretion.

Subsequently, anion-sensitive ATPase activity has been demonstrated in microsomal fractions of many tissues (Table 1), in which either bicarbonate or hydrogen transport occurs, supporting a role of the enzyme in this transport. However, the fact that the ATPase of mitochondria is also anion-sensitive, has led to a debate in the literature whether there really exists a plasma membrane-bound anion-sensitive ATPase, and if so, whether this plays a role in hydrogen or bicarbonate transport (see Section 2h).

As the name of the enzyme indicates, it is sensitive to anions. For the assay of this enzyme activity the assay medium should minimally contain ATP, Mg^{2+}, buffer and the anion(s) to be tested. Since a buffer usually contains one or more anions and ATP is an anion at physiological pH, it is difficult to have a control medium without anions. This has led us to introduce an assay method using a medium, which contains in principle, in addition to ATP, only one specific anion, which is part of the buffer [9,12]. This method has the advantage that the ionic strength is kept constant, in contrast to assay systems in which the anion to be tested is introduced additionally.

We prefer to determine the enzyme activity in three media: one with a stimulating anion (mostly HCO_3^-), one with a neutral anion (mostly Cl^-) and one with an inhibitory anion (mostly SCN^-). The enzyme activity in each of these media should be given. Some authors report only the difference between a stimulating and a neutral medium whereas others give the ratio of the activities. The differential method operates on the assumption that the activity in the medium with the neutral anion is due to other ATPase activities in the enzyme preparation. The ratio method assumes that the activity in the medium with the neutral anion is completely due to a low-activity form of the anion-sensitive ATPase. Since both assumptions are contradictory and probably not true for the relatively crude enzyme preparations, it is better not to give only the difference or the ratio of the activities in these media.

(b) Effects of substrate, cations and pH

The preferred substrate for the anion-sensitive ATPase is ATP. The K_m value reported for various preparations ranges from 0.1 to 0.6 mM [5,6,17,37,38]. In most experiments ATP concentrations between 2 and 5 mM have been used. Replacement of ATP by other trinucleotides generally leads to much lower enzyme activities. In

TABLE 1

Tissues in which anion-sensitive ATPase activity has been reported

Tissue	Species	References
Gastric mucosa	Frog	4, 5
	Dog	6, 7
	Necturus	8
	Lizard	9
	Rat	10
	Cattle	11
	Rabbit	12
Midgut	Hyalophora	13
Kidney	Rat	14, 15, 16
	Rabbit	12, 17, 18
Pancreas	Cat	19, 20
	Dog	20
	Rat	21
Salivary gland	Rabbit	22
	Dog	23
	Rat	24
Seminiferous tubules	Rat, hamster, gerbil	25
Liver	Rat	26
Brain	Rat, monkey	27
Gill	Rainbow trout	12, 28, 29
	Goldfish	30
Erythrocytes	Rabbit	31–34
Ascites tumour cells	Mouse	35
Intestine	Eel	36
	Rat	37
Uterus	Rat	38
Pancreatic islets	Rat	39
Malphigian tubules	Locust	40
Placenta	Man	41

rabbit gastric mucosa the activity for GTP and ITP is only 50 to 60% of that with ATP [12]. Other trinucleotides give only minor activities. Similar results have been obtained for anion-sensitive ATPases from dog gastric mucosa [6], cat pancreas [19] and eel intestinal mucosa [36], although in the latter two preparations UTP gives approximately the same activity as ITP. For ADP and AMP no significant activities have been reported in these tissues.

The anion-sensitive ATPase activities in brush-border preparations from rabbit kidney [17] and rat small intestine [37] have a much lower substrate specificity. In the kidney preparation the activities with GTP and ITP are the same as with ATP, whereas with UTP a lower and with CTP no activity is found. In brush-border preparations from rat ileum and jejunum the activities with GTP, UTP and XTP (xanthosine triphosphate) are about equal to that with ATP, and in these preparations 30–50% of the activity with ATP is obtained with AMP and p-nitrophenylphosphate.

Divalent cations are required for the anion-sensitive ATPase. The highest activity has been found with Mg^{2+}, while with Mn^{2+} the activity is nearly the same in gastric mucosal preparations [6,12], but considerably lower in cat pancreas preparations [19]. In the latter preparation Ca^{2+} is a relatively good substitute for Mg^{2+}. The activity with Ca^{2+} is generally much lower than with Mg^{2+} [12,19], and Ca^{2+} is inhibitory in the presence of Mg^{2+} [4].

The optimal Mg^{2+} concentration varies between 0.5 and 2 mM, while higher concentrations inhibit the activity partially [4,12] or even completely [37,40]. The optimal Mg^{2+}/ATP ratio varies between 0.2 and 2. It is unclear whether Mg^{2+}-ATP is the true substrate of the enzyme or whether a divalent cation is necessary for the enzyme activity as such. Monovalent cations have little or no effect on the activity [4,5].

The pH dependence of the anion-sensitive ATPase is generally not very pronounced. Without stimulating anion there is either no pH dependence [9,12,26], or there is a slight increase with increasing pH between 6 and 9 [6,17,19,30]. In the presence of a stimulating anion (bicarbonate, sulfite, sulfate) the activity is generally higher over the entire pH range up to pH 8.5–9. The enzyme in rat intestinal brush border, on the contrary, shows a steep pH dependence in the absence of bicarbonate and only a slight stimulation by this anion at pH 8.5 or below, whereas it inhibits at higher pH values [37].

(c) Anion dependence

Most authors have studied the anion dependence of the ATPase by adding increasing amounts of the anion to be tested to the medium. Because of its possible physiological role, bicarbonate has most often been investigated. This ion always stimulates the ATPase, the maximal activity being reached at 20–50 mM [4,19,26,28,36]. Half maximal activity is reached at bicarbonate concentrations varying between 6 and 17 mM. The enzymes from brush border preparations show a considerably higher half-maximal activating concentration of 36 mM [17,37]. When supraoptimal bicarbonate concentrations are used, a decrease in activity has sometimes been reported [4,26], whereas in other cases the activity stays at its maximal level [28,36].

Bicarbonate as activating anion can be replaced by several other anions. Particularly other oxy-anions, like selenite, borate, sulfite, sulfate, arsenate and arsenite, stimulate the enzyme [6,13,20,24]. Blum et al. [6] reported a positive correlation between the degree of stimulation and the pK value and concentration of the stimulating anion. This led them to conclude that a proton transfer step is involved in the activation mechanism. There is, however, a large variation between the stimulation by the different anions in various preparations, which makes the general applicability of this assumption unlikely. The half-maximal stimulating concentration of sulfite is generally lower than that of bicarbonate (1.4 vs. 12 mM in rainbow trout gill [29] and 0.8 vs. 36 mM in renal brush border [17]. The maximal stimulation reached with both anions is equal in the gill preparation [29], whereas in renal brush border with sulfite a five times higher activity is reached than with bicarbonate [17]. In intestinal brush-border membranes, however, 25 mM sulfite stimulates, whereas 100 mM sulfite leads to a decrease of the ATPase activity [37].

The anion-sensitive ATPase can generally be inhibited by thiocyanate [4,5]. Since this anion inhibits gastric acid secretion, this finding has been used as an argument in favour of a role of the enzyme in proton transport. The concentration of thiocyanate required for maximal inhibition usually amounts to 5–10 mM. The residual activity, which sometimes occurs, can be attributed to an anion-insensitive ATPase. The enzyme from brush border shows again a deviating behaviour in that it is relatively insensitive towards thiocyanate, the inhibition being less than 30% [17,37].

De Pont et al. [9] have investigated the anion sensitivity of the enzyme in lizard gastric mucosa in a completely different way. They used an incubation medium in which besides ATP in principle only one anion was present. The level of the enzyme activity thus obtained can be attributed to the anion present. They found that thiocyanate inhibits the enzyme activity, but the effectiveness of this lipophilic anion as an inhibitor depends on the other anion present. E.g. with glucuronate as the other anion only 0.35 mM thiocyanate was necessary to obtain 50% inhibition, whereas with nitrate 20 mM thiocyanate was needed. From the half inhibition concentrations the relative affinities of the various anions for the enzyme could be determined. The same relative affinities were found when chloride and another anion were used in combination and the 50% intermediate activities were determined. These findings could most simply be explained by assuming the presence on the enzyme of a single anion binding site with different affinities for the various anions. The affinity of the enzyme was high for thiocyanate and very low for glucuronate. Also in rainbow trout gill [29] a single anion site was found with a relative affinity for thiocyanate and chloride of 35, which value is slightly higher than the value of 21 for the lizard gastric mucosa enzyme [9].

De Pont et al. [9] observed, moreover, that when the logarithm of the relative affinity is plotted as a function of the relative activity obtained with the anion (both relative to chloride) an inverse relationship is obtained (Fig. 1). This indicates that the tighter the anion is bound, the lower the enzyme activity is. Only bicarbonate showed twice the activity expected from its affinity.

Using this constant ion concentration method, Van Amelsvoort et al. [12] measured the relative anion sensitivity of the enzyme from rabbit gastric mucosa for 13 different anions (Table 2). Of these anions sulfite was the maximally stimulating anion, followed by oxalate and bicarbonate. Besides thiocyanate, nitrate and perchlorate and particularly azide are also inhibitory anions relative to chloride.

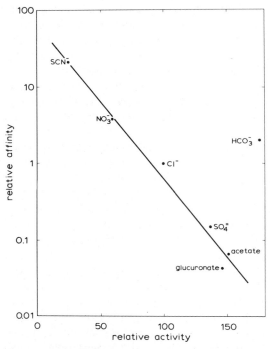

Fig. 1. Relationship between the relative anion affinity of anion-sensitive ATPase of lizard gastric mucosa and the relative enzyme activity in the presence of each anion alone. (From [9].)

Since in the latter method the activity with chloride was set at 1, the absolute affinity of this anion for the enzyme could not be measured. Bornancin et al. [29] recently measured the ATPase activity in a microsomal fraction from rainbow trout gill using Mg-ATP as substrate and a glycine–Tris buffer without any other anions

TABLE 2

Effects of various anions on "peak II" HCO_3^--stimulated ATPase

Relative activities (activity in Cl^- medium set at 1.00) are presented with S.E. for four experiments. Mean specific activity in Cl^- medium is 58.9 μmol ATP/h/mg protein. Peak II represents the 40–44% (w/v) subfraction of the microsomal fraction of rabbit gastric mucosa. From [12].

Major anion	Relative activity	Major anion	Relative activity
Azide	0.10 ± 0.01	Citrate	1.18 ± 0.04
Thiocyanate	0.28 ± 0.02	Sulfate	1.29 ± 0.03
Perchlorate	0.41 ± 0.01	Acetate	1.38 ± 0.06
Nitrate	0.65 ± 0.03	Bicarbonate	1.54 ± 0.04
Chloride	≡ 1.00	Oxalate	1.63 ± 0.05
Iodide	1.05 ± 0.05	Sulfite	2.59 ± 0.16
Formate	1.13 ± 0.10		

present. Addition of chloride led to a stimulation of the activity, maximally to one-third of that obtained with bicarbonate. The K_m for this effect of chloride was 3.3 mM. When buffers containing sulfonate groups, (PIPES or HEPES) were used, the stimulating effect of chloride did not occur. This suggests a competitive effect between these sulfonate buffers and chloride. These authors ascribe the lack of chloride stimulation reported by most other investigators either to the use of sulfonate buffers, or to the presence of already saturating chloride concentrations in the assay medium, e.g. by using $MgCl_2$ or Tris–HCl. In the constant ion-concentration method [23] a stimulating effect of chloride is found, when the activity is first measured with a less stimulating anion such as nitrate, which is gradually replaced by chloride [9].

These properties of the anion-sensitive ATPase from microsomal fractions are strikingly similar to those of the mitochondrial ATPase in several tissues [42–45].

(d) Other properties of the enzyme

In several tissues, like gastric mucosa [6,8,46,47], pancreas [19,20], submandibular gland [22] and intestinal mucosa [36], the anion-sensitive ATPase can be solubilized by Triton X-100, a detergent/protein ratio of 3:1 being most effective [6,8,47]. The solubilized enzyme from gastric mucosa [46] and pancreas [20] could be partially purified by gel filtration and sucrose density gradient centrifugation. The properties of the solubilized enzyme do not differ very much from those of the particulate enzyme.

Except for the inhibitory effects of thiocyanate and other anions, not many specific inhibitors have been reported. The enzyme from gastric mucosa is inhibited by OH-group-directed inhibitors, such as diisopropylfluorophosphate and methanesulfonyl chloride and by the modifying solvent D_2O [6]. Sulfhydryl reagents, like *p*-chloromercuribenzoate [4,20] and $HgCl_2$ [5], inhibit the anion-sensitive ATPase from gastric mucosa and pancreas.

Some studies have been carried out on the role of phosphorylated intermediates and the role of partial reactions, subunit structure and involvement of phospholipids. All these studies have been done with crude preparations from gastric mucosa, which means that the preparation contains $(K^+ + H^+)$-ATPase activity (see Section 3). This makes it difficult to relate these findings to the anion-sensitive ATPase.

(e) Localization

The localization of the anion-sensitive ATPase has generally been investigated by measuring the activity of the enzyme in fractions obtained after differential centrifugation. After removal of nuclei and cell debris by centrifugation at $10^4 \times g/\text{min}$, a heavy mitochondrial fraction is obtained by centrifugation at $10^5 \times g/\text{min}$, a light mitochondrial fraction at $4-7 \cdot 10^5 \times g/\text{min}$ and a microsomal fraction at $6 \cdot 10^6 \times g/\text{min}$. In some cases only one mitochondrial fraction is isolated by omitting the

$10^5 \times g$/min step. The anion-sensitive ATPase is always present in both mitochondrial fractions and mostly also in the microsomal fraction. In liver of rat [26], and pancreas of rat [10] and rabbit [21] no significant activity could be found in the microsomal fraction. The specific activity of the anion-sensitive ATPase in both mitochondrial fractions of rabbit gastric mucosa, rabbit kidney and rainbow trout gill was higher than in the microsomal fractions of these tissues [12]. The postmicrosomal supernatant is always free of any ATPase activity, indicating that the enzyme is particulate. The mitochondrial and microsomal fractions are sometimes further fractionated by isopycnic centrifugation over discontinuous or linear sucrose gradients.

It is now clear that the anion-sensitive ATPase present in the mitochondrial fractions originates, mainly if not completely, from the mitochondrial inner membrane where it forms the so called F_1 part of the mitochondrial ATPase complex [43]. The crucial question is now: What is the origin of the anion-sensitive ATPase in the microsomal fraction and its subfractions? Most investigators have studied this problem by comparing the distribution of the anion-sensitive ATPase with that of a mitochondrial marker enzyme, such as succinate dehydrogenase or cytochrome c oxidase or with that of a plasma-membrane marker enzyme like 5'-nucleotidase or $(Na^+ + K^+)$-ATPase. From this type of experiments a plasma-membrane localisation of the anion-sensitive ATPase has been postulated for dog gastric mucosa [6,7], dog pancreas [20], rat submandibular gland duct [24], rainbow trout gill [28,29] and goldfish gill [30].

A different conclusion has been reached by Van Amelsvoort et al. [12,18,21,47,48]. They have investigated the localisation of the anion-sensitive ATPase in four different tissues. After subfractionation of both the light-mitochondrial and the microsomal fractions by linear sucrose-gradient centrifugation one to three ATPase peaks appeared in each fraction. In the light-mitochondrial and the microsomal fractions of rabbit gastric mucosa [12] the three peaks were designated I, II and III in order of increasing density (Fig. 2). The ATPase activity of peak I (d = 1.11 g/l), which was most prominent in the microsomal fraction, was anion-insensitive. Since this ATPase peak coincided with the $(Na^+ + K^+)$-ATPase and 5'-nucleotidase peaks and showed little or no cytochrome c oxidase activity, it apparently represents a plasma membrane fraction. Peaks II and III both contained anion-sensitive ATPase and cytochrome c oxidase activity. Peak III (d = 1.18 g/l) was most prominent in the mitochondrial fraction and showed high cytochrome c oxidase activity, indicating its mitochondrial origin. Peak II (d = 1.16 g/l) had a higher anion-sensitive ATPase/cytochrome c oxidase ratio than in peak III. At first glance this would suggest that the anion-sensitive ATPase activity present in this fraction could not be completely attributed to mitochondrial contamination. Comparable fractions could be obtained from the mitochondrial fraction of rabbit pancreas [21] and the light mitochondrial and microsomal fractions of rabbit kidney [18]. The question now was the origin of the "peak II" fraction in these tissues.

To answer this question, various properties of peak II (unknown) and peak III (mitochondrial) subfractions were compared. Typical mitochondrial ATPase inhibi-

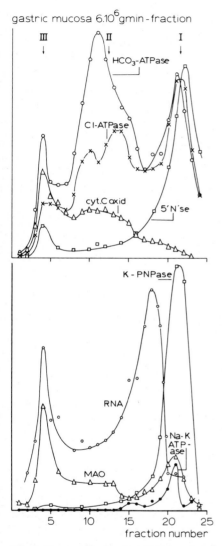

Fig. 2. Enzyme distribution patterns after density gradient centrifugation of a microsomal fraction of rabbit gastric mucosa. (From [12].)

tors, like oligomycin, quercetin, aurovertin and DCCD had virtually equal inhibitory effects on the anion-sensitive ATPase activities of the two peaks. The same was the case for the effects of liver mitochondrial inhibitor protein (Table 3). The earlier reported oligomycin-insensitivity of the anion-sensitive ATPase in the pancreas [20] was shown to be an artifact of proteolytic or lipolytic degradation during fractiona-

TABLE 3
Effect of inhibitors of anion ATPase of different tissue fractions

Inhibitor	pI_{50} [a]	pI_{99}	Rest activ. (%)	pI_{50}	pI_{99}	Rest activ. (%)
	Gastric mucosa [12]					
	peak II			peak III		
Oligomycin	7.2	5.0	15	7.1	4.9	20
Aurovertin D	6.1	4.9	46	6.2	5.1	40
Olig. 20 μM + Aurovertin 35 μM	–	–	9	–	–	7
Quercetin	4.2	3.7	6	4.1	3.7	8
DCCD	6.6	5.8	16	6.7	6.0	16
Mitoch. inhib. protein (g/U)	3.7	–	–	3.4	–	–
Acetazolamide (10 mM)	–	–	93	–	–	93
	Pancreas (1 mM EDTA present) [21]					
	peak II			peak III		
Oligomycin	7.8	5.8	16	7.7	5.6	11
Auroverin D (16 μM)	–	–	44	–	–	42
Mitoch. inhib. protein (g/U)	3.4	–	–	3.1–3.3	–	–
	Kidney [18]					
	microsomal fraction			mitochondrial fraction		
Oligomycin	7.8	4.8	26	7.0	4.7	10
Aurovertin D	6.3	5.1	48	6.2	5.0	51
DCCD	6.6	5.6	29	6.5	5.7	18

[a] pI_{50} is the negative logarithm of the inhibitor concentration (molar unless otherwise indicated) at half-maximal inhibition; pI_{99} is that giving 99% of the maximal inhibition; Rest activ. is the % activity remaining at maximal inhibition. Activity represents total ATPase activity in the HCO_3^- medium. The figures are averages for 2–4 experiments.

tion [21]. The phospholipid patterns of peak II and III from gastric mucosa microsomes were closely similar [12]. They differed from that of peak I in having high levels of cardiolipin and low levels of sphingomyelin, again suggesting that peaks II and III are from the same (mitochondrial) origin. Solubilisation of peak II from gastric mucosa by Triton X-100, followed by sodium dodecylsulfate gel electrophoresis, shows two dominant bands of 64 200 and 61 600 M_r and a lesser band of 28 500, similar to the pattern given by mitochondrial F_1 ATPase [47]. Electronmicroscopically peak II showed small mitochondria, mostly lacking an outer membrane, lysosomes and rough and smooth endoplasmic reticulum, while peak III consisted of small intact mitochondria and rough endoplasmic reticulum [18]. Peak I, on the other hand, consisted mainly of vesicles, probably deriving from plasma membrane. All these findings strongly suggest that the anion-sensitive ATPase activity in the peak II is of mitochondrial origin. The lower cytochrome *c* oxidase activity in this subfraction as compared to peak III could be due to a specific inactivation of this enzyme activity or to an activation of the anion-sensitive ATPase activity, caused by the loss of endogenous mitochondrial inhibitor protein [49]. Since the ATPase in peak I, which is probably mainly of plasma membrane origin, is anion-insensitive, these findings disfavour the presence of a plasma membrane-bound anion-sensitive ATPase in the four tissues investigated in this study.

The contrasting findings of other investigators [6,20,24,28–30,46] may be artefactual due to more drastic homogenisation and to non-equilibrium gradient centrifugation. Too drastic homogenisation may inactivate the mitochondrial anion-sensitive ATPase by release of the mitochondrial inhibitor protein. Moreover the fragments obtained by this method may become so small that they do not reach their equilibrium position in the time used for gradient centrifugation. Van Amelsvoort et al. [12] showed for gastric mucosa that completely different gradient-density patterns were obtained when either shorter centrifugation periods were used or when a drastic sonication procedure was applied instead of mild homogenisation.

(f) Brush-border membranes

The apical plasma membrane of epithelial cells of small intestinal and renal proximal tubules is characterised by the presence of many microvilli (brush border). These membranes can be isolated relatively easily by centrifugation and free flow electrophoresis techniques. Kinne-Saffran and Kinne [15] found that after free-flow electrophoresis of a rat kidney-cortex membrane preparation, the anion-sensitive ATPase co-migrated with the alkaline phosphatase activity but was separated from the $(Na^+ + K^+)$-ATPase activity, which is assumed to be a marker of basolateral plasma membranes. This suggests that the brush-border membrane of the proximal tubule contains an anion-sensitive ATPase. The same conclusion was reached by Liang and Sacktor [17] for a brush-border preparation from rabbit kidney.

Van Amelsvoort et al. [18] found in a crude brush border preparation from rabbit kidney an Mg-ATPase activity which could be stimulated by bicarbonate by 51%, which value is comparable to that reported by Liang and Sacktor [17]. The enzyme activity was relatively insensitive towards oligomycin and also towards DCCD and aurovertin D. Further purification of the crude brush-border fraction by density-gradient centrifugation resulted in a preparation with an ATPase activity which could only be 12% stimulated by HCO_3^-. Freezing and thawing raised the stimulation to 28%. The purification of the brush border was accompanied by a decrease of the cytochrome *c* oxidase activity and a reduction in the number of enclosed mitochondria observed electronmicroscopically [18]. It was concluded that the brush-border preparation contained two ATPase activities, one rather anion-insensitive and an inherent property of these membranes and one due to mitochondrial contamination. The increased bicarbonate stimulation after freezing and thawing was explained by assuming an increased accessibility of the substrate or stimulating anion to the enzyme.

The question whether or not there is an anion-sensitive ATPase in brush border membranes was reinvestigated by Kinne-Saffran and Kinne [16]. They compared the properties of Mg-ATPases in a mitochondrial and a brush-border fraction from rat kidney cortex. The mitochondrial ATPase could be stimulated 90% by bicarbonate, and was very sensitive towards oligomycin, aurovertin, carboxyactryloside and the mitochondrial inhibitor protein. The brush-border Mg-ATPase could only be stimulated 28% by bicarbonate and was inhibited at most 18% by these mitochondrial inhibitors. On the other hand, the antibiotic filipin, which reacts with cholesterol in membranes, inhibits the ATPase activity in the cholesterol-rich brush-border membranes but not that in the cholesterol-poor mitochondria. The filipin treatment, however, increased the bicarbonate sensitivity of the residual ATPase activity in the brush-border preparation, suggesting that there was still some mitochondrial contamination in this preparation.

In brush border of intestine also a relatively anion-insensitive ATPase could be found [37]. Previous negative findings [50] could be explained by the fact that the chelator EDTA, which inactivates this enzyme [51], was present in the incubation medium. This inactivation can be overcome by addition of excess Zn^{2+} [51]. This finding and the fact that the anion-sensitive ATPase in intestinal brush-border membranes can be inhibited by L-phenylalanine and L-cysteine suggests that the alkaline phosphatase activity and the anion-sensitive ATPase activity originates from a single enzyme [37,41]. The same conclusion has been reached for the anion-sensitive ATPase of brush border from human placenta [41], but not for the enzyme from rat kidney [16], where the alkaline phosphatase activity is inhibited by 1-p-bromotetramisole, whereas the anion-sensitive ATPase activity is not affected.

These findings indicate that brush-border membranes contain an Mg-ATPase, which can be stimulated by anions, but the level of stimulation is rather low as compared to that obtained in mitochondrial preparations. Some properties of this enzyme (e.g. substrate and anion dependence) are different from those of the mitochondrial enzyme and the activity might be a property of another enzyme such as alkaline phosphatase.

(g) Erythrocytes

Erythrocytes do not contain mitochondria and any anion-sensitive ATPase activity in a particulate fraction must therefore be attributed to the plasma membrane. An Mg-ATPase, which could be stimulated by bicarbonate was found in rabbit erythrocyte membranes [31–33] and should thus be plasma-membrane bound. The properties of the enzyme in rabbit erythrocyte membranes are, however, completely different from those of the anion-sensitive ATPase from other tissues. Whereas sulfite is a good stimulant for the Mg-ATPase in mitochondrial and microsomal fractions of various tissues, it only slightly stimulates [32] or even inhibits [33] the erythrocyte enzyme. Other oxy-anions inhibit the ATPase of erythrocyte membranes [32]. The substrate dependence of the enzyme is greatly different from that of the enzyme from other tissues [32,33]. The half maximal inhibitory concentration of oligomycin is approx. 100 times higher than that in microsomal or mitochondrial preparations [33]. Triton X-100 inhibits the enzyme at concentrations, where it effectively solubilises the gastric microsomal ATPase [6,7,33].

These findings suggest that the erythrocyte membrane is different from that from other tissues. Van Amelsvoort et al. [33] investigated whether this enzyme activity could be a property of other proteins present in the membrane. No effects of ouabain or of 4,4'-diisothiocyano-dihydrostilbene-2,2'-disulfonic acid (H_2DIDS) were found, indicating that the activity is neither due to the ($Na^+ + K^+$)-ATPase nor to the anion exchange protein present in erythrocyte membrane. On the other hand, substances like ruthenium red, chlorpromazine and EGTA, which are inhibitors of the ($Ca^{2+} + Mg^{2+}$)-ATPase, were found to inhibit the anion-sensitive ATPase activity. Addition of Ca^{2+} stimulates the ATPase activity. This suggests that the anion-sensitivity is an inherent property of the ($Ca^{2+} + Mg^{2+}$)-ATPase present in the erythrocyte membranes.

Recent findings of Au [34] support this conclusion. He showed that erythrocyte ($Ca^{2+} + Mg^{2+}$)-ATPase and anion-sensitive ATPase activity are both stimulated by calmodulin from various sources and are inhibited by an inhibitory protein from pig erythrocytes. Also after separation of rabbit erythrocytes on the basis of their density the two activities varied in parallel in the fractions.

(h) Transport function

Arguments in favour of a role of the anion-sensitive ATPase in transport are mainly based on an observed parallelism between the effect of certain inhibitors on the enzyme activity and the transport process. However, in nearly all cases the inhibitor used is the lipophilic anion thiocyanate, which might inhibit the transport process by other means. Moreover, inhibition of the mitochondrial ATPase activity led to a reduction in the amount of ATP available for other transport processes (e.g. ($Na^+ + K^+$)-ATPase) causing indirect inhibition of anion transport.

A prerequisite for a role of the anion-sensitive ATPase in transport would be the localisation of the enzyme in plasma membranes, since a direct role of a mitochondrial enzyme in membrane transport can be excluded. As discussed in Section 2e, it is very likely that the anion-sensitive ATPase present in microsomal fractions of various tissues from different species is due to mitochondrial contamination.

Only in brush-border membranes of some specialised cells and in erythrocyte membranes there seems to be an Mg-ATPase, which can be stimulated by anions, but only to a minor degree. Moreover, the properties of the enzyme in these membranes differ considerably from those of the activities in the microsomal and mitochondrial fractions of most other tissues. The anion sensitivity of the ATPase activity in the brush-border membranes of placenta and small intestine is a property of the alkaline phosphatase [37,41] and that in erythrocytes is part of the ($Ca^{2+} + Mg^{2+}$)-ATPase activity [33,34]. Although this does not definitely exclude a role of the enzyme in anion transport, no valid arguments in favour of a role of this enzyme in anion or proton transport have been advanced.

Other arguments against such a role are the fact that the enzyme is not activated, but is rather non-specifically inhibited by the binding of anions to a single anion-binding site [9] (see Section 2c). Moreover, the molar ratio between proton transport and ATP hydrolysis in lizard gastric mucosa lies between 0.06 and 0.17 [9], which is far below that of the 3:1 Na^+/ATP ratio for $(Na^+ + K^+)$-ATPase and the 2:1 Ca^{2+}/ATP ratio for $(Ca^{2+} + Mg^{2+})$-ATPase.

In conclusion, at the moment the arguments against a role of the anion-sensitive ATPase in anion or proton transport are much stronger than those in favour of such a role.

3. $(K^+ + H^+)$-ATPase

(a) Introduction

The gastric mucosa is able to generate a proton gradient of $10^6:1$. The energy required for this transport can in principle be delivered by an ATPase and a search for such an enzyme was indicated. The earlier suggestion that a membrane-bound anion-sensitive ATPase in cooperation with carbonic anhydrase would be responsible, has been considered in Section 2 and found to be unlikely.

Forte et al. [52] found a K^+-stimulated phosphatase activity, which in contrast to such an activity in many other tissues could not be inhibited by ouabain, indicating that the activity was not due to the $(Na^+ + K^+)$-ATPase system. Later studies of this group showed first that the appearance of the phosphatase activity ontogenetically coincided with the development of acid secretion [53]. Later they found that a K^+-stimulated ATPase activity was closely associated with the K^+-stimulated phosphatase activity [54]. The K^+-stimulated ATPase activity was shown to be stimulated by K^+-sensitive ionophores and by membrane disruption procedures [55]. After membrane disruption the ionophore had no further effect, suggesting that the ATPase activity is present in the membranes of vesicles with low K^+-permeability. The K^+-sensitive site of the enzyme would then have to be located at the inside of the vesicular membrane.

The relation between the K^+-stimulated ATPase and gastric acid secretion was greatly supported by the finding of Lee et al. [56] that addition of ATP to gastric mucosal vesicles leads to alkalinisation of the medium, indicating proton uptake by the vesicles. This suggests that the ATPase might drive a $K^+ - H^+$ exchange. The enzyme has since then generally been designated as $(K^+ + H^+)$-ATPase, although a direct effect of H^+ on the ATPase activity has not been shown. Several reviews on this interesting enzyme have appeared [57–61].

(b) Purification

Several purification procedures have been described. Generally scrapings of the fundic mucosa of frog, rabbit or pig are homogenised in isotonic buffer [62–64].

Either the post-mitochondrial or the microsomal fraction is taken for further purification by continuous [63] or discontinuous [62,64] density-gradient centrifugation. A fraction thus obtained is often used as enzyme preparation. This fraction has a density of 1.10–1.15 and the $(K^+ + H^+)$-ATPase in this fraction can usually be stimulated by K^+-specific ionophores (valinomycin, nigericin).

Further purification of this fraction is possible by a second (continuous) gradient centrifugation [64]. Alternatively, free-flow electrophoresis in the presence of Mg-ATP has been applied [63], which leads to a fraction with an activity of 110 mol/mg protein/h in the presence of valinomycin. These latter purification steps often lead to a reduced recovery, and also to a decrease or loss of the effect of ionophores on the activity, indicating that the vesicular structures have been opened by the purification procedure. Less purified preparations have, therefore, mostly been used for the study of vesicular transport.

Upon purification, the K^+-stimulated phosphatase activity is always copurified with the $(K^+ + H^+)$-ATPase activity [63–65]. Mitochondrial markers, such as cytochrome c oxidase, succinate dehydrogenase, monoamino-oxidase, and the ribosomal marker RNA are largely removed by the purification procedure. The same is true for the anion-sensitive ATPase and 5′nucleotidase activities, but some $(Na^+ - K^+)$-ATPase activity is still present in highly purified $(K^+ + H^+)$-ATPase preparations. Purification is also characterised by a lowering of the K^+-insensitive Mg ATPase activity, but even in the purest preparations some Mg^{2+}-ATPase activity (4% of $(K^+ + H^+)$-ATPase activity) is still present. This may represent an impurity or an inherent property of the enzyme.

(c) Structural and chemical properties

Electronmicroscopic studies show that the purified ATPase preparation consists almost entirely of smooth-surfaced vesicles with a diameter of 0.1–0.2 μm [63,66,67] and a volume of 2 μl/mg protein [67,68]. The general organisation of intramembranous particles seen after freeze-fracture suggests that the vesicles may originate from the tubulovesicular system of the intact parietal cells [67]. This system is abundant in the apical region of these cells and appears to be directly involved in the acid secretion process [69].

The same conclusion for the origin of $(K^+ + H^+)$-ATPase-rich membranes has been reached from immunological studies. Fluorescent antibodies against the purified enzyme stain the apical cell surface and the secretory canaliculi, but not the basolateral membrane of parietal cells [69,70].

The purified enzyme preparation contains 28 g carbohydrate/100 g protein, characterised by a high amount of glucose and by a lack of neuraminic acid [71]. Sodium dodecyl gel electrophoresis shows that in purified preparations a protein with an apparent $M_r = 100\,000$ comprises more than 75% of the total amount of protein [63,66]. The 100 000 protein can be phosphorylated by $[\gamma-^{32}P]ATP$ [66], suggesting similarities with the catalytic subunit of $(Na^+ + K^+)$-ATPase [1] and $(Ca^{2+} + Mg^{2+})$-ATPase [2]. The 100 000 band seems, however, to be heterogeneous, since upon isoelectrofocussing a number of proteins appears [7]. Tryptic digestion studies also lead to the conclusion that the 100000 band contains, in addition to the catalytic subunit, a glycoprotein and another protein with approx. the same molecular weight [78].

(d) General enzymatic properties

The enzyme has a high specificity for ATP as substrate. A much lower specific activity is found with deoxyATP (62% [64]) and with CTP (15% [66]; 17% [64]). No activity is found with ADP, AMP, AMPPNP, AMPPCP and ITP [64,66]. With GTP no activity [64] or a slight activity (12%) [66] is found. Values reported for the K_m for ATP vary between 20 and 100 μM [56,62,66] but recently Sachs et al. [71] reported the presence of a low affinity ($K_m = 1$ mM) site in addition to a high affinity site ($K_m = 74$ μM). These authors reported that K^+ decreases the affinity of ATP for the enzyme.

The ATPase requires Mg^{2+} [54] and can be stimulated by monovalent cations in the following order of affinity: $Tl^+ > K^+ > Rb^+ > Cs^+$, NH_4^+. Little or no activity is found with Na^+ and Li^+ [62,64,66]. The maximal rate obtained with these monovalent cations is nearly independent of the cation used [62,64]. There is a large variation in the values reported for the K_m for K^+ (0.2–2.7 mM), which is probably due to the fact that the K_m value has been shown to increase with decreasing pH [67] and with increasing ATP concentration [71]. K^+ concentrations, which are more than 10 times the K_m value, tend to decrease the enzyme activity, but this effect also decreases with decreasing pH [67].

The pH optimum of $(K^+ + H^+)$-ATPase has previously been reported to be 7.5 [65] but more recent reports give values of 6.7–7.0 [62,64]. It thus seems that ATP, monovalent cations and protons influence the ATPase activity in a complex way, the values obtained by varying one of these substances being affected by the choice or concentration of the other two compounds.

(e) Partial reactions

The enzyme can be phosphorylated by ATP [62] to a maximal level of 1.6 nmol/mg protein [71]. The rate of phosphorylation is markedly increased when enzyme and ATPase are first mixed, followed by addition of Mg^{2+}, as compared to the situation when ATP, Mg^{2+} and enzyme are mixed simultaneously [74]. This suggests that binding of ATP precedes the phosphorylation step. The binding step as such cannot easily be measured with ATP, since phosphorylation immediately follows its binding.

Van de Ven et al. [75] have, therefore, used the non-hydrolysable analogue adenylyl imidodiphosphate (AMPPNP) as a substitute for ATP and have measured the binding characteristics of this compound. A dissociation constant of 53 μM was obtained and the maximal AMPPNP binding was 2.8 nmol/mg protein. The amount of binding was decreased by monovalent cations in the same order of potency as for their stimulation of the enzyme. The effect of Mg^{2+} was first investigated by the complexing agent CDTA. Addition of CDTA to the enzyme, which contained rather firmly bound Mg^{2+} (20–100 nmol/mg protein), decreased the amount of bound AMPPNP. This appeared not to be due to chelation of Mg^{2+}, but rather to direct binding of CDTA or to removal of stabilising cations other than Mg^{2+}. Addition of extra Mg^{2+} also decreased the amount of AMPPNP binding. At low Mg^{2+} concentrations (<0.5 mM) the apparent number of binding sites was reduced, whereas at higher Mg^{2+} concentrations (0.5 mM) the binding of AMPPNP was inhibited in a competitive way. They concluded, in agreement with previous suggestions of Sachs and coworkers [59], that the enzyme is a dimer, which can bind 2 mol AMPPNP/mol enzyme (Fig. 3). In the presence of Mg^{2+} only 1 mol of AMPPNP could be bound, the binding of which is further competitively decreased by additional Mg^{2+}. This explains also why the amount of phosphorylation is only about half of that of the maximal AMPPNP binding. These findings would fit a half of the site-reactivity model for the enzyme [58].

The rate of phosphorylation is much faster than the overall turnover rate of the enzyme, suggesting that the phosphorylation step is not rate-limiting [71]. Surprisingly, Wallmark and Mårdh [74] found that at low substrate concentration (5 μM ATP) Na^+ decreases the rate of phosphorylation, the steady-state level of phosphorylation and the $(K^+ + H^+)$-ATPase activity. In the absence of K^+, Na^+ has a stimulating effect on the ATPase activity at this ATP concentration. Since at ATP concentrations normally used in the ATPase assay (1–5 mM) Na^+ does not affect the ATPase activity, these findings are difficult to interpret.

Both rate and level of phosphorylation are markedly reduced by K^+ and other monovalent cations [62,74]. This is due to the fact that these cations stimulate the dephosphorylation reaction. The order of potency of this effect of the monovalent cations is again the same as for their effect on the overall ATPase activity [62]. The kinetics of the dephosphorylation reaction is complex [71,74]. At least two components can be distinguished in the dephosphorylation reaction, a slow and a fast one. Only the slow component appears to be rate-limiting in the overall ATPase reaction.

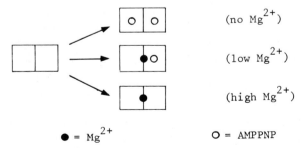

Fig. 3. Schematic presentation of the binding of AMPPNP to a dimeric model of $(K^+ + H^+)$-ATPase and the effect of Mg^{2+} on this process. Symbol ▯▯ represents a dimeric enzyme molecule, ○ stands for AMPPNP, ● for Mg^{2+}.

ADP has little or no stimulatory effect on the rate of dephosphorylation [71,74], which finding is in contrast with the fact that the enzyme catalyses an ATP–ADP exchange reaction [71]. Dephosphorylation can also occur with hydroxylamine [74,76], suggesting that the intermediate is an acylphosphate as is also the case for $(Na^+ + K^+)$-ATPase [1] and $(Ca^{2+} + Mg^{2+})$-ATPase [2].

As previously discovered [52], the enzyme shows a cation-stimulated phosphatase activity, the cation specificity of which is the same as that of the ATPase and which is copurified with the enzyme [63]. The specific activity of the phosphatase reaction is 60–80% of that of the $(K^+ + H^+)$-ATPase activity [63,77,78]. This is much higher than for $(Na^+ + K^+)$-ATPase, where the activity of the phosphatase activity is only 10–20% of the ATPase activity [1].

(f) Activators and inhibitors

Ray [79] has presented evidence that in the soluble supernatant fraction of rabbit fundic mucosal cells there is a heat-labile, non-dialysible, and protease-sensitive factor, which is able to activate the $(K^+ + H^+)$-ATPase and K^+-stimulated phosphatase activity of frog, rabbit and pig mucosal microsomes by a factor 1.5–3.5. Ca^{2+} in a concentration above 20 μM abolishes the activation by this factor. Since the activity of the preparation used in this study is only 10–15% of the highest activity reported, further investigations on this activator are warranted.

Various inhibitors of the $(K^+ + H^+)$-ATPase activity have been described, for some of which the point of attack is known. Sulfhydryl reagents, such as *p*-chloromercuribenzene sulfonate [66], N-ethylmaleimide [54,62] and 5,5'-dithiobis-(2-nitrobenzoic acid) (DTNB) [80] inhibit the $(K^+ + H^+)$-ATPase activity. N-ethylmaleimide also inhibits the phosphatase activity and abolishes the stimulating effect of K^+ on the dephosphorylation reaction [62]. DTNB also inhibits the K^+-stimulated phosphatase activity, but some of the other effects of this reagent depend on the ligands present during the inactivation reaction [80]. The rate of inactivation by DTNB, which is linear in the absence of Mg^{2+}, is enhanced by monovalent cations and is decreased by ATP and ADP. From these experiments a dissociation constant of 44 μM could be deduced for the enzyme-K^+ complex. Since the dissociation constant of the enzyme-DTNB complex is not affected by ATP, it is concluded that the sulfhydryl group essential for enzyme activity is not involved in the binding of ATP. In the presence of Mg^{2+}, the inactivation by DTNB does not obey pseudo-first-order kinetics. Under these circumstances the phosphorylation level is decreased, but the stimulating effect of K^+ on the dephosphorylation step is not impaired. In the absence of Mg^{2+} the phosphorylation level is not affected, but DTNB reacts with a sulfhydryl group which is involved in the stimulation of the dephosphorylation reaction by K^+, as previously observed for the effect of N-ethylmaleimide [62].

From inactivation studies with the arginine-modifying reagent butanedione it was concluded that such a residue is involved in the ATP-binding centre [64,81]. The $(K^+ + H^+)$-ATPase is inactivated by butanedione in a second-order process [64] and the inactivation is protected by ATP [64,81]. In the absence of ATP the rate of inactivation is decreased by monovalent cations, whereas in the presence of ATP these cations have the reverse effect [64]. Moreover, K^+ decreases the apparent ATP-enzyme dissociation constant deduced from these experiments. These experiments can be interpreted by assuming that ATP affects the conformational state of enzyme, leading to a reduced binding of monovalent cations and opposite. Mg^{2+} also increases the rate of inactivation by DTNB, which is in agreement with the assumption that this ion also induces a change in the conformational state.

The amino group reagent 2-methoxy-2,4-diphenyl-3-dihydrofuranone also inhibits the $(K^+ + H^+)$-ATPase activity [81], ATP protects against this effect. This suggests that amino groups are also important for the enzyme activity. Modification of histidine residues by the reagent diethylpyrocarbonate also inactivates the $(K^+ + H^+)$-ATPase activity, but the K^+-stimulated phosphatase activity is not impaired [71]. The inactivation of the $(K^+ + H^+)$-ATPase activity can be partially prevented by addition of ATP. Carboxyl group reagents such as N,N'-dicyclohexylcarbodiimide [66] and N,N'-ethoxycarbonyl-2-ethoxy-1,2-dihydroquinoline (EEDQ) [82] also inhibit the $(K^+ + H^+)$-ATPase activity. The inactivation caused by the latter reagent is enhanced by ATP, ADP, and/or Mg^{2+} but decreased by K^+. The incorporation of [^{14}C]glycine ethyl ester in the enzyme protein, which is stimulated by EEDQ, is also reduced by K^+. This suggests that a carboxyl group would be involved in the K^+ binding centre. From the pH dependence of the inactivation reaction Saccomani et al. [82] conclude that this carboxyl group would have a rather high pK value (above 6.5).

The enzyme can also be inhibited by some other agents, but the mechanism of their action has not been elucidated. These agents are Zn^{2+} [54,62], fluoride [54,62], vanadate [83] and dipicrylamine [66]. Other agents such as thiocyanate and ouabain, which are inhibitors of the anion-sensitive ATPase and $(Na^+ + K^+)$-ATPase, respectively, have no effect on the $(K^+ + H^+)$-ATPase [54]. Two out of five antibody preparations against purified $(K^+ + H^+)$-ATPase inhibit the ATPase activity for 80% and the K^+-phosphatase activity for 35%, while the three other preparations have no effect on the enzyme activity [70], suggesting a heterogeneity of these preparations.

Inactivation of the enzyme is also possible by a mild treatment with trypsin [73]. The loss of both the $(K^+ + H^+)$-ATPase and the K^+-stimulated phosphatase activity shows a biphasic pattern, the slow phase of which can be abolished by ATP and ADP. The rate of inactivation of both activities is also reduced by Na^+ and by K^+. The phosphoenzyme level is first increased by a factor 2, suggesting that a step leading to dephosphorylation is blocked by the proteolytic treatment. Several tryptic peptides were formed in the absence of ATP, two of which could be labelled by [γ-^{32}P]ATP. The two peptides formed after tryptic digestion in the presence of ATP could not be phosphorylated, suggesting that a protein distinct from the catalytic protein was digested. A glycoprotein, which comprises 30% of the total 100 000 band protein, was not digested by trypsin. This suggests that the purified enzyme still contains three different proteins in the 100 000 band.

(g) Phospholipid dependence

The membrane preparation, which is enriched in $(K^+ + H^+)$-ATPase, contains in addition to proteins a considerable amount of cholesterol and phospholipids. The content and composition of this lipid fraction (high cholesterol and sphingomyelin content, Table 4) is characteristic for a plasma membrane fraction as is also found for a purified $(Na^+ + K^+)$-ATPase preparation [1].

The requirement of certain phospholipids for the activity of an enzyme can be investigated by means of specific phospholipases [84]. The degree of breakdown of the various phospholipids depends on the substrate specificity of the phospholipase, but also on the accessibility of the phospholipids. Phospholipids located on the inside of vesicles or in close connection with membrane proteins may not be available to the phospholipases.

Saccomani et al. [85] treated a vesicular $(K^+ + H^+)$-ATPase preparation from pig gastric mucosa with phospholipase A_2, resulting in a breakdown of 50% of the phospholipids. This treatment also results in partial loss of ATPase activity, but the residual activity is still 25% of the original activity. The K^+-stimulated phosphatase activity is not affected by this treatment. Schrijen et al. [86] used two phospholipases with slight difference in substrate specificity, alone and in combination. With each of these phospholipases approx. 50% of the phospholipids could be hydrolysed resulting in a 50% loss in enzyme activity. When the two phospholipases were used successively 70% of the phospholipids were hydrolysed and the loss of activity was also 70%. This represents a striking parallelism between residual phospholipid content and ATPase activity. In this case the K^+-stimulated phosphatase activity

TABLE 4

Effect of phospholipase treatment on phospholipid composition and activity of $(K^+ + H^+)$-ATPase preparation

The phospholipid content of the control preparation was 0.79 mg/mg protein, the cholesterol content 0.19 mg/mg protein. From [86].

	Untreated	PL-ase C (*B. cereus*)	PL-ase C (*C. welchii*)	Both phospholipases
	% of total phospholipids in control preparation			
Phospholipid composition				
Total	100	51	50	35
Sphingomyelin	28	27	17	15
Phosphatidylcholine	30	9	6	3
Phosphatidylserine	11	4	11	5
Phosphatidylinositol	4	5	6	8
Phosphatidylethanol	27	6	10	4
	% of control activity			
$(K^+ + H^+)$-ATPase activity	100	43	45	29

showed the same behaviour as the $(K^+ + H^+)$-ATPase activity. From these experiments it can be concluded that phospholipids are in some way important for the $(K^+ + H^+)$-ATPase activity. Whether there is some specificity in the interaction between phospholipids and the enzyme protein has not yet been established. From reactivation studies, showing that all phospholipids except phosphatidylinositol can restore the enzyme activity [85], no firm conclusions could be drawn either. Schrijen et al. [86] found that the properties of the phospholipase C-treated enzyme are the same as that of the untreated $(K^+ + H^+)$-ATPase, except that the apparent affinity for K^+ is doubled.

Membrane enzymes often show a transition in the Arrhenius plot of the enzyme activity, which is usually attributed to a change in the lipid configuration. $(K^+ + H^+)$-ATPase preparations show such a transition at 27–28°C [71,86]. However, a transition in the polarisation signal of a lipid viscosity probe (diphenylhexatriene), was not observed in this region [71], indicating that the transition in the ATPase activity is either due to a change in the rate-limiting step of the reaction or to the presence of an annulus of specific lipids with physical properties different from those of the bulk lipids.

(h) Vesicular transport

The finding of Lee et al. [56] that addition of Mg-ATP to membrane vesicles containing $(K^+ + H^+)$-ATPase, incubated in a KCl medium, leads to alkalinisation of the medium, was a breakthrough in gastric physiology. This finding indicated that the vesicles are able to accumulate protons and thus that the ATPase could be involved in gastric acid secretion. The proton uptake depends on an intact vesicular structure, since detergent treatment of the vesicles abolishes proton uptake, and on preincubation of the vesicles in a KCl solution. Lee et al. [56] suggested that the ATPase would catalyse an exchange process of K^+ for H^+.

Strong evidence for this assumption has been provided by Sachs and coworkers [58,66,78,87]. They found that the rate of proton uptake depends on the nature of the cation present and that the sequence of the stimulating effect of these cations is the same as for the ATPase reaction. The only exception is Tl^+, which strongly stimulates the ATPase but inhibits proton transport [71]. The substrate specificity for the proton transport is also the same as for the $(K^+ + H^+)$-ATPase activity. Most inhibitors of the enzyme reaction, described in Section 3g, also inhibit the proton transport process.

By means of an ionophore specific for potassium (e.g. valinomycin) or for protons (tetrachlorsalicylanilide, TCS), Sachs et al. showed that the exchange process is electroneutral. The proton gradient generated by addition of ATP to the vesicles slowly dissipates due to the ion permeability of the vesicle. Addition of TCS leads to some increase in the rate of proton efflux, but a much faster proton release is observed upon addition of valinomycin (Fig. 4). This means that the permeability of the vesicle for K^+ is lower than for H^+ [66,78]. The ionophore nigericin, which exchanges H^+ for K^+, completely abolishes the proton gradient [58].

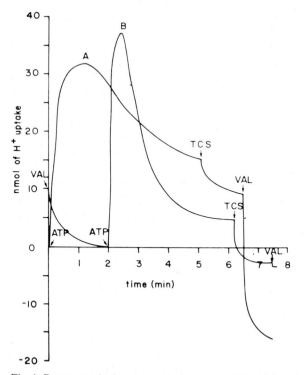

Fig. 4. Proton uptake by pig gastric mucosal vesicles. In curve A vesicles were incubated in a medium containing 150 mM KCl, followed by the addition of $1.7 \cdot 10^{-5}$ M ATP. At the indicated times tetrachlorsalicylanilide (TCS) and valinomycin (VAL) were added. In curve B valinomycin was added immediately followed by the addition of the vesicles, then $1.7 \cdot 10^{-5}$ M ATP. (From [78].)

It is unlikely that the effect of valinomycin is due to a conductance change which would stimulate an electrogenic pump, since protonophores like TCS have much less effect than K^+ ionophores and lipid-permeable cations do not substitute for these ionophores [58]. The effect of K^+-ionophores must be attributed to a requirement of the pump enzyme for internal K^+. Under certain conditions H^+ transport could be shown to be driven by a large K^+ gradient in the absence of ATP [66].

The gastric vesicles showed also a temperature-sensitive ^{86}Rb and ^{36}Cl uptake with approximately equal half times, indicating that these uptake processes are coupled [66,68]. Addition of ATP to the vesicles resulted in a transient release of ^{86}Rb, but the ^{36}Cl content did not change. This indicates that the ^{86}Rb efflux is accompanied by proton uptake. The effect of ATP on the ^{86}Rb efflux is rather specific for this nucleotide [68]. The pH dependence of this efflux process corresponds to that of the $(K^+ + H^+)$-ATPase, while some of the ATPase inhibitors also inhibit the ^{86}Rb efflux. The effect of ATP on the ^{86}Rb efflux is not due to a potential difference, since neither valinomycin nor lipid-permeable cations affect the

^{86}Rb efflux [66,68]. Calculation of the ratio between ^{86}Rb efflux and H^+-influx measured under identical circumstances, leads to a value close to one. These findings are all in agreement with the assumption that $(K^+ + H^+)$-ATPase catalyses an electroneutral $K^+ - H^+$ exchange process.

The use of potential-sensitive probes, like [^{14}C]thiocyanate and anilinonaphthosulfonic acid (ANS) confirm the electroneutrality of the $(K^+ + H^+)$-ATPase [66]. In the absence of ionophores no uptake of [^{14}C]thiocyanate and no change in ANS fluorescence were observed upon addition of ATP. Only in the presence of valinomycin does the inward K^+ gradient result in a positive potential, which leads to enhanced thiocyanate uptake and ANS fluorescence. This indicates that the absence of an effect when ionophores are present is due to the electroneutrality of the ATPase and not to the insensitivity of the method.

The proton uptake method described above has the disadvantage that with 2 mM Mg^{2+} and an ionic strength of 0.2 it can only be used at pH 6.1, since otherwise the ATP hydrolysis as such would cause a pH change. The ^{86}Rb efflux method, on the other hand, is rather cumbersome and slow. Lee and Forte [86] have investigated some fluorescent amines, which accumulate in acid compartments and lead to quenching of fluorescence. They showed that 9-amino-acridine behaves as would theoretically be expected. Addition of ATP to vesicles incubated in a medium containing Mg^{2+}, K^+ and 9-amino-acridine leads to a quenching of the fluorescence in the vesicles. At an external pH of 7.0 the ATP-generated pH difference between inside and outside was calculated to be 4–4.5 pH units. With this method Lee et al. [67] clearly showed that the proton uptake depends on the permeability of the anion accompanying K^+. The anion permeability decreases in the order:

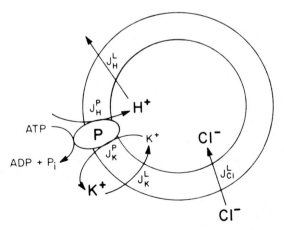

Fig. 5. Schematic model for the ion movements across gastric microsomal vesicles. The J values are the ionic fluxes with the superscripts designating pump flux (P) or leak pathway (L). The model consists of an ATP-driven H^+-K^+ exchange pump, $(K^+ + H^+)$-ATPase, and the passive leak pathways for the principal ions K^+, H^+ and Cl^-. (From [67].)

$NO_3^- > Br^- > Cl^- > I^- \simeq$ acetate $\simeq$ isethionate. The same is true for the valinomycin-stimulated ATPase reaction. The vesicular gradient for H^+ was abolished by a combination of a K^+-ionophore and a protonophore or a K^+-H^+ exchange ionophore (nigericin). Under these circumstances the ATPase is uncoupled and does not generate proton transport.

From these experiments a pump-leak model for the vesicular transport system with a $(K^+ + H^+)$-ATPase pump and passive conductances for K^+, Cl^- and H^+ ions has been postulated (Fig. 5).

(i) Role in gastric acid secretion

The gastric vesicles described above originate from the tubulovesicular system of the parietal cell. It has been established that upon stimulation of acid secretion the tubulovesicular system disappears with a simultaneous increase in the microvillar area of the apical membrane [89]. This phenomenon has generally been interpreted as a fusion process of the vesicles with the apical plasma membrane. Stimulants of the acid secretion process would thus primarily stimulate the fusion process. Berglindh et al. [69] have postulated an alternative explanation, viz. that the tubulovesicular system consists of collapsed tubules, which would osmotically expand by a KCl flux into the tubular lumen. The stimulants of the acid secretion process (histamine, gastrin, acetylcholine) would increase the KCl permeability of the apical membrane. Cyclic AMP or Ca^{2+} [90] is thought to be the second messenger of the stimulants of the acid secretion process.

In both models the $(K^+ + H^+)$-ATPase acts by pumping H^+ to the lumen in exchange for K^+. The K^+ ions must originate from the cytoplasm and be recycled by the pump enzyme. It is not yet known whether this ion in fact reaches the luminal side or remains inside the membrane, where it is bound by the $(K^+ + H^+)$-ATPase. Several other problems need to be elucidated before the vesicular $(K^+ + H^+)$-ATPase can be accepted as the system fully responsible for gastric acid secretion [60].

References

1. Schuurmans Stekhoven, F.M.A.H. and Bonting, S.L. (1981) This volume, Chapter 6.
2. Hasselbach, W. (1981) This volume, Chapter 7.
3. Durbin, R.P. and Kasbekar, D.K. (1965) Fed. Proc. 24, 1377–1381.
4. Kasbekar, D.K. and Durbin, R.P. (1965) Biochim. Biophys. Acta 105, 472–482.
5. Sachs, G., Mitch, W.E. and Hirschowitz, B.I. (1965) Proc. Soc. Exp. Biol. Med. 119, 1023–1027.
6. Blum, A.L., Shah, G., St. Pierre, T., Helander, H.F., Sung, C.P., Wiebelhaus, V.D. and Sachs, G. (1971) Biochim. Biophys. Acta 249, 101–113.
7. Sachs, G., Shah, G., Strych, A., Cline, G. and Hirschowitz, B.I. (1972) Biochim. Biophys. Acta 266, 625–638.
8. Wiebelhaus, V.D., Sung, C.P., Helander, H.F., Shah, G., Blum, A.L. and Sachs, G. (1971) Biochim. Biophys. Acta 241, 49–56.
9. De Pont, J.J.H.H.M., Hansen, T. and Bonting, S.L. (1972) Biochim. Biophys. Acta 274, 189–200.
10. Soumarmon, A., Lewin, M., Cheret, A.M. and Bonfils, S. (1974) Biochim. Biophys. Acta 339, 403–414.

11 Hegner, D. and Anika, S. (1975) Comp. Biochem. Physiol. 50B, 339–343.
12 Van Amelsvoort, J.M.M., de Pont, J.J.H.H.M. and Bonting, S.L. (1977) Biochim. Biophys. Acta 466, 283–301.
13 Turbeck, B.O., Nedergaard, S. and Kruse, H. (1968) Biochim. Biophys. Acta 163, 354–361.
14 Katz, A.I. and Epstein, F.H. (1971) Enzyme 12, 499–507.
15 Kinne-Saffran, E. and Kinne, R. (1974) Proc. Soc. Exp. Biol. Med. 146, 751–753.
16 Kinne-Saffran, E. and Kinne, R. (1979) J. Membr. Biol. 49, 235–251.
17 Liang, C.T. and Sacktor, B. (1976) Arch. Biochem. Biophys. 176, 285–297.
18 Van Amelsvoort, J.M.M., de Pont, J.J.H.H.M., Stols, A.L.H. and Bonting, S.L. (1977) Biochim. Biophys. Acta 471, 78–91.
19 Simon, B. and Thomas, L. (1972) Biochim. Biophys. Acta 288, 434–442.
20 Simon, B., Kinne, R. and Sachs, G. (1972) Biochim. Biophys. Acta 282, 293–300.
21 Van Amelsvoort, J.M.M., Jansen, J.W.C.M., de Pont, J.J.H.H.M. and Bonting, S.L. (1978) Biochim. Biophys. Acta 512, 296–308.
22 Simon, B., Kinne, R. and Knauf, H. (1972) Pflügers Arch. 337, 177–184.
23 Izutsu, K.T. and Siegel, I.A. (1972) Biochim. Biophys. Acta 284, 478–484.
24 Wais, U. and Knauf, H. (1975) Pflügers Arch. 361, 61–64.
25 Setchell, B.P., Smith, M.W. and Munn, E.A. (1972) J. Reprod. Fert. 28, 413–418.
26 Izutsu, K.T. and Siegel, I.A. (1975) Biochim. Biophys. Acta 382, 193–203.
27 Kimelberg, H.K. and Bourke, R.S. (1973) J. Neurochem. 20, 347–359.
28 Kerstetter, T.H. and Kirschner, L.B. (1974) Comp. Biochem. Physiol. 48B, 581–589.
29 Bornancin, M., De Renzis, G. and Naon, R. (1980) Am. J. Physiol. 238, R251–R259.
30 De Renzis, G. and Bornancin, M. (1977) Biochim. Biophys. Acta 467, 192–207.
31 Duncan, C.J. (1975) Life Sci. 16, 955–966.
32 Izutsu, K.T., Madden, P.R., Watson, E.L. and Siegel, J.A. (1977) Pflügers Arch. 369, 119–124.
33 Van Amelsvoort, J.M.M., van Hoof, P.M.K.B., de Pont, J.J.H.H.M. and Bonting, S.L. (1978) Biochim. Biophys. Acta 507, 83–93.
34 Au, K.S. (1979) Int. J. Biochem. 10, 687–689.
35 Ivanshchenko, A.I., Zhubanova, A.A., Balmukhanov, B.S. and Ryskulova, S.T. (1975) Biokhimiya 40, 629–633.
36 Morisawa, M. and Utida, S. (1976) Biochim. Biophys. Acta 445, 458–463.
37 Humphreys, M.H. and Chou, L.Y.N. (1979) Am. J. Physiol. 236, E70–E76.
38 Iretani, N. and Wells, W.W. (1976) Biochim. Biophys. Acta 436, 863–868.
39 Sener, A., Valverde, I. and Malaisse, W.J. (1979) FEBS Lett. 105, 40–42.
40 Anstee, J.H. and Fathpour, H. (1979) Insect. Biochem. 9, 383–388.
41 Boyd, C.A. and Chipperfield, A.R. (1980) J. Physiol. 303, 63P.
42 Racker, E. (1962) Fed. Proc. 21, A54.
43 Mitchell, P. and Moyle, J. (1971) Bioenergetics 2, 1–11.
44 Lambeth, D.O. and Lardy, H.A. (1971) Eur. J. Biochem. 22, 355–363.
45 Ebel, R.E. and Lardy, H.A. (1975) J. Biol. Chem. 250, 191–196.
46 Sachs, G., Wiebelhaus, V.D., Blum, A.L. and Hirschowitz, B.I. (1972) in G. Sachs, E. Heinz and K.J. Ulrich (Eds.), Gastric Secretion, pp. 321–343, Academic Press, New York.
47 Bonting, S.L., Van Amelsvoort, J.M.M. and de Pont, J.J.H.H.M. (1978) Acta Physiol. Scand. Spec. Suppl. 329–340.
48 Bonting, S.L., de Pont, J.J.H.H.M., Van Amelsvoort, J.M.M. and Schrijen, J.J. (1980) Ann. N.Y. Acad. Sci. 341, 335–356.
49 Chan, S.H.P. and Barbour, R.L. (1976) Biochim. Biophys. Acta 430, 426–433.
50 Van Os, C.H., Mircheff, A.K. and Wright, E.M. (1977) J. Cell. Biol. 73, 257–260.
51 Kaysen, G.A., Chou, L.Y. and Humphreys, M.H. (1979) J. Cell. Biol. 82, 780–782.
52 Forte, J.G., Forte, G.M. and Saltman, P. (1967) J. Cell. Physiol. 69, 293–304.
53 Limlomwangse, L. and Forte, J.G. (1970) Am. J. Physiol. 219, 1717–1722.
54 Ganser, A.L. and Forte, J.G. (1973) Biochim. Biophys. Acta 307, 169–180.
55 Ganser, A.L. and Forte, J.G. (1973) Biochem. Biophys. Res. Commun. 54, 690–696.

56 Lee, J., Simpson, G. and Scholes, P. (1974) Biochem. Biophys. Res. Commun. 54, 690–696.
57 Forte, J.G. and Lee, H.C. (1977) Gastroenterology 73, 921–926.
58 Sachs, G. (1977) Rev. Physiol. Biochem. Pharmacol. 79, 133–162.
59 Sachs, G., Spenney, J.G. and Lewin, M. (1978) Physiol. Rev. 58, 106–173.
60 Forte, J.G., Machen, T.E. and Öbrink, K.J. (1980) Annu. Rev. Physiol. 42, 111–126.
61 Schuurmans Stekhoven, F.M.A.H. and Bonting, S.L. (1981) Physiol. Rev. 61, 1–76.
62 Ray, T.K. and Forte, J.G. (1978) Biochim. Biophys. Acta 443, 451–467.
63 Saccomani, G., Stewart, H.B., Shaw, O., Lewin, M. and Sachs, G. (1977) Biochim. Biophys. Acta 465, 311–330.
64 Schrijen, J.J., Luyben, W.A.H.M., de Pont, J.J.H.H.M. and Bonting, S.L. (1980) Biochim. Biophys. Acta 597, 331–344.
65 Forte, J.G., Ganser, A., Beesley, R. and Forte, T.M. (1975) Gastroenterology 69, 175–189.
66 Sachs, G., Chang, H.H., Rabon, E., Schackman, R., Lewin, M. and Saccomani, G. (1976) J. Biol. Chem. 251, 7690–7698.
67 Lee, H.-C., Breitbart, H., Berman, M. and Forte, J.G. (1979) Biochim. Biophys. Acta 553, 107–131.
68 Schackman, R., Schwartz, A., Saccomani, G. and Sachs, G. (1977) J. Membrane Biol. 32, 361–381.
69 Berglindh, T., Dibona, D.R., Ito, S. and Sachs, G. (1980) Am. J. Physiol. 238, G165–G176.
70 Saccomani, G., Helander, H.F., Crago, G., Chang, H.H., Daily, D.W. and Sachs, G. (1979) J. Cell. Biol. 83, 271–283.
71 Sachs, G., Berglindh, T., Rabon, E., Stewart, H.B., Barcellona, M.L., Wallmark, B. and Saccomani, G. (1980) Ann. N.Y. Acad. Sci. 341, 312–334.
72 Sachs, G., Rabon, E. and Saccomani, G. (1979) in Y. Muhahata and L. Packer (Eds.), Cation Flux Across Biomembranes, pp. 53–66, Academic Press, New York.
73 Saccomani, G., Daily, D.W. and Sachs, G. (1979) J. Biol. Chem. 254, 2821–2827.
74 Wallmark, B. and Mårdh, S. (1979) J. Biol. Chem. 254, 11899–11902.
75 Van de Ven, F.J.M., Schrijen, J.J., de Pont, J.J.H.H.M. and Bonting, S.L. (1981) Biochim. Biophys. Acta 640, 487–499.
76 Tanisawa, A.S. and Forte, J.G. (1971) Arch. Biochem. Biophys. 147, 165–175.
77 Forte, J.G., Ganser, A.L. and Tanisawa, A.S. (1974) Ann. N.Y. Acad. Sci. 242, 255–267.
78 Chang, H., Saccomani, G., Rabon, E., Schackmann, R. and Sachs, G. (1977) Biochim. Biophys. Acta 464, 313–327.
79 Ray, T.K. (1978) FEBS Lett. 92, 49–52.
80 Schrijen, J.J., van Groningen-Luyben, W.A.H.M., de Pont, J.J.H.H.M. and Bonting, S.L. (1981) Biochim. Biophys. Acta 640, 473–486.
81 Lee, H.-C. and Forte, J.G. (1980) Biochim. Biophys. Acta 598, 595–605.
82 Saccomani, G., Barcellona, M.L., Rabon, E. and Sachs, G. (1980) in I. Schulz et al. (Eds.), Hydrogen Ion Transport in Epithelia, Elsevier/North-Holland, Amsterdam, pp. 175–183.
83 O'Neal, S.G., Rhoads, D.B. and Racker, E. (1979) Biochim. Biophys. Res. Commun. 89, 845–850.
84 Bonting, S.L. and de Pont, J.J.H.H.M. (1980) Biochem. Soc. Transact. 8, 40–42.
85 Saccomani, G., Chang, H.H., Spisni, A., Helander, H.H., Spitzer, H.L. and Sachs, G. (1979) J. Supramol. Struct. 11, 429–444.
86 Schrijen, J.J., Omachi, A., Van Groningen Luyben, W.A.H.M., de Pont, J.J.H.H.M. and Bonting, S.L. (1981) in preparation.
87 Sachs, G., Chang, H., Rabon, E., Schackmann, R., Sarou, H.M. and Saccomani, G. (1977) Gastroenterology 73, 931–940.
88 Lee, H.C. and Forte, J.G. (1978) Biochim. Biophys. Acta 508, 339–356.
89 Helander, H.F. and Hirschowitz, B.I. (1972) Gastroenterology 63, 951–961.
90 Proverbio, F. and Michelangeli, F. (1978) J. Membrane Biol. 42, 301–315.

CHAPTER 9

Mitochondrial ion transport

A.J. MEIJER and K. VAN DAM

*Laboratory of Biochemistry, B.C.P. Jansen Institute,
University of Amsterdam, Plantage Muidergracht 12,
1018 TV Amsterdam, The Netherlands*

1. Mitochondrial metabolite transport

(a) Introduction

Since many metabolic processes in the cell require the participation of both intra- and extramitochondrial enzyme reactions, transport of certain metabolites across the mitochondrial membrane is obligatory. Developments in the field of mitochondrial metabolite transport have been rapid in the last 15–20 years. Whereas most of this work during the first decade was carried out with isolated mitochondria (see [1–3] for review), during the last decade research gradually moved in the direction of elucidation of the role that these transport systems (translocators, carriers) play in the regulation of metabolism in the intact cell [4–7].

Since the reviews mentioned above have described the history of mitochondrial metabolite transport in great detail and since space is limited, in the present review we will highlight recent developments with emphasis on the possible regulatory role that the translocators play in integrating mitochondrial and cytosolic processes.

(b) Survey of the mitochondrial metabolite translocators

Fig. 1 gives a schematic presentation of the known mitochondrial metabolite translocators together with the main metabolic processes in which they are involved.

Transport of phosphate, pyruvate and glutamate via their respective translocators is electroneutral and H^+-coupled. Although in Fig. 1 their transport has been written as cotransport of the anion with H^+ (symport), it is also possible that the actual transport mechanism involves an exchange of the anion with OH^- (antiport). Experimentally, these two mechanisms cannot be distinguished. The translocators for α-oxoglutarate and malate (the latter is also called the dicarboxylate translocator) also mediate electroneutral exchanges.

The tricarboxylate translocator mediates an electroneutral exchange between citrate (or isocitrate) and malate. In this case H^+ is involved since citrate^{3-} must be converted to citrate^{2-}. This translocator also transports phosphoenolpyruvate,

Inhibitors [a]	out	in	Function in [b]	Driving force
Atractylate * Carboxyatractylate * Bongkrekate	ADP^{3-} → ← ATP^{4-}		oxidative phosphorylation	$\Delta\psi$
N-Ethylmaleimide Mersalyl	P_i^- ←→ H^+ ←→		oxidative phosphorylation	ΔpH
α-Cyanocinnamates *	Pyr^- ←→ H^+ ←→		citric acid cycle gluconeogenesis from lactate	ΔpH
Trimethylamino-acylcarnitine	acylcarn ←→ ←→ carn		long-chain fatty acid oxidation	–
Benzene 1,2,3-tricarboxylate	mal^{2-} ←→ ←→ cit^{3-}; $isoc^{3-}$; PEP H^+		lipogenesis; gluconeogenesis (PEP); isocitrate-α-oxoglutarate shuttle	ΔpH
Phthalonate	αOG^{2-} ←→ → mal^{2-}		gluconeogenesis from lactate; malate-aspartate shuttle; isocitrate-α-oxoglutarate shuttle; nitrogen metabolism	–
n-Butylmalonate * Mersalyl	P_i^{2-} ←→ → mal^{2-}		gluconeogenesis from lactate, pyruvate, amino acids; urea synthesis	–
N-Ethylmaleimide Bromocresolpurple	Glu^- ←→ H^+ ←→		urea synthesis; nitrogen metabolism	ΔpH
Glisoxepide	GluH → ← asp^-		urea synthesis; malate-aspartate shuttle; gluconeogenesis from lactate	$-Z\Delta pH$ $+\Delta\psi$
	Gln —(?)→		glutamine degradation	?
Lysine ?	$Orn.H^+$ → H^+ ↓ H^+		urea synthesis	ΔpH
	←(?)— citrull		urea synthesis	?
	neutral amino acids (?)→		degradation	?

Fig. 1. Mitochondrial metabolite translocators.
[a] Inhibitors with * may be used in metabolic studies with intact cells.
[b] For metabolic pathways, see [4,8].
Abbreviations: P_i, phosphate; pyr, pyruvate; acylcarn, acylcarnitine; carn, carnitine; mal, malate; cit, citrate; isoc, isocitrate; PEP, phosphoenolpyruvate; α-OG, α-oxoglutarate; glu, glutamate; asp, aspartate; gln, glutamine; orn, ornithine; citrull, citrulline.

which is important in tissues where phosphoenolpyruvate synthesis for gluconeogenesis occurs inside the mitochondria.

Transport of adenine nucleotides is presumably fully electrophoretic, and therefore driven by the mitochondrial membrane potential, because ADP^{3-} exchanges with ATP^{4-} [9]. Aspartate transport is both electrophoretic and H^+-coupled because undissociated glutamate exchanges with the aspartate anion [10]. Since in the intact cell the mitochondrial membrane potential is positive outside and the mitochondrial matrix slightly alkaline with respect to the cytosol, it follows that in vivo movement of ATP and of aspartate is unidirectional, out of the mitochondria.

The recently discovered carnitine carrier [11,12] involves an exchange between two electroneutral species and is not affected by the mitochondrial energy state.

Transport of glutamine into the mitochondria, important in kidney and intestine, has been a very controversial issue. At present, the mechanism of glutamine transport has not been solved [6].

Transport of ornithine and citrulline, important for urea synthesis in liver, has been a controversial issue too [6]. In Fig. 1 transport of ornithine and citrulline is depicted as two separate pathways, but this has not yet been settled. It is also possible that ornithine exchanges with citrulline.

(c) The use of mitochondrial transport inhibitors in metabolic studies

For most mitochondrial transport systems inhibitors are available and these have been very useful in studies with isolated mitochondria, not only in the study of their kinetic properties but also for the isolation of some of the translocator proteins. In addition, some of the inhibitors can be used in studying the role of mitochondrial transport systems in metabolic processes as they occur in the intact cell. Fig. 1 gives a list of inhibitors commonly used. For a more complete list the reader is referred to [2,3,5–7].

Problems may arise when the transport inhibitors are used for metabolic studies in intact cells, for two reasons. Firstly, the intact cell is metabolically far more complex than an isolated mitochondrial preparation, so that the inhibitor specificity becomes important. Evidently, the −SH binding reagents mersalyl and N-ethylmaleimide cannot be used in intact cells because they will attack all kinds of −SH containing proteins. Benzene-1,2,3-tricarboxylate inhibits fatty acid synthesis in isolated hepatocytes of neonatal chicks [13]. Although mitochondrial citrate efflux is an intermediate step in this process [14], interpretation of this inhibition is difficult because the inhibitor also affects acetyl-CoA carboxylase, the key enzyme in lipogenesis [13]. Phthalonate is a powerful inhibitor of the α-oxoglutarate translocator. However, its use in studies with intact cells is limited because it also inhibits lactate dehydrogenase and phosphoenolpyruvate carboxykinase [15]. Although α-cyanocinnamate, which inhibits pyruvate transport, has been successfully applied in many metabolic studies in intact cells, it may not be useful under all conditions. In studies on lipogenesis with pyruvate as the precursor interpretation of the results is complicated by the fact that α-cyanocinnamate also inhibits acetyl-CoA carboxylase (W.J. Vaartjes, personal communication).

Secondly, the transport inhibitor must be able to pass the cell membrane. The inability of benzene-1,2,3-tricarboxylate to inhibit gluconeogenesis from lactate in perfused pigeon liver, a tissue in which mitochondrial efflux of phosphoenolpyruvate is obligatory for glucose synthesis, is presumably due to lack of penetration through the plasma membrane [16]. This observation is interesting since it shows that the ability of an inhibitor to penetrate the cell membrane may vary from tissue to tissue; benzene-1,2,3-tricarboxylate does inhibit lipogenesis in hepatocytes of neonatal chicks, as discussed above. Another example is the apparent relative impermeability of the plasma membrane of isolated foetal rat hepatocytes, as compared with that from adult rats, for atractyloside, the inhibitor of the adenine nucleotide translocator [17].

(d) Kinetic properties of the individual translocators

For a proper understanding of the role of the translocators in the regulation of metabolism, knowledge of their kinetic constants is indispensable. Table 1 summarises these parameters. Because of technical difficulties, in most cases it has not been possible to determine these parameters at 37°C. It is also important to stress that the kinetic constants have been determined in isolated mitochondria. It is likely that the kinetic constants in the intact cell are different, one reason being that there is an inhibitory interaction of cytosolic anions with the various translocators. Some of these effects are given in Table 2. With the exception of a few cases (the α-oxoglutarate translocator in heart [18] and the carnitine and aspartate translocators; see Sections 1f, iii and iv), little is known about the K_m values of the metabolites to be transported from the matrix side of the mitochondrial membrane. In the case of citrate and ATP transport such information is difficult to obtain because most of the intramitochondrial citrate and ATP is chelated with Mg^{2+} and only the free anions are transported. Likewise, little is known about possible competition between metabolites present in the matrix for export out of the mitochondria. The complexity of the complete kinetic analysis of a translocator, in which both the external and internal concentrations have been taken into account, is illustrated by the studies of Sluse et al. [18] on α-oxoglutarate transport in heart mitochondria.

It must be realised that the kinetic properties of the translocators may vary from species to species and even from tissue to tissue in the same species. One example is the relatively high capacity of the tricarboxylate translocator in liver compared to that in heart [1,19], which is related to the absence of extramitochondrial fatty acid synthesis in heart. Another example is the high activity of the dicarboxylate translocator in liver compared to that in heart [19] which is related to the absence of gluconeogenesis in heart.

(e) Studies on the distribution of metabolites across the mitochondrial membrane in the intact cell

To study the role of the mitochondrial transport systems in the regulation of metabolic pathways as they occur in the intact cell, knowledge of the mitochondrial and cytosolic metabolite concentrations in the intact cell is essential. At present, three methods have been developed for the determination of these parameters:

(1) Fractionation of freeze-clamped, freeze-dried tissue in organic solvents [31].

(2) Rapid fractionation of hepatocytes by brief treatment with digitonin, which damages the plasma membrane but not the inner mitochondrial membrane [32].

(3) Rapid fractionation of hepatocytes by mechanical shearing forces [33].

For an extensive discussion of the technical details of each of these methods the reader is referred to [34].

Most of the studies on the intracellular distribution of metabolites across the mitochondrial membrane have been carried out with perfused liver (method 1) or isolated hepatocytes (methods 2 and 3), mainly because the metabolic properties of these preparations had been well characterised in the past. It is important that, in general, all fractionation methods give comparable values for the anion distribution across the mitochondrial membrane under comparable conditions although they differ in detail [34].

One of the conclusions that can be drawn at present is that previous calculations of mitochondrial and cytosolic metabolite concentrations from total cell amounts (for a review of the calculation methods, see [5]) are incorrect, presumably because the assumptions on which these calculations were based (near-equilibrium of certain enzyme reactions) appear to be invalid [33–35].

One of the questions that can in principle be answered by the fractionation techniques is whether the anion transport systems play a regulatory role in metabolism or whether they are sufficiently active to maintain near-equilibrium. In equilibrium, the distribution of anions whose transport is electroneutral and either directly or indirectly coupled to H^+ movement is given by the equation:

$$\log \frac{[A^{n-}]_{mit}}{[A^{n-}]_{cyt}} = n\Delta pH$$

in which $\Delta pH = pH_{mit} - pH_{cyt}$ and n is the number of charges that the anion carries at the particular pH [36]. Anions like citrate, malate and α-oxoglutarate distribute according to this equation only when all the translocators involved (tricarboxylate, α-oxoglutarate, dicarboxylate and phosphate translocators) are sufficiently active to maintain equilibrium. Although this situation can be easily met in isolated mitochondria under non-metabolising conditions, with zero net flux through the translocators [36], this is not necessarily the case for intact cells.

Most studies with isolated hepatocytes or perfused liver indicate the existence of a ΔpH across the mitochondrial membrane, ranging from 0–0.6 [33,34,37–39], but under some conditions the calculated ΔpH depends on the metabolite chosen. This indicates disequilibrium in one or more of the transport steps. Under most conditions it is unlikely that the phosphate translocator is out of equilibrium because of its very high V_{max} (Table 1). Problems do arise, however, when the dicarboxylate translocator is out of equilibrium because it connects the movement of citrate, malate and α-oxoglutarate with that of H^+. In that case the above equation cannot

TABLE 1
Kinetic properties of translocators [a]

Translocator	K_m for external metabolite			V_{max} (nmol/min/mg/ mitochondrial protein) (°C)
ADP, ATP [7]	ADP:	1–12 μM		600 (30°)
	ATP:	150 μM (energised mitochondria)		
		1–2 μM (deenergised mitochondria)		
P_i [20]	1.6 mM			205 (0°)
Pyruvate [21]	0.2–0.6 mM			40–100 (37°)
Carnitine [22]	carnitine: 4–5 mM			500 [b] (37°)
Tricarboxylate [23]	citrate	0.2 mM		22 (9°)
	isocitrate	0.2 mM		22 (9°)
	malate	0.7 mM		22 (9°)
Dicarboxylate [24]	malate:	0.2 mM		70 (9°)
	P_i:	1.5 mM		70 (9°)
α-Oxoglutarate [25]	α-oxoglutarate:	0.05 mM		43 (9°)
	malate:	0.1 mM		43 (9°)
Aspartate [6,26]	glutamate 6 mM			200–250 (30°)
Glutamate [7,27]	glutamate 4 mM			23–30 (25°)
Glutamine [28] (kidney)	2.7 mM			150–300 (23°)
Ornithine [29]	1.0 mM			8 [c] (20°)

[a] All values shown are for rat-liver mitochondria with the exception of glutamine. In order to convert V_{max} values to rates per g dry weight of liver tissue one has to multiply with 220, since 1 g dry weight of liver contains about 220 mg mitochondrial protein.
[b] At infinite intramitochondrial carnitine concentration.
[c] This rate is probably underestimated, since citrulline synthesis in liver mitochondria is faster.

be used to calculate ΔpH even when the α-oxoglutarate-malate and citrate-malate exchanges are at equilibrium.

Equilibrium of the α-oxoglutarate-malate exchange implies that (α-oxoglutarate/malate)$_{mit}$/(α-oxoglutarate/malate)$_{cyt}$ = 1, since H$^+$ is not involved in this exchange. From data published by Siess et al. [35,37] it can be calculated that this condition is approximately fulfilled in hepatocytes incubated with lactate.

At equilibrium of the citrate-malate exchange system log[(citrate/malate)$_{mit}$/(citrate/malate)$_{cyt}$] = ΔpH, since one H$^+$ accompanies citrate in its exchange for malate. This value must be equal to the logarithm of the pyruvate and glutamate concentration gradients if both the pyruvate and glutamate translocators are also at equilibrium.

In most of the studies published so far the actual ΔpH across the mitochondrial membrane in the intact cell has not been checked by using an independent method,

TABLE 2
Inhibition of mitochondrial metabolite transport by extramitochondrial metabolites

Exchange	Inhibiting anion	Type of inhibition	K_i (mM)
ADP_{out} – ATP_{in} [30]	ATP	competitive	0.2
$Citrate_{out}$ – $malate_{in}$ [23]	malate	competitive	0.7
	phosphoenolpyruvate	competitive	0.11
$Malate_{out}$ – $phosphate_{in}$ [24]	phosphate	largely non-competitive	4
α-$Oxoglutarate_{out}$ – $malate_{in}$ [25]	malate	competitive	0.12
	citrate	competitive	3.6
$Glutamate_{out}$ – $aspartate_{in}$ [26]	aspartate	non-competitive	4

in conjunction with the mitochondrial metabolite distributions, so that no definite conclusions can be drawn at present. However, in a recent study by Soboll et al. [39] the subcellular metabolite distribution in the perfused liver was compared with the mitochondrial and cytosolic pH values, as calculated from the distribution of the weak acid 5,5-dimethyl-2,4-oxazolidinedione. It was found that in liver from fed rats citrate, malate and α-oxoglutarate were distributed across the mitochondrial membrane according to the pH gradient indicating that the translocators were at equilibrium. This was not the case in the fasted state, presumably because the translocators were kinetically limited by the low metabolite concentrations under these conditions.

The main weakness in these studies is the assumption that the cytosolic and mitochondrial activity coefficients are equal, which may not be true. As we will see later, experiments with isolated mitochondria have indicated that under some experimental conditions diffusion gradients of certain metabolites within the mitochondrial matrix may exist. It remains to be established how such gradients can be maintained in the steady state. Although it is unknown whether this also occurs under physiological conditions, the possibility must certainly be considered. Another problem is a possible difference in concentration of cytosolic and mitochondrial free Mg^{2+} with consequent differential binding of certain metabolites (citrate, ATP, ADP) to Mg^{2+} in both compartments.

Despite the problems involved in the interpretation of measured metabolite concentration gradients across the mitochondrial membrane, knowledge of the approximate magnitude of mitochondrial and cytosolic metabolite concentrations has already provided important information with regard to the regulation of key enzymes by certain effector metabolites in each of the two compartments.

(f) Recent developments in mitochondrial metabolite transport

In this chapter we shall discuss some specific issues that have received a relatively great deal of attention in recent years.

(i) Transport of adenine nucleotides
The role of adenine nucleotide transport in metabolism.

In energised mitochondria the 1:1 exchange of ADP for ATP is unidirectional. This is due not only to the higher affinity of the translocator for extramitochondrial ADP (Table 1) but also to a higher affinity for intramitochondrial ATP [40,41]. Only in deenergised mitochondria does the reverse exchange become possible. Exchange between extramitochondrial ADP and intramitochondrial ATP is efficient from the point of view of physiological functioning of the mitochondria since most ATP-requiring reactions occur outside the mitochondria.

During the $ADP_{out}^{3-} - ATP_{in}^{4-}$ exchange one net negative charge moves across the mitochondrial membrane. Consequently, the extramitochondrial ATP/ADP ratio exceeds the intramitochondrial ratio, the difference depending on the magnitude of the mitochondrial membrane potential under conditions where there is no net flux through the translocator [9]. Continuous exchange of ADP^{3-} for ATP^{4-} requires that one H^+ must be pumped out of the mitochondria per ATP. This H^+ can reenter the mitochondria together with phosphate so that the net result is that for extramitochondrial ATP synthesis one extra H^+ ion must be pumped by the respiratory chain. Since it is conceivable that under flux conditions the ADP-ATP exchange is not fully electrophoretic, the number of extra H^+ ions per ATP may be less than one. The H^+ movement linked to substrate movement may explain why intramitochondrial P/O ratios exceed extramitochondrial P/O ratios [42,43].

The difference in extramitochondrial and intramitochondrial ATP/ADP ratio is observed not only in isolated mitochondria but also in intact cells [44–46]. It allows the symbiosis of the intra- and extramitochondrial ATP systems which have to operate at different phosphate potentials [40].

Recently attempts have been made to ascertain whether adenine nucleotide transport across the mitochondrial membrane in intact cells can be a rate-determining factor for extramitochondrial ATP-utilising processes.

Work with isolated liver mitochondria has suggested that in respiratory states between state 3 and 4 (as defined by Chance and Williams [47]), a condition thought to exist in the cell, respiration is controlled by the extramitochondrial ATP/ADP ratio [48–50]. Since a relatively high extramitochondrial ATP concentration would inhibit ADP entry into the mitochondria [51], it has been concluded that the adenine nucleotide translocator limits respiration under these conditions. This was supported by the finding that even very low concentrations of carboxyatractyloside decreased the rate of respiration in the presence of glucose and limiting amounts of hexokinase, in contrast to the situation in state 3 (with excess of hexokinase) where a sigmoidal titration curve for carboxyatractyloside was obtained [52].

According to Holian et al. [53], however, respiration in tightly coupled mitochondria is controlled by the extramitochondrial ATP/ADP $\times P_i$ ratio (the phosphate potential) rather than by ATP/ADP. Since P_i is not involved in ADP transport it was concluded that ADP transport could not be rate-limiting. However, since under flux conditions both P_i and ADP have to enter to allow exit of ATP, it will be impossible to uncouple the effect of P_i and adenine nucleotide transport. This is illustrated by the findings of Lemasters and Sowers [54] who showed that P_i still exerted its stimulatory effect on respiration in the presence of partially inhibitory concentrations of atractyloside.

In isolated hepatocytes, Wilson et al. [55] consider the first two sites of the respiratory chain to be in near-equilibrium with the cytosolic phosphate potential. Since ADP transport is part of this system the translocator must also operate at near-equilibrium and therefore cannot be a rate-limiting step according to this view. On the basis of two kinetic arguments Vignais and coworkers [7,56,57] also consider the translocator to operate at near-equilibrium. Firstly, the V_{max} of the adenine nucleotide translocator exceeds the estimated flux through the translocator in hepatocytes by 30% [56] (which is not very much). Secondly, the calculated cytosolic free ADP concentration (115 μM, according to the data of Akerboom et al. [44], based on a cytosolic free Mg^{2+} concentration of 0.4 mM) is 10–100 times higher than its K_m for transport (Table 1). However, the possibility that part of the cytosolic ADP is protein-bound cannot be excluded [44].

Incompatible with the view of near-equilibrium of the ADP–ATP exchange are the observations of Akerboom et al. [44]. Isolated hepatocytes, incubated with various gluconeogenic substrates, were titrated with atractyloside. Since significant inhibition of gluconeogenesis was already obtained at concentrations of atractyloside in the order of its K_i value for the translocator [58], it was concluded that in hepatocytes the transport of ADP is, in fact, a rate-limiting step. This view has been challenged by Vignais and coworkers [7,56,57] who consider the translocator as part of a cyclic multi-enzyme system in which the bulk of ADP arising in the cytosol as the result of ATP-consuming processes is transported in exchange for mitochondrial ATP to feed the oxidative phosphorylation system. According to these authors in a cyclic system any inhibition of a component enzyme is expected to depress the cyclic rate to some extent. This argument is, however, incorrect since an equilibrium enzyme, even though it is part of a cycle, should have sufficient activity to maintain equilibrium and thus cannot be rate-limiting.

Van der Meer et al. [59] have described the regulation of mitochondrial respiration on the basis of the principles of irreversible thermodynamics. According to these authors, and thus in contrast to the view of Holian et al. [53], the entire system of oxidative phosphorylation in the hepatocyte is displaced from equilibrium. They observed a linear relationship between the rate of oxygen consumption and the affinity of the entire oxidative phosphorylation system, a term which includes the cytosolic phosphate potential, the mitochondrial $NADH/NAD^+$ ratio and the partial pressure of oxygen. They concluded that the adenine nucleotide translocator may be a rate-limiting step in cellular oxidative phosphorylation.

Compartmentation of adenine nucleotides in the mitochondrial matrix?

According to Klingenberg and Rottenberg [9] the ratio $(ATP/ADP)_{out}/(ATP/ADP)_{in}$ in isolated liver mitochondria, incubated under conditions of zero net flux through the adenine nucleotide translocator, varies linearly with the mitochondrial membrane potential. According to their results a membrane potential of 160 mV, which is the value that presumably exists in hepatocytes [60], corresponds with an $(ATP/ADP)_{out}/(ATP/ADP)_{in}$ ratio of 100 (at an external free Mg^{2+} concentration of 4 mM). In isolated hepatocytes this ratio is 3–6 for cells from starved rats [44,45]. With regard to cells from fed rats conflicting values have been reported: either 3–6 [61] or 50 [45]. From this one may conclude that in the intact hepatocyte either the adenine nucleotide translocator is not in equilibrium (with a possible exception for hepatocytes from fed rats) or that the activity coefficients of intra- and extramitochondrial ATP (ADP) are not equal. Apart from the possibility of binding of ATP and ADP to cytosolic proteins and Mg^{2+}, discussed in the previous section, there is also the possibility of compartmentation by binding in the mitochondrial matrix.

Some evidence to support this last possibility comes from experiments with isolated mitochondria. Thus it was shown that, at low temperature (0–4°C), [^{14}C]ADP added to state-3 mitochondria in the steady state did not readily mix with intramitochondrial unlabelled ADP during [^{14}C]ATP synthesis [62,63] (contrast [64]). Compartmentation of intramitochondrial adenine nucleotides would have important consequences for intramitochondrial ATP-utilising reactions like citrulline synthesis and pyruvate carboxylation. It must be realised, however, that compartmentation at 4°C does not necessarily mean that it also exists at 37°C. According to Raijman and Bartulis [65] inhibition of citrulline synthesis by added ADP in isolated mitochondria is not accompanied by a decreased intramitochondrial ATP/ADP ratio. It was concluded that carbamoyl-phosphate synthetase (ammonia) is not in equilibrium with ATP at the concentrations of the major matrix pool which the measurements reflect. On the other hand, several other groups do find significant decreases in mitochondrial ATP/ADP going from state 4 to state 3 [66–69]. This apparent discrepancy is most likely explained by technical difficulties involved in the separation of mitochondria from their suspending medium, which lead to overestimation of mitochondrial ATP/ADP in state 3 [45,70].

Inhibition of adenine nucleotide transport by fatty acyl-CoA esters.

It has been proposed that fatty acyl-CoA esters, which are potent inhibitors of ADP transport in isolated mitochondria, also control flux through the translocator in vivo and hence the production of ATP (see [4] for literature). However, as pointed out by Stubbs [7], such inhibition makes little physiological sense in situations like starvation or long-term exercise, when fatty acids are an important fuel; especially under the latter conditions the translocator must operate at high capacity to provide the cytosol with ATP. Indeed, in isolated hepatocytes incubated with various concentrations of fatty acids no inhibition of ATP transport by fatty acyl-CoA could be observed [71]. Possibly this inhibition is prevented by a low-molecular weight cytosolic protein with a high affinity for fatty acyl-CoA [72].

(ii) Transport of pyruvate
After the discovery of the pyruvate translocator in 1971 by Papa et al. [73] and the introduction of the α-cyanocinnamates as specific inhibitors of this transport system

[74] studies on pyruvate transport have given conflicting results, mainly because of technical problems (see [6,21,75] for a discussion of these problems). The following facts are, however, now established. Pyruvate transport is electroneutral and H^+-coupled. The K_m for pyruvate is 0.2–0.6 mM and the V_{max} is 40–100 nmol/min/mg mitochondrial protein at 37°C (Table 1). At high, unphysiological concentrations (above 2 mM) pyruvate is no longer only transported via the translocator but passive diffusion into the mitochondrial matrix becomes important [76]. The pyruvate translocator also mediates transport of β-hydroxybutyrate and acetoacetate [76–79] but the ketone bodies can also pass the mitochondrial membrane by diffusion if present at high concentrations [76,79]. Entry of pyruvate may be accelerated by exchange with ketone bodies which could be important for gluconeogenesis under ketogenic conditions [79,80].

α-Cyano-4-hydroxy-cinnamate has been successfully applied in metabolic studies with intact cell preparations. In epididymal fat-pads it inhibits fatty acid synthesis from glucose and fructose, but not from acetate [81]. In perfused rat heart it inhibits oxidation of glycolytic pyruvate [81]. In kidney-cortex slices gluconeogenesis from pyruvate and lactate, but not from succinate, is inhibited [81]. In isolated hepatocytes, gluconeogenesis from lactate but not from alanine is inhibited by α-cyano-4-hydroxy-cinnamate [82]; it was concluded that alanine enters the mitochondria before transamination with α-oxoglutarate, despite the fact that the bulk of the cellular alanine aminotransferase is cytosolic. The importance of mitochondrial alanine aminotransferase in alanine degradation has also been stressed by other studies [83]. Hepatic gluconeogenesis from serine is also inhibited by α-cyano-4-hydroxy-cinnamate [82,83]. It was concluded that serine enters the gluconeogenic pathway at the level of cytosolic pyruvate via serine dehydratase, a cytosolic enzyme. However, in this case the conclusion is not justified because the alternative pathway of serine degradation, via serine aminotransferase, a mitochondrial enzyme, also requires the participation of the pyruvate translocator, not only for pyruvate entry into the mitochondria to allow transamination [84], but also for transport of hydroxypyruvate, the product of transamination, to the cytosol [85].

Of practical importance in all these metabolic studies is the fact that the effectiveness of the α-cyanocinnamates decreases in the presence of albumin, presumably because they are bound to it [81].

An important question is whether mitochondrial pyruvate transport can regulate pyruvate metabolism. One way to approach this problem is to carry out careful titrations of pyruvate-dependent processes with the transport inhibitors. So far, such experiments have not been done. Inspection of the kinetic properties of the pyruvate translocator, however, shows that limitation of pyruvate metabolism by its transport into the mitochondria is possible. For liver the average reported V_{max} is 70 nmol/min/mg mitochondrial protein, which is 900 μmol/g dry weight of liver tissue/h (Table 1). The maximum rate of glucose synthesis from lactate in hepatocytes is about 430 μmol/g dry weight/h [86], so that flux through the pyruvate translocator under these conditions is $2 \times 430 = 860$ μmol/g dry weight/h, which is close to its V_{max}. In the presence of lactate plus ethanol mitochondrial pyruvate transport will certainly slow down because the cytosolic pyruvate concentration then falls to low values due to the high cytosolic reduction state with ethanol present. It may be speculated that this is, in fact, the reason for the generally observed inhibition by ethanol of gluconeogenesis from lactate.

(iii) Transport of acylcarnitine
Transport of long-chain fatty acids into the mitochondria occurs via the 1:1 exchange of fatty acylcarnitine for mitochondrial carnitine [11,12]. During exchange the total mitochondrial carnitine pool remains constant [11,12]. The exchange rate shows saturation kinetics with respect to the extramitochondrial substrate. In heart mitochondria, the K_m for extramitochondrial carnitine, in its homologous exchange with intramitochondrial carnitine, is about 5 mM [22]. The K_m of the carnitine esters is lower and decreases with increasing chain length of the fatty acid [87]. The K_m of intramitochondrial carnitine for export is also about 5 mM in heart [87] and somewhat lower in liver mitochondria (1.8 mM [88]). Since the carnitine concentration in freshly isolated mitochondria is 1–2 mM (heart [11,12]) or 0.2–0.4 mM (liver [87,89]) the acylcarnitine translocator is subsaturated with respect to the matrix carnitine concentration in these preparations. In addition, the acylcarnitine$_{out}$–carnitine$_{in}$ exchange may be influenced by competition between acylcarnitine and carnitine for binding to the mitochondrial matrix side of the translocator. Such competition (not yet shown experimentally) would prevent depletion of intramitochondrial carnitine and thereby protect intramitochondrial CoASH from the inhibitory effect of excessive acylation on β-oxidation of fatty acids, which requires both acyl-CoA and free CoA [87].

The pool-size of intramitochondrial carnitine and carnitine esters can be altered by a leak reaction in which uncompensated carnitine flux across the mitochondrial membrane occurs. This leak is presumably mediated by the carnitine translocator and is 100–200 times slower than the exchange rate [22,87,89].

Of great interest is the fact that under ketogenic conditions (fasting, alloxan diabetes, glucagon treatment) liver mitochondrial carnitine transport is accelerated because of an increased mitochondrial carnitine pool under these conditions, so that the ability of the liver to transport fatty acids into the mitochondria is increased [89].

(iv) The glutamate-aspartate translocator
Mitochondrial aspartate efflux is important in processes like urea synthesis [90], gluconeogenesis from lactate and the transport of cytosolic reducing equivalents to the mitochondria via the so-called malate-aspartate shuttle, e.g. during ethanol oxidation and gluconeogenesis from reduced substrates like glycerol, sorbitol, and xylitol [4,5]. During gluconeogenesis from lactate oxaloacetate is transported to the cytosol as aspartate to circumvent the low permeability of the mitochondrial membrane for oxaloacetate.

Since the glutamate-aspartate exchange is both electrophoretic and H^+ coupled

(Fig. 1) at equilibrium the following equation should hold [91]:

$$RT \ln \frac{(\text{glutamate/aspartate})_{in}}{(\text{glutamate/aspartate})_{out}} = -F\Delta\psi + 2.3\,RT\Delta pH$$

(in which $\Delta\psi$ is the mitochondrial membrane potential, inside minus outside, $\Delta pH = pH_{in} - pH_{out}$). Because in the intact hepatocyte under the conditions studied so far this relationship is not fulfilled it is likely that mitochondrial aspartate efflux in the intact cell is under kinetic control [92]. However, kinetic studies on the aspartate translocator have led to conflicting interpretations with regard to the activity coefficient of mitochondrial aspartate.

Murphy et al. [93] used an integrated rate equation to analyse their data on aspartate efflux from isolated mitochondria. It was assumed that the carrier is unidirectional and that intramitochondrial glutamate does not affect its activity. The observed rate of aspartate efflux was independent of pH_{in}, pH_{out} or ΔpH (see also [94]), but varied linearly with $\Delta\psi$ over a range from 70–160 mV. The K_m for intramitochondrial aspartate was 4–5 mM and was independent of the extramitochondrial glutamate concentration (which only affected V_{max}), ΔpH and $\Delta\psi$. The K_m for extramitochondrial glutamate was 6 mM (cf. [26]). Using the kinetic parameters found, together with a K_i of 4 mM for non-competitive inhibition of the glutamate$_{out}$–aspartate$_{in}$ exchange by extramitochondrial aspartate, mitochondrial aspartate efflux was calculated for hepatocytes synthesising urea in the presence of lactate, NH_3, and ornithine. This value agreed well with aspartate efflux calculated from the sum of accumulated urea and aspartate.

Whereas these results indicate that mitochondrial microcompartmentation of aspartate in the intact hepatocyte does not occur, data obtained by LaNoue et al. [94,95] led to the opposite view. According to these authors aspartate efflux is first order with respect to intramitochondrial aspartate, so that the carrier does not appear to be saturated with it. Moreover, in mitochondria treated with N-ethylmaleimide to inhibit glutamate entry via the glutamate translocator (Table 1), at 4°C, newly synthesised unlabelled aspartate (from added oxaloacetate and glutamate) did not readily mix with a preexisting small (1 nmol/mg protein) intramitochondrial pool of [^{14}C]aspartate. In submitochondrial particles with an inside-out membrane orientation, the measured K_m for external aspartate in its exchange for internal glutamate was 38 μM, which according to the authors, represents the true K_m for aspartate on the matrix side of the carrier. From these and other data it was concluded that in the mitochondrial matrix there is a diffusion gradient of aspartate with the lowest concentration near the membrane where the carrier resides. Thus, according to this view measured intramitochondrial aspartate vastly overestimates the concentration of free aspartate near the carrier [95].

Obviously, both views are incompatible. One possibility is that the diffusion gradient of matrix aspartate, observed at 4°C, does not exist at 37°C. Also, the assumption that the K_m for matrix aspartate can be measured in submitochondrial particles may be incorrect because of alterations in properties of the carrier during particle preparation. In fact, the V_{max} of aspartate transport in these particles was drastically decreased as compared to intact mitochondria [94]. On the other hand, although the equations derived by Murphy et al. [93] satisfactorily describe mitochondrial aspartate efflux under one particular set of conditions, it remains to be seen whether they also fit the data under other conditions.

It is clear that more experiments need to be done to clarify the apparent disagreement.

(v) Hormones and mitochondrial metabolite transport

Recent studies strongly indicate that mitochondrial anion translocators may be targets for hormone action.

In 1975 Garrison and Haynes [96] showed that mitochondria isolated from glucagon-treated hepatocytes metabolised added pyruvate at higher rates than normal. Later studies demonstrated that transport of pyruvate into the mitochondria was accelerated [97,98]. This is not due to a direct stimulation of the pyruvate translocator but rather due to an increased matrix pH [98]. This, in turn, is caused by activation of mitochondrial electron transport, which results in the generation of a higher proton-motive force [98].

An increase in pH gradient across the mitochondrial membrane would also affect the mitochondrial/cytosolic distribution of pyruvate and of all other anions that are H^+ coupled. Some information on this point may be extracted from data published by Siess et al. [35] who studied the effect of glucagon in hepatocytes incubated with lactate. In their experiments glucagon increased both the ratio $(pyruvate)_{mit}/(pyruvate)_{cyt}$ (from 1.68 to 2.69) and $(citrate/malate)_{mit}/(citrate/malate)_{cyt}$ (from 4.79 to 8.56), as anticipated. Interestingly, the ratio $(\alpha\text{-oxoglutarate}/malate)_{mit}/(\alpha\text{-oxoglutarate}/malate)_{cyt}$ changed little (from 1.15 to 1.39) which is consistent with the fact that the electroneutral α-oxoglutarate-malate exchange is not H^+-coupled. Contrary to the expectation the ratio $(glutamate/aspartate)_{mit}/(glutamate/aspartate)_{cyt}$ decreased slightly (from 3.03 to 2.46) by addition of glucagon. With the reservation that there is no microcompartmentation of aspartate in the mitochondria (see above), this may indicate that the aspartate translocator was under kinetic control. This is supported by the fact that glucagon caused a fall in cytosolic glutamate from 7.7 to 2.2 mM, just in the region of the K_m for the aspartate translocator, with no effect on mitochondrial aspartate [35].

According to Halestrap [98] activation of mitochondrial electron transport not only increases the proton-motive force but also the intramitochondrial ATP concentration which is important for intramitochondrial ATP utilising reactions like pyruvate carboxylation and citrulline synthesis, processes known to be activated by glucagon. Siess et al. [35], however, showed that in hepatocytes glucagon not only increased mitochondrial ATP, but also the sum of ATP, ADP and AMP at the expense of the cytosolic pool of adenine nucleotides, a phenomenon to which no attention has been paid in the literature. Exchange between ADP and ATP cannot increase the mitochondrial adenine nucleotide pool. Possibly net influx of adenine nucleotides can occur via exchange between ADP and mitochondrial phosphoenolpyruvate (see [4] for literature).

Changes in the intramitochondrial adenine nucleotide pool in rat liver are also found in the period around birth. The large increase in the rate of ADP translocation immediately after birth appears to be related to the intramitochondrial ATP content which increases about 4-fold in this period [99,100]. It is believed that at birth a short burst of hormones (adrenaline, glucagon) induces glycogenolysis, followed by enhanced glycolytic flux. This causes a spike in cytosolic ATP production which leads to an increased mitochondrial adenine nucleotide content and hence to increased rates of ADP transport so that the capacity of the mitochondria to produce ATP increases [101].

The adenine nucleotide translocator is also subject to activity changes under other conditions. According to Babior et al. [102] its activity in rat liver mitochondria depends on the thyroid status of the animal. The activity is relatively low in mitochondria from thyroidectomised rats and increases several-fold in the hyperthyroid state with intermediate values in the euthyroid state. It has been suggested that these changes are due to a change in the lipid environment of the translocator [103].

(vi) Isolation of translocators

In recent years several attempts have been made to isolate some of the translocator proteins. Major progress has been achieved with the purification of the adenine nucleotide translocator. This protein has been isolated in its native form and its molecular weight and immunological properties have been characterised [57,104]. The carrier protein from beef heart and rat liver is a dimer. The M_r of each of the subunits is 30000. In heart, the carrier protein makes up 10% of the total mitochondrial protein. The amino acid composition of the yeast protein has been determined recently [105].

The carrier has been incorporated in liposomes and its properties in the reconstituted system closely resemble those observed in the intact mitochondria [106,107].

In studies performed by Wohlrab [108] and by Hofmann and Kadenbach [109] a mitochondrial inner membrane protein with an M_r of 31000–32000 with either N-[^{3}H]ethylmaleimide [108] or [^{203}Hg]mersalyl [109] bound to it was isolated, which was claimed to be the phosphate translocator. Proof of its identity, however, can only be obtained if the protein is able to mediate phosphate transport in a reconstituted system.

Attempts to isolate proteins involved in the translocation of citrate and of glutamate have been less successful so far [6].

2. Mitochondrial cation transport

(a) H^+

The active transport of protons through the mitochondrial inner membrane is one of the central events in aerobic energy transduction. According to the chemiosmotic theory this mitochondrial membrane contains a number of enzyme complexes that catalyse (reversibly) the coupled movement of electrons from low to high potential redox components and of H^+ ions from inside to outside [110,111]. In this way chemical free energy of oxidation is converted into osmotic free energy of a proton gradient. One way to utilise the latter is by the dehydration of ADP and P_i to form ATP coupled to the influx of H^+ ions, a reaction that is catalysed (reversibly) by the H^+-ATPase complex [110,111]. Thus, the transmembrane proton gradient is an intermediate in transduction of free energy of oxidation into free energy of hydrolysis (of ATP).

As has become clear in the foregoing the free energy contained in a proton gradient across the mitochondrial membrane can also be used to drive other transport processes. In this respect, it is useful to remember that the proton gradient has a chemical and an electrical component. This is formally represented by the equation:

$$\Delta \tilde{\mu}_H = \Delta \psi - Z \Delta pH$$

in which $\Delta \tilde{\mu}_H$ denotes the difference in electrochemical potential of protons (inside-outside), $\Delta \psi$ the electrical potential across the membrane (inside-outside) and $Z\Delta pH$ the pH difference (inside-outside) normalised in such a way that each of the terms is expressed in V ($Z \approx 0.06$ at $300°K$). Changes in either the membrane potential or the pH gradient will result in changes in $\Delta \tilde{\mu}_H$. Conversely, a $\Delta \tilde{\mu}_H$ that is generated by respiration may influence processes via the pH term or via the electrical term. Examples of the different possibilities are given in Fig. 1.

To evaluate the coupling between different processes in the mitochondrial membrane it is obviously of importance to know the stoichiometry of this coupling. Thus, many attempts have been made to determine the number of H^+ ions transported per redox event or per ATP hydrolysed. Unfortunately the results of these experiments do not as yet allow unambiguous conclusions to be drawn as to the stoichiometries.

The earliest experiments to determine the H^+/O ratio (the number of protons ejected per oxygen consumed) made use of the fact that the sudden addition of small amounts of oxygen to an anaerobic suspension of mitochondria leads to rapid reversible pH changes in the medium [112]. To prevent the build-up of a large $\Delta \psi$ in such experiments valinomycin plus K^+ ions (or Ca^{2+} ions) were added. It turned out that the H^+/O ratio depended on the available substrate: it was 6 with NAD-linked substrates, 4 with succinate and 2 if electrons were fed into the respiratory chain at the level of cytochrome c. More recently, several authors have challenged these earlier results using slightly modified experimental conditions. They claim that the true H^+/O ratios may be as high as 12, 8 and 4, respectively ([113–115], but see also [116]). Especially for the part of the respiratory chain between cytochrome c and oxygen a H^+/O of greater than 2 now seems plausible [114].

Knowledge of the H^+/O ratios is of great importance for an understanding of the mechanism by which proton ejection occurs. The chemiosmotic theory proposes that the redox carriers are arranged loop-wise in the mitochondrial membrane in such a way that each time two electrons move to the external surface they take two protons with them. Since the reduction of one oxygen atom requires two electrons, a H^+/O ratio of 12 would require the presence of six loops. With the presently known components of the respiratory chain it would be hard to construct so many loops. In fact, the search for feasible loops in the terminal region of the respiratory chain led to the formulation of the so-called "Q-cycle" [117].

The H^+/ATP ratio (the number of protons ejected per ATP hydrolysed) was originally also determined by a pulse method and found to be 2 [118]. Again, later experiments indicated a higher value of 3 or 4 [119]. As discussed in Section 1f (i), in the case of ATP synthesis or hydrolysis the transport of substrates and products may be accompanied by net movement of a proton across the membrane, so that more protons are involved than those transported via the ATPase complex itself. H^+/ATP ratios have also been calculated via equations containing thermodynamic parameters; a stoichiometry of around 2.5–3 was deduced [120–122].

The implications of the different stoichiometries for the overall P/O ratios are obvious. If the proton gradient is the only and necessary intermediate in oxidative phosphorylation, the "theoretical" P/O ratio can be calculated from the H^+/O and H^+/ATP ratios. The original experiments agreed with the long-accepted P/O ratios of 3, 2 and 1 for NAD-linked substrates, succinate and cytochrome c, respectively. The more recent findings may not always agree with these accepted values [123].

(b) Ca^{2+}

Mitochondria from most sources have the capacity to accumulate relatively large amounts of Ca^{2+}, especially in the presence of permeant acids [124]. This energy-dependent uptake even takes precedence over oxidative phosphorylation. If the permeant acid forms an insoluble salt with Ca^{2+}, as for instance phosphate does, precipitates can be formed inside the mitochondria.

The energy for Ca^{2+} uptake can be derived either from substrate oxidation or from ATP hydrolysis. Since adenine nucleotides stabilise calcium phosphate precipitates the presence of adenine nucleotides stimulates Ca^{2+} uptake, even if it is respiration-driven [125]. It is generally found that 2 Ca^{2+} ions enter the mitochondrion at the expense of 1 ATP or its equivalent [124]. It is, however, less clear how many electrical charges cross the membrane during this process. Most experiments indicate that free Ca^{2+}, carrying 2 positive charges, is the mobile species [126,127]. Some experiments suggest, however, that a complex of the form $(CaX)^+$, in which X^- is a monovalent anion (for instance phosphate), moves [128]. The distinction between these two possibilities plays an important role in the discussion on the stoichiometry of the mitochondrial H^+ pump.

Physiologically, the uptake of Ca^{2+} by mitochondria may regulate the cytosolic Ca^{2+} concentration, which in turn modulates processes such as membrane permeability, muscle contraction, glycogenolysis, etc. [129]. Since the mitochondria can only function as a storage depot, it is important to know also what factors determine the efflux of Ca^{2+} from mitochondria. In this respect there appear to be several modes of Ca^{2+} efflux, different from the uptake pathway. The main evidence for such a difference lies in the insensitivity of the efflux process towards ruthenium red, an inhibitor of the uptake [130].

One mode of Ca^{2+} release is manifested in a $Ca^{2+}-nH^+$ exchange [130–132]. Experimentally this system can function bidirectionally, but presumably in the cell the extramitochondrial pH is always lower than the intramitochondrial pH so that in vivo this exchange system catalyses Ca^{2+} efflux. Similarly, the Na^+-dependent Ca^{2+} flux found in mitochondria from some tissues [129,133] will be outwardly directed in the cell, since extramitochondrial Na^+ is higher than intramitochondrial Na^+ due to the presence of a Na^+-H^+ exchange system. It is tempting to think that the Na^+-dependent pathway for Ca^{2+} flux is actually a combination of the Na^+-H^+ and the $Ca^{2+}-nH^+$ exchange systems, but this does not appear to be likely [129].

If the Ca^{2+} uptake and release systems were simultaneously operative, an energy-dissipating cycle would be set up. It appears that this is prevented through modulation of the activity of the efflux pathway. A high reduction state of the intramitochondrial nicotinamide nucleotides might inhibit and an oxidised state might activate the $Ca^{2+}-nH^+$ exchange system [126,134]. Since the redox state of the nicotinamide nucleotides is in turn regulated by the overall energetic condition of the cell, a feed-back loop would be set up in which the cytosolic phosphate potential determines the cytosolic Ca^{2+} concentration. Interestingly the distribution of Ca^{2+} between mitochondrial and cytosolic is also influenced by hormones [135,136].

(c) Monovalent cations

Mitochondria have the ability to accumulate and retain K^+ ions; indeed, in mitochondria isolated in sucrose media K^+ is the main intramitochondrial cation. In the cell, however, the main cytosolic cation is also K^+ so that concentrative uptake of this ion in vivo is impossible for osmotic reasons. A continuous slow uptake of K^+ may be required to compensate for the efflux catalysed by the presence of a low K^+-H^+ exchange activity [111]. The Na^+-H^+ exchange activity of mitochondria is much more active, as indicated by the rapid spontaneous swelling of mitochondria in isotonic sodium phosphate solution [137]. There is evidence that the activity of the intrinsic monovalent cation-transporting systems depends on the energetic state of the membrane [138]. This could result under highly energised conditions in an energy-dissipating cyclic ion transport, preventing development of extremely high proton gradients [139].

References

1 Chappell, J.B. (1968) Br. Med. Bull. 25, 150–157.
2 Klingenberg, M. (1970) in P.N. Campbell and F. Dickens (Eds.), Essays of Biochemistry, Vol. 6, pp. 119–159, Academic Press, London.

3 Fonyo, A., Palmieri, F. and Quagliariello, E. (1976) in E. Quagliariello, F. Palmieri and T.P. Singer (Eds.), Horizons of Biochemistry and Biophysics, Vol. 2, pp. 60–105, Addison-Wesley, Reading, MA.
4 Meijer, A.J. and Van Dam, K. (1974) Biochim. Biophys. Acta 346, 213–244.
5 Williamson, J.R. (1976) in R.W. Hanson and M.A. Mehlman (Eds.), Gluconeogenesis, pp. 165–220, Wiley, London.
6 LaNoue, K.F. and Schoolwerth, A.C. (1979) Annu. Rev. Biochem. 48, 871–922.
7 Stubbs, M. (1979) Pharmacol. Ther. 7, 329–349.
8 Williamson, J.R. (1976) in J.M. Tager, H.D. Söling and J.R. Williamson (Eds.), Use of Isolated Liver Cells and Kidney Tubules in Metabolic Studies, pp. 79–95, Elsevier/North-Holland, Amsterdam.
9 Klingenberg, M. and Rottenberg, H. (1977) Eur. J. Biochem. 73, 125–130.
10 LaNoue, K.F., Meijer, A.J. and Brouwer, A. (1974) Arch. Biochem. Biophys. 161, 544–550.
11 Pande, S.V. (1975) Proc. Natl. Acad. Sci. USA 72, 883–887.
12 Ramsay, R.R. and Tubbs, P.K. (1975) FEBS Lett. 54, 21–25.
13 Goodridge, A.G. (1973) J. Biol. Chem. 248, 4318–4326.
14 Lowenstein, J.M. (1968) in T.W. Goodwin (Ed.), Metabolic Roles of Citrate, pp. 61–86, Biochem. Soc. Symp. Vol. 27, Academic Press, London.
15 Meijer, A.J., Van Woerkom, G.M. and Eggelte, T.A. (1976) Biochim. Biophys. Acta 430, 53–61.
16 Söling, H.D., Kleineke, J., Willms, B., Janson, G. and Kuhn, A. (1973) Eur. J. Biochem. 37, 233–244.
17 Van Lelyveld, P.H. and Hommes, F.A. (1978) Biochem. J. 174, 527–533.
18 Sluse, F.E., Duyckaerts, C. and Liébecq, C. (1979) Eur. J. Biochem. 100, 3–17.
19 Sluse, F.E., Meijer, A.J. and Tager, J.M. (1971) FEBS Lett. 18, 149–151.
20 Coty, W.A. and Pedersen, P.L. (1974) J. Biol. Chem. 249, 2593–2598.
21 Denton, R.M. and Halestrap, A.P. (1979) Essays Biochem. 15, 37–77.
22 Ramsey, R.R. and Tubbs, P.K. (1976) Eur. J. Biochem. 69, 299–303.
23 Palmieri, F., Genchi, G., Stipani, I. and Quagliariello, E. (1977) in K. van Dam and B.F. van Gelder (Eds.), Structure and Function of Energy-Transducing Membranes, pp. 251–260, Elsevier/North-Holland, Amsterdam.
24 Palmieri, F., Prezioso, G., Quagliariello, E. and Klingenberg, M. (1971) Eur. J. Biochem. 22, 66–74.
25 Palmieri, F., Quagliariello, E. and Klingenberg, M. (1972) Eur. J. Biochem. 29, 408–416.
26 Tischler, M.E., Pachence, J., Williamson, J.R. and LaNoue, K.F. (1976) Arch. Biochem. Biophys. 173, 448–462.
27 Hoek, J.B. and Njogu, R.M. (1976) FEBS Lett. 71, 341–346.
28 Goldstein, L. and Boylan, J.M. (1978) Am. J. Physiol. 234, F514–521.
29 McGivan, J.D., Bradford, N.M. and Beavis, A.D. (1977) Biochem. J. 162, 147–156.
30 Souverijn, J.H., Huisman, L.A., Rosing, J. and Kemp, A. (1973) Biochim. Biophys. Acta 305, 185–198.
31 Elbers, R., Heldt, H.W., Schmucker, P., Soboll, S. and Wiese, H. (1974) Hoppe S. Z. Physiol. Chem. 355, 378–393.
32 Zuurendonk, P.F., Akerboom, T.P.M. and Tager, J.M. (1976) in J.M. Tager, H.D. Söling and J.R. Williamson (Eds.), Use of Isolated Liver Cells and Kidney Tubules in Metabolic Studies, pp. 17–27, Elsevier/North-Holland, Amsterdam.
33 Tischler, M.E., Hecht, P. and Williamson, J.R. (1977) Arch. Biochem. Biophys. 181, 278–292.
34 Akerboom, T.P.M., Van der Meer, R. and Tager, J.M. (1979) in Techniques in Metabolic Research, B205, pp. 1–33, Elsevier/North-Holland Biomedical Press, Shannon.
35 Siess, E.A., Brocks, D.G., Lattke, H.K. and Wieland, O.H. (1977) Biochem. J. 166, 225–235.
36 Palmieri, F., Quagliariello, E. and Klingenberg, M. (1970) Eur. J. Biochem. 17, 230–238.
37 Siess, E.A., Brocks, D.G. and Wieland, O.H. (1978) Hoppe S. Z. Physiol. Chem. 339, 785–798.
38 Soboll, S., Scholz, R., Freisl, M., Elbers, R. and Heldt, H.W. (1976) in J.M. Tager, H.D. Söling and J.R. Williamson (Eds.), Use of Isolated Liver Cells and Kidney Tubules in Metabolic Studies, pp. 29–40, Elsevier/North-Holland, Amsterdam.
39 Soboll, S., Elbers, R., Scholz, R. and Heldt, H.W. (1980) Hoppe S. Z. Physiol. Chem. 361, 69–76.
40 Klingenberg, M. (1979) in Y. Mukohota and L. Packer (Eds.), Cation Flux across Biomembranes, pp. 387–397, Academic Press, New York.

41 Duszynski, J., Savina, M.V. and Wojtczak, L. (1978) FEBS Lett. 86, 9–13.
42 Duszynski, J., Bogucka, K., Letko, H.G., Küster, U. and Wojtczak, L. (1979) in E. Quagliariello, S. Papa, F. Palmieri and M. Klingenberg (Eds.), Function and Molecular Aspects of Biomembrane-Transport, pp. 309–312, Elsevier/North-Holland, Amsterdam.
43 Brand, M.D. (1979) Biochem. Biophys. Res. Commun. 91, 592–598.
44 Akerboom, T.P.M., Bookelman, H., Zuurendonk, P.F., Van der Meer, R. and Tager, J.M. (1978) Eur. J. Biochem. 84, 413–420.
45 Soboll, S., Scholz, R. and Heldt, H.W. (1978) Eur. J. Biochem. 87, 377–390.
46 Soboll, S., Werdan, K., Bozsik, M., Müller, M., Erdmann, E. and Heldt, H.W. (1979) FEBS Lett. 100, 125–128.
47 Chance, B. and Williams, G.R. (1956) Adv. Enzymol. 17, 65–134.
48 Davis, E.J. and Lumeng, L. (1975) J. Biol. Chem. 250, 2275–2282.
49 Davis, E.J. and Davis-van Thienen, W.I.A. (1978) Biochem. Biophys. Res. Commun. 83, 1260–1266.
50 Küster, U., Bohnensack, R. and Kunz, W. (1976) Biochim. Biophys. Acta 440, 391–402.
51 Slater, E.C., Rosing, J. and Mol, A. (1973) Biochim. Biophys. Acta 292, 534–553.
52 Kunz, W., Bohnensack, R., Küster, U., Letko, G. and Schönfeld, P. (1979) in E. Quagliariello, S. Papa, F. Palmieri and M. Klingenberg (Eds.), Function and Molecular Aspects of Biomembrane-Transport, pp. 313–316, Elsevier/North-Holland, Amsterdam.
53 Holian, A., Owen, C.S. and Wilson, D.F. (1977) Arch. Biochem. Biophys. 181, 164–171.
54 Lemasters, J.J. and Sowers, A.E. (1979) J. Biol. Chem. 254, 1248–1251.
55 Wilson, D.F., Stubbs, M., Veech, R.L., Erecinska, M. and Krebs, H.A. (1974) Biochem. J. 140, 57–64.
56 Stubbs, M., Vignais, P.V. and Krebs, H.A. (1978) Biochem. J. 172, 333–342.
57 Vignais, P.V. and Lauquin, G.J.M. (1979) TIBS 4, 90–92.
58 Akerboom, T.P.M. (1979) Compartmentation of adenine nucleotides in rat hepatocytes, Ph.D. Thesis, University of Amsterdam.
59 Van der Meer, R., Akerboom, T.P.M., Groen, A.K. and Tager, J.M. (1978) Eur. J. Biochem. 84, 421–428.
60 Hoek, J.B., Nicholls, D.G. and Williamson, J.R. (1980) J. Biol. Chem. 255, 1458–1464.
61 Sies, H., Akerboom, T.P.M. and Tager, J.M. (1977) Eur. J. Biochem. 72, 301–307.
62 Kemp, A. and Out, T.A. (1974) Biochem. Soc. Trans. 2, 516–517.
63 Vignais, P.V., Vignais, P.M. and Doussiere, J. (1975) Biochim. Biophys. Acta 376, 219–230.
64 Klingenberg, M. (1979) in K. van Dam and B.F. van Gelder (Eds.), Structure and Function of Energy-Transducing Membranes, pp. 275–282, Elsevier/North-Holland, Amsterdam.
65 Raijman, L. and Bartulis, T. (1979) Arch. Biochem. Biophys. 195, 188–197.
66 Klingenberg, M. (1972) in S.G. Van den Bergh, P. Borst, L.L.M. van Deenen, J.C. Riemersma, E.C. Slater and J.M. Tager (Eds.), Mitochondria/Biomembranes, pp. 147–162, Elsevier/North-Holland, Amsterdam.
67 Stucki, J.W., Brawand, F. and Walter, P. (1972) Eur. J. Biochem. 27, 181–191.
68 Söling, H.D. and Kleineke, J. (1976) in R.W. Hanson and M.A. Mehlman (Eds.), Gluconeogenesis, pp. 369–462, Wiley, London.
69 Bryła, J. and Niedźwiecka, A. (1979) Int. J. Biochem. 10, 235–239.
70 Hansford, R.G. (1977) J. Biol. Chem. 252, 1552–1560.
71 Akerboom, T.P.M., Bookelman, H. and Tager, J.M. (1977) FEBS Lett. 74, 50–54.
72 Barbour, R.L. and Chen, S.H.P. (1979) Biochem. Biophys. Res. Commun. 89, 1168–1175.
73 Papa, S., Francavilla, A., Paradies, G. and Meduri, B. (1971) FEBS Lett. 12, 285–288.
74 Halestrap, A.P. and Denton, R.M. (1974) Biochem. J. 138, 313–316.
75 Vaartjes, W.J., Geelen, M.J.H. and Van den Bergh, S.G. (1979) Biochim. Biophys. Acta 548, 38–47.
76 Pande, S.V. and Parvin, R. (1978) J. Biol. Chem. 253, 1565–1573.
77 Mowbray, J. (1975) Biochem. J. 148, 41–47.
78 Paradies, G. and Papa, S. (1975) FEBS Lett. 52, 149–152.
79 Halestrap, A.P. (1978) Biochem. J. 172, 377–387.

80 Scholz, R., Olson, M.S., Schwab, A.J., Schwabe, U., Noell, C. and Braun, W. (1978) Eur. J. Biochem. 86, 519–530.
81 Halestrap, A.P. and Denton, R.M. (1975) Biochem. J. 148, 97–106.
82 Mendes-Mourão, J., Halestrap, A.P., Crisp, D.M. and Pogson, C.I. (1975) FEBS Lett. 53, 29–32.
83 Dieterle, P., Brawand, F., Moser, U.K. and Walter, P. (1978) Eur. J. Biochem. 88, 467–473.
84 Snell, K. (1975) FEBS Lett. 55, 202–205.
85 Kitagawa, Y. and Sugimoto, E. (1979) Biochim. Biophys. Acta 582, 276–282.
86 Krebs, H.A., Lund, P. and Stubbs, M. (1976) in R.W. Hanson and M.A. Mehlman (Eds.), Gluconeogenesis, pp. 269–291, Wiley, London.
87 Tubbs, P. and Ramsay, R. (1979) in E. Quagliariello, S. Papa, F. Palmieri and M. Klingenberg (Eds.), Function and Molecular Aspects of Biomembrane-Transport, pp. 279–286, Elsevier/North-Holland, Amsterdam.
88 Pande, S.V. and Parvin, R. (1979) in E. Quagliariello, S. Papa, F. Palmieri and M. Klingenberg (Eds.), Function and Molecular Aspects of Biomembrane-Transport, pp. 287–290, Elsevier/North-Holland, Amsterdam.
89 Parvin, R. and Pande, S.V. (1979) J. Biol. Chem. 254, 5423–5429.
90 Meijer, A.J., Gimpel, J.A., DeLeeuw, G., Tischler, M.E., Tager, J.M. and Williamson, J.R. (1978) J. Biol. Chem. 253, 2308–2320.
91 Rottenberg, H. (1976) Arch. Biochem. Biophys. 173, 461–462.
92 Williamson, J.R. and Viale, R.O. (1979) in S. Fleischer and L. Packer (Eds.), Methods in Enzymology, Vol. 56, pp. 252–278, Academic Press, New York.
93 Murphy, E., Coll, K.E., Viale, R.O., Tischler, M.E. and Williamson, J.R. (1979) J. Biol. Chem. 254, 8369–8376.
94 Duszynski, J., Mueller, G. and LaNoue, K.F. (1978) J. Biol. Chem. 253, 6149–6157.
95 LaNoue, K.F., Duszynski, J., Watts, J.A. and McKee, E. (1979) Arch. Biochem. Biophys. 195, 578–590.
96 Garrison, J.C. and Haynes, Jr., R.C. (1975) J. Biol. Chem. 250, 2769–2777.
97 Titheradge, M.A. and Coore, H.G. (1976) FEBS Lett. 63, 45–50.
98 Halestrap, A.P. (1978) Biochem. J. 172, 389–398.
99 Nakazawa, T., Asami, K., Suzuki, H. and Yukawa, O. (1973) J. Biochem. (Tokyo) 73, 397–406.
100 Pollak, J.K. (1975) Biochem. J. 150, 477–488.
101 Sutton, R. and Pollak, J.K. (1980) Biochem. J. 186, 361–367.
102 Babior, B.M., Creagan, S., Ingbar, S.H. and Kipnes, R.S. (1973) Proc. Natl. Acad. Sci. USA 70, 98–102.
103 Hoch, F.L. (1977) Arch. Biochem. Biophys. 178, 535–545.
104 Klingenberg, M. (1979) TIBS 4, 249–252.
105 Brandolin, G., Lauquin, G.J.M., Jollès, J., Jollès, P. and Vignais, P.V. (1979) FEBS Lett. 98, 161–164.
106 Kramer, R. and Klingenberg, M. (1977) FEBS Lett. 82, 363–367.
107 Schertzer, H.G. and Racker, E. (1976) J. Biol. Chem. 251, 2446–2452.
108 Wohlrab, H. (1979) Biochemistry 18, 2098–2102.
109 Hofmann, H.D. and Kadenbach, B. (1979) Eur. J. Biochem. 102, 605–613.
110 Mitchell, P. (1961) Nature 191, 144–148.
111 Mitchell, P. (1968) in Chemiosmotic Coupling and Energy Transduction, Glynn Research, Bodmin.
112 Mitchell, P. and Moyle, J. (1967) Biochem. J. 105, 1147–1162.
113 Brand, M.D. (1979) Biochem. Soc. Trans. 7, 874–880.
114 Wikström, M. and Krab, K. (1979) Biochem. Soc. Trans. 7, 880–887.
115 Rossi, E. and Azzone, G.F. (1969) Eur. J. Biochem. 7, 418–426.
116 Mitchell, P. and Moyle, J. (1979) Biochem. Soc. Trans. 7, 887–894.
117 Mitchell, P. (1977) Annu. Rev. Biochem. 46, 996–1005.
118 Mitchell, P. and Moyle, J. (1965) Nature 208, 147–151.
119 Brand, M.D. and Lehninger, A.L. (1977) Proc. Natl. Acad. Sci. USA 74, 1955–1959.

120 Van Dam, K., Wiechmann, A.H.C.A. and Hellingwerf, K.J. (1977) Biochem. Soc. Trans. 5, 485–487.
121 Nicholls, D.G. (1977) Biochem. Soc. Trans. 5, 200–203.
122 Azzone, G.F., Pozzan, T. and Massari, S. (1978) Biochem. Biophys. Acta 501, 307–316.
123 Hinkle, P.C. and Yu, M.L. (1979) J. Biol. Chem. 254, 2450–2455.
124 Lehninger, A.L., Carafoli, E. and Rossi, C.S. (1967) Adv. Enzymol. 9, 259–320.
125 Carafoli, E., Rossi, C.S. and Lehninger, A.L. (1965) J. Biol. Chem. 240, 2254–2261.
126 Fiskum, G., Reynafarje, B. and Lehninger, A.L. (1979) J. Biol. Chem. 254, 6288–6295.
127 Deana, R., Arrabaca, J.D., Mathieu-Shire, Y. and Chappell, J.B. (1979) FEBS Lett. 106, 231–234.
128 Moyle, J. and Mitchell, P. (1977) FEBS Lett. 77, 136–140.
129 Nicholls, D.G. (1978) Biochem. J. 170, 511–522.
130 Fiskum, G. and Lehninger, A.L. (1979) J. Biol. Chem. 254, 6236–6239.
131 Fiskum, G. and Cockrell, R.S. (1978) FEBS Lett. 92, 125–128.
132 Bernardi, P. and Azzone, G.F. (1979) Eur. J. Biochem. 102, 555–562.
133 Crompton, M., Moser, R., Lüdi, H. and Carafoli, E. (1978) Eur. J. Biochem. 82, 25–31.
134 Lehninger, A.L., Vercesi, A. and Bababunmi, E.A. (1978) Proc. Natl. Acad. Sci. USA 75, 1690–1694.
135 Babcock, D.F., Chen, J-L.J., Yip, B.P. and Lardy, H.A. (1979) J. Biol. Chem. 254, 8117–8120.
136 Blackmore, P.F., Dehaye, J.-P. and Exton, J.H. (1979) J. Biol. Chem. 254, 6945–6950.
137 Mitchell, P. and Moyle, J. (1969) Eur. J. Biochem. 9, 149–155.
138 Brierley, G.P., Settlemire, C.T. and Knight, V.A. (1968) Arch. Biochem. Biophys. 126, 276–288.
139 Nicholls, D.G. (1977) Eur. J. Biochem. 77, 349–356.

CHAPTER 10

Transport across bacterial membranes

WIL N. KONINGS, KLAAS J. HELLINGWERF and GEORGE T. ROBILLARD *

*Department of Microbiology, Biological Centre, University of Groningen, Kerklaan 30, 9751 NN Haren, and * Department of Physical Chemistry, University of Groningen, Nijenborgh 16, 9747 AG Groningen, The Netherlands*

1. Introduction

The cytoplasm of bacteria is surrounded by a cell envelope which is composed of several layers. The inner layer, the cytoplasmic membrane, is in direct contact with the cytoplasm. It consists of a liquid-crystalline bilayer of phospholipids in which proteins are embedded.

Exterior of the cytoplasmic membrane lies the cell wall which is mainly composed of peptidoglycan, a network of polysaccharides cross-linked by short peptides (Fig. 1).

In Gram-negative organisms a third layer, the outer membrane, surrounds the cell wall. This outer membrane consists of phospholipids and lipopolysaccharides in which proteins are embedded.

The functional properties of the three layers of the cell envelope of Gram-negative bacteria are distinctly different. The cytoplasmic membrane is the main borderline between the cytoplasm and the cell's surroundings. It forms a diffusion barrier through which most solutes can pass only via specific translocation systems. Energy-transducing systems such as electron transfer systems and the ATPase complex are also located in this cytoplasmic membrane. The cell wall is a rigid layer which determines the volume and the shape of the cell. Due to the rigid properties of the cell wall the cell can resist the osmotic pressure of the cytoplasm which is several-fold higher than of the environment. Most solutes and ions which are used by the bacteria can freely diffuse through the pores of the network of the cell wall. The outer membrane of Gram-negative bacteria forms, just like the cytoplasmic membrane, a barrier for solutes. This membrane differs from the cytoplasmic membrane by the presence of proteins which form non-specific hydrophilic pores through which solutes up to 600 M_r can penetrate. The area located between the cytoplasmic membrane and the outer membrane is termed the periplasmic space. It contains several proteins, some involved in the perception of chemotactic stimuli and others in the transport of solutes across the cell envelope. Because the cytoplasmic membrane is the most important diffusion barrier of the cell envelope, insight into

Bonting/de Pont (eds.) Membrane transport
© *Elsevier/North-Holland Biomedical Press, 1981*

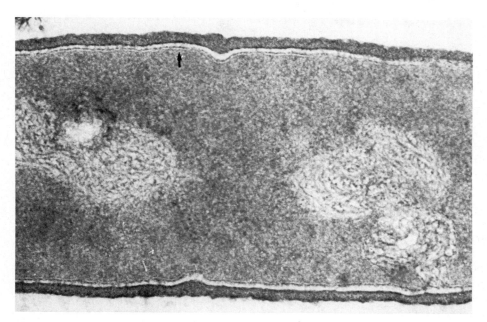

Fig. 1. Structure of the envelope of a Gram-positive bacterium. *Bacillus subtilis* cells were fixed in glutaraldehyde and OsO_4^-, embedded in epon and thin-sectioned according to standard procedures. The outermost structure is the cell wall; the cell membrane is indicated by the arrow. Magnification: approx. 100 000-fold.

the mechanism of solute transport by bacteria requires knowledge about the mechanisms of solute translocation across the cytoplasmic membrane. Many solutes are translocated against their concentration gradient for which metabolic energy is needed. An important aspect of solute transport, therefore, is its coupling to the cell's energy-transducing machinery. In the last decade significant progress has been made in this field of research. This progress has been initiated by two major developments. Firstly, the isolation of cytoplasmic membrane preparations which form closed structures (membrane vesicles), retaining the functional properties of the cytoplasmic membrane in vivo [1]. The availability of these vesicles opened the possibility to study translocation processes without the interference of cytoplasmic (or periplasmic) constituents. Secondly, experimental support for the chemiosmotic hypothesis [2] led to the realisation that energy-transducing electron transfer and ATPase systems are special forms of specific transport systems. This strongly influenced the perception of the transport phenomena and consequently led to new experimental approaches. In this chapter we shall discuss these new developments without trying to review the experimental data. These have been reviewed extensively in recent years and the reader is directed to these reviews for more detailed information [3–5].

2. The chemiosmotic concept

A very elegant explanation for the mechanism of coupling between energy-generating and energy-consuming processes in the cytoplasmic membrane has been offered by the chemiosmotic theory [5,7]. It was postulated that the energy-transducing systems like electron-transfer systems, ATPase complexes and solute transport systems, which are incorporated anisotropically in the membrane, act as electrogenic proton pumps. In the cytoplasmic membrane of bacteria these proton pumps translocate protons from the cytoplasmic side of the membrane to the external medium. The cytoplasmic membrane is practically impermeable to ions and in particular to protons and hydroxyl ions. Proton translocation by these electrogenic pumps will therefore result in the generation of two gradients. A pH-gradient (chemical gradient of protons; ΔpH) is formed as a result of alkalinization of the cytoplasm due to removal of protons and acidification of the external medium due to accumulation of protons. An electrical potential ($\Delta\psi$) is formed because positive charges are removed from the cytoplasm and accumulated in the external medium.

Both gradients exert an inwardly directed force on the protons, the proton-motive force ($\Delta\tilde{\mu}_{H^+}$):

$$\Delta\tilde{\mu}_{H^+} = \Delta\psi - Z\Delta pH$$

in which $Z = 2.3\ RT/F$; R is the gas constant, T the absolute temperature and F the Faraday constant. Z has a numerical value of about 60 mV at 25°C. The chemiosmotic hypothesis thus visualizes the energy-transducing systems in the membrane as primary transport systems which convert chemical or light energy into electrochemical energy.

A second aspect of the chemiosmotic coupling theory postulates that the proton-motive force (pmf) drives energy-consuming processes in the membrane by a reversed flow of protons [8] (Fig. 2). The energy of $\Delta\tilde{\mu}_{H^+}$ is thus either converted into ATP by a reversed action of the ATPase complex, or drives osmotic work such as the formation of solute gradients by secondary transport or drives mechanical work such as flagellar movements.

According to the chemiosmotic concept therefore the central intermediate between the energy-transducing processes in the cytoplasmic membrane is the electrochemical potential gradient of protons. pmf-producing and pmf-consuming processes found in bacteria are listed below.

pmf-generating processes	pmf-driven processes
cytochrome-linked electron-transfer	transhydrogenase reaction
ATP-hydrolysis	ATP-synthesis
bacteriorhodopsin–light interaction	secondary solute transport
efflux of metabolic end products	flagellar movement

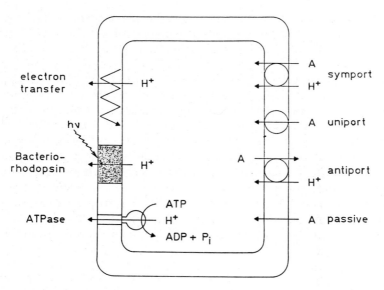

Fig. 2. Schematic presentation of primary and secondary transport systems in bacteria.

3. Energy-transducing systems

Aspects of four energy-transducing systems which can play an important role in bacterial energy metabolism will be discussed: electron-transfer systems, the Ca^{2+}, Mg^{2+}-activated ATPase complex, bacteriorhodopsin and secondary solute transport systems.

(a) Cytochrome-linked electron-transfer systems

A wide variety of different cytochrome-linked electron-transfer systems is encountered in bacteria: respiratory chains with oxygen, nitrate or sulphate as electron acceptors, fumarate reductase systems and light-driven cyclic electron-transfer systems (Fig. 3). All these systems are composed of several electron-transfer carriers, the nature of which varies considerably in different organisms. Electron carriers which are most common in bacterial electron-transfer systems are flavoproteins (dehydrogenases), quinones, non-heme iron centres, cytochromes and terminal oxidases and reductases. One common feature of all electron-transfer systems is that they are tightly incorporated in the cytoplasmic membrane. Another important general property of these systems is that electron transfer results in the translocation of protons from the cytoplasm into the external medium. Electron transfer therefore

Transport across bacterial membranes

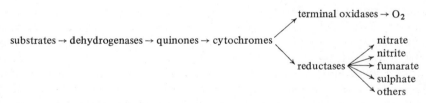

Fig. 3. Generalized scheme of electron-transfer systems in bacteria.

leads to the generation of a proton-motive force, internally alkaline and negative.

The mechanism of proton translocation by the electron-transfer systems is at this moment hardly understood. Mitchell [9] postulated the redox loop model. The electron carriers are located alternately at the inner surface and the outer surface of the cytoplasmic membrane thus forming loops in the electron transfer chain (Fig. 4A). Each loop corresponds to a coupling site as postulated by the chemical hypothesis of energy coupling [10]. In each loop two hydrogen atoms move from the inner side of the membrane outward through a hydrogen carrier (for instance a quinone) (Fig. 4A). Subsequently two electrons pass back to the inside through an electron carrier (for instance a cytochrome), and two protons are liberated into the external medium. Per redox loop therefore two protons are translocated from the cytoplasm into the external medium per two electrons transferred. The total number of protons that will be translocated per two electrons transferred by the electron transfer chain will therefore depend on the number of redox loops. The loop concept offers a simple explanation for electron-transfer coupled proton translocation.

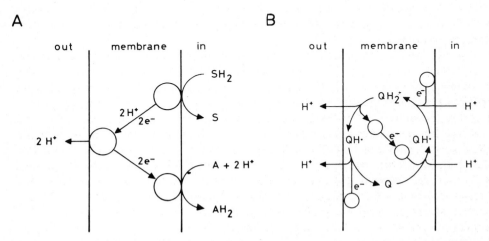

Fig. 4. Scheme of proton translocation during electron transfer via (A) a loop mechanism; (B) a proton-motive Q cycle. SH_2 and A represent the electron donor and acceptor, respectively. 1, 2 and 3 are components of the electron-transfer system. Q represents quinone, circles represent electron carriers.

However, some of the properties of electron carriers (such as their observed redox potentials) do not fit in such a simple loop model. This has led Mitchell [11] to propose a modified mechanism, the so-called proton-motive Q cycle (Fig. 4B). In this model quinones function in two separate reactions in the $QH_2/QH\cdot$ and the $Q/QH\cdot$ couple. These couples have different midpoint redox potentials and would operate at the reducing and the oxidizing site of cytochrome b, respectively. During these reactions proton translocation is supposed to occur by diffusion of the quinones in the fully oxidized (Q) and fully reduced (QH_2) forms through the hydrophobic environment between their successive reaction sites at both sides of the membrane. Recently some experimental support for such a role of quinones has been obtained. Alternative models which will not be discussed here, have been postulated by Papa [12] and Williams [13]. Currently there is no conclusive support for a specific model.

The number of protons that are translocated across the membrane per 2 electrons transferred (the $H^+/2e$ stoicheiometry) is the subject of turbulent discussion and evidence for stoichiometries of 3 or 4 have been presented [14].

The maximal value that the proton-motive force can reach is limited by the redox potential difference ($\Delta E_0'$) between the electron donor and acceptor and by the total number of protons (n) translocated per two electrons transferred:

$$\Delta \tilde{\mu}_{H^+} \leq \frac{2}{n} \cdot \Delta E_0'$$

This relationship holds when all electron carriers of the electron-transfer chain participate in proton translocation and when the redox potential difference of all proton translocating sites is the same. This equation allows us to calculate the maximum proton-motive force that can be generated. In Table 1 this proton-motive force has been calculated for three different electron transfer systems and for three different n-values.

The available information about the proton-motive force in aerobically grown cells indicates that this value is usually between -200 and -250 mV. Proton-motive force values of around -200 mV have also been found during nitrate respiration. In cells carrying out fumarate reduction the values of the proton-motive force appear to be lower (around -100 mV). The number of protons which are translocated during electron transfer are not known with certainty. In all three systems translocation of 6 protons per two electrons during NADH-oxidation would be sufficient to generate the observed proton-motive force. The proton-motive force that can be maintained in a cell will depend on the activities of proton-motive force-generating and proton-motive force-dissipating processes. Of these processes are especially the passive secondary transport processes (Fig. 2) beyond control of the organism. The uptake of cations (H^+, K^+, Na^+) and the efflux of anions (such as anionic pool constituents) will increase with increasing proton-motive force. In order to maintain the internal environment as constant as possible bacteria have developed transport systems which extrude cations such as Na^+ and Ca^{2+} by a countertransport

TABLE 1

Theoretically maximal values of the proton-motive force that can be generated with NADH as electron donor

Electron transfer system	$\Delta E'_0$ (mV)	Maximal value of the proton-motive force for n is:		
		2	4	6
NADH to oxygen	1138	1138	569	379
NADH to nitrate	753	753	376	251
NADH to fumarate	353	353	176	118

(antiport) with protons. In a scheme as shown in Fig. 5 the uptake of one proton is required for every sodium ion removed. Bacteria have also developed secondary transport systems which transport metabolic intermediates (such as glucose-6-phosphate, succinate, etc.) from the medium back to the cytoplasm. These and other processes will result in the uptake of protons and/or charge. As a consequence the steady state proton-motive force will seldom be in thermodynamic equilibrium with the redox potential difference or with the phosphate potential (see below).

This also explains why respiratory control, i.e. control of electron flow by the proton-motive force, is rarely observed in bacteria. In contrast, respiratory control is found in organelles (such as mitochondria, chloroplasts) which are surrounded by an almost constant environment.

(b) The ATPase complex

The ATPase complex is a primary proton pump present in energy-transducing membranes of many different bacteria. This complex can convert the free energy of the proton-motive force into free energy necessary for ATP synthesis in a coupled process in which protons are translocated from the external medium into the cytoplasm. Alternatively it catalyzes the reversed process in which ATP is hydrolyzed and protons are extruded from the cytoplasm into the external medium (Fig. 6). The ATPase complex consists of two distinct groups of polypeptides. One group (F_1) can be removed from the membrane easily by a low ionic strength washing; the other group (F_0) forms an integral part of the membrane and can only be isolated after solubilization of the membrane [15]. The F_1 part of the ATPase complex is composed of at least 5 different subunits, has an M_r of about 360000 and contains the catalytically active site. The F_0 part is composed of 3 to 5 subunits and functions as a proton-conducting pathway through the membrane [15]. This proton channel can be blocked by specific inhibitors (like DCCD). ATP hydrolysis is also subjected to the action of these inhibitors in the membrane-bound intact ATPase complex. Many ATPase complexes contain a small subunit which acts as an inhibitor of the catalytic activity. It has been suggested that this inhibitor functions as a check valve, to prevent wasteful splitting of ATP.

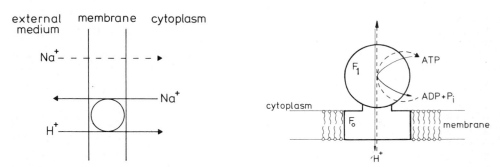

Fig. 5. Entrance of Na$^+$ by passive secondary transport and extrusion by Na$^+$/H$^+$-antiport.

Fig. 6. Schematical drawing of a membrane-bound energy-transducing ATPase complex.

Much of the information on the structure and function of ATPase complexes has been derived from reconstitution studies of isolated complexes [15] and their subunits and from studies on bacterial unc (uncoupled) mutants which lack a functional ATPase complex [16]. During ATP hydrolysis by the intact ATPase complex, protons are extruded from the complex via the F$_0$ part. If the complex is properly embedded in a membrane ATP hydrolysis will lead to the generation of a proton-motive force. Two protons are probably extruded per ATP hydrolysed but H$^+$/ATP stoicheiometries of 4 have also been reported. The mechanism by which the high-energy phosphate bond between ADP and P$_i$ is formed from the free energy of proton transfer is still completely unclear. Two mechanisms have been proposed [17]: the first, advanced by Mitchell, postulates that the protons attack one of the phosphate oxygens, to form a molecule of water thus converting the phosphate group to a high-energy form which can react directly with ADP. The second, proposed by Boyer, envisions the role of protons as a means of changing the conformation of the F$_1$. Due to this change in conformation ATP, spontaneously formed from ADP and P$_i$ in an hydrophobic environment, can be released.

The ATPase complex catalyzes the following reaction:

$$\text{ATP}^{4-} + \text{H}_2\text{O} + n\text{H}^+_{\text{in}} \rightleftharpoons \text{ADP}^{3-} + \text{P}_i^- + n\text{H}^+_{\text{out}}$$

The maximal $\Delta\tilde{\mu}_{\text{H}^+}$ that thermodynamically can be generated by this reaction is determined by the phosphate potential ($\Delta G'_{\text{ATP}}$) and the value of n:

$$\Delta\tilde{\mu}_{\text{H}^+} = \frac{1}{n}\Delta G_{\text{ATP}}$$

in which

$$\Delta G'_{\text{ATP}} = \Delta G'_0 - 2.3\frac{RT}{F}\log\frac{(\text{ATP})}{(\text{ADP})\cdot(\text{P}_i)}\text{ mV}$$

($\Delta G_0'$ = free energy under standard conditions; (P_i) = free inorganic phosphate concentration in cell). The free inorganic phosphate concentration is usually around 0.01 M. At 25°C and an internal pH of 8, the $\Delta G_0' = -340$ mV. Under these conditions $\Delta G'_{ATP}$ will be:

$$\Delta G'_{ATP} = -460 - 60 \log (ATP)/(ADP) \text{ mV}$$

When the H^+/ATP stoichiometry equals 2 the equilibrium value of $\Delta \tilde{\mu}_{H^+}$ that can be reached will be: $\Delta \tilde{\mu}_{H^+} = -230 - 30 \log(ATP)/(ADP)$. This relationship is shown in Fig. 7. This figure shows that under conditions of low $\Delta \tilde{\mu}_{H^+}$ and high (ATP)/(ADP) a $\Delta \tilde{\mu}_{H^+}$ can be generated by ATP-hydrolysis, while under conditions of low (ATP)/(ADP) and high $\Delta \tilde{\mu}_{H^+}$ ATP-synthesis may occur. An equilibrium between the phosphate potential and the proton-motive force will only be reached when no other factors regulate the ATPase activity. Such a regulating factor (ATPase inhibitor protein) has been reported for *E. coli* but it is not clear at this moment whether such regulation mechanisms are present in other organisms too.

(c) Bacteriorhodopsin

Bacteriorhodopsin is found in differentiated regions of the cytoplasmic membrane of several *Halobacteria*. These so-called purple membranes are composed of two major lipid species (25% by weight) and one protein component which contains retinal (the aldehyde form of vitamin A) as a covalently bound prosthetic group. The retinal is bound via a protonated Schiff-base linkage to a lysine residue of the protein moiety of bacteriorhodopsin and absorbs light with a maximum absorbance at 570 nm [18].

Reconstitution of the proton pump function of bacteriorhodopsin by incorpora-

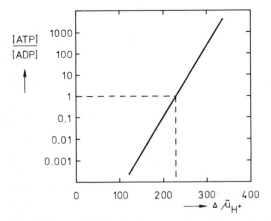

Fig. 7. Theoretical relation between the proton-motive force and the ATP/ADP ratio in a cell under conditions where the reaction catalyzed by the ATPase complex is in equilibrium.

tion of this protein into phospholipid liposomes has been performed and the results of these studies convincingly demonstrated the proton-pump function of bacteriorhodopsin [19].

At this moment a large amount of information is available on the structure of bacteriorhodopsin [20] and it is very likely that this proton pump will be the first for which a detailed molecular description of its catalytic activity can be given. The amino acid sequence, the folding of the polypeptide chain and the arrangement of the protein molecules in the purple membrane are known [21,22]. The protein consists of 7 α-helices, each spanning the membrane. The retinal group is located in a hydrophobic pocket, about 15 Å from the extracellular surface of the membrane.

Upon light absorption a cyclic sequence of reactions takes place during which the transmembrane transfer of one proton is catalyzed. At least 6 intermediates with different absorption maxima can be discriminated during this photocycle.

The Schiff-base proton plays an important role in the mechanism of proton translocation; during the cycle a deprotonation and reprotonation of the nitrogen atom of the Schiff-base linkage occurs.

Various models for the mechanism of proton translocation by bacteriorhodopsin have been postulated. The essential features of these models are: in the ground state the C_6-ring of the all-*trans* retinal group is tightly bound in a hydrophobic pocket

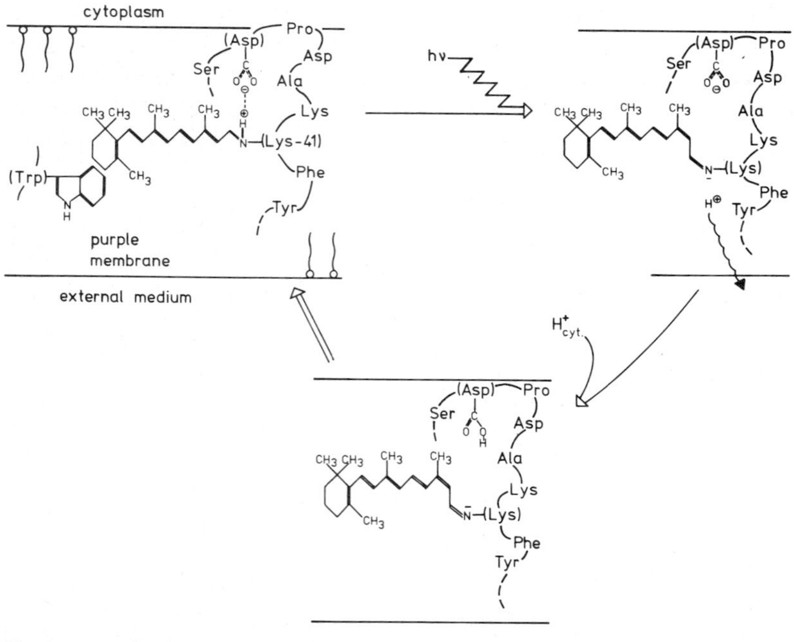

Fig. 8. Tentative scheme of the mechanism of the catalytic activity of bacteriorhodopsin. H_{cyt}^+ indicates that reprotonation during the photocycle of bacteriorhodopsin occurs from the cytoplasmic side. A proton is released first at the extracellular side of the membrane.

and the "tail" of the retinal group is approximately linear. The protonated Schiff-base, linking the retinal to lysine-41 of the flexible part of the polypeptide backbone, is positively charged and forms a salt bridge with a negatively charged amino residue. Upon light absorption the retinal group isomerizes (most probably to the 13-*cis* form) and consequently the Schiff-base moves a few Ångströms to the extracellular surface. As a result the salt-bridge is disrupted, the protonated Schiff-base dissociates its proton to the extracellular surface and the negatively charged amino acid is reprotonated from the cytoplasmic side (Fig. 8). In its deprotonated form, the retinal group is able to re-isomerize to the all-*trans* form and the catalytic cycle is completed by a transfer of the proton from the protonated amino acid residue to the Schiff-base linkage and a reformation of the salt-bridge. Many details of the catalytic mechanism of the bacteriorhodopsin still have to be resolved (like the assignment of distinct conformational states to the different intermediates). Possibly, the electrical measurements on purple membranes [24] which show charge displacements in the protein during the occurrence of the different photointermediates can be of use in this assignment. The information obtained from the bacteriorhodopsin system may also be useful in the elucidation of the mechanism of other proton pumps.

4. Solute transport

Solute transport across the cytoplasmic membrane of bacteria occurs by two major mechanisms: (i) Secondary transport systems; transport by these systems is driven by electrochemical gradients and will lead to the translocation of solute in unmodified form; (ii) Group translocation: solute is substrate for a specific enzyme system in the membrane; the enzyme reaction results in a chemical modification of the solute and release of the products at the cytoplasmic side. The only well-established group translocation system is the phosphoenolpyruvate phospho-transferase system (PTS) (see below).

Most solutes are translocated across the cytoplasmic membrane by secondary transport either passively, without the involvement of specific membrane proteins or facilitated by specific carrier proteins. This latter mechanism is often termed "active transport" (see Fig. 2).

Mitchell [7] visualized three different systems for facilitated secondary transport. "Uniport": only one solute is translocated by the carrier protein. "Symport": two or more different solutes are translocated in the same direction by the carrier protein. "Antiport": two or more different solutes are translocated by one carrier in opposite directions.

The driving forces for solute transport will depend on the overall charge and on the protons which are translocated, as well as on the solute gradient. The driving force is therefore composed of components of the proton-motive force and of the solute gradient, and translocation of solute will proceed until the total driving force is zero. At that stage a steady-state level of accumulation is reached. Fig. 9 shows

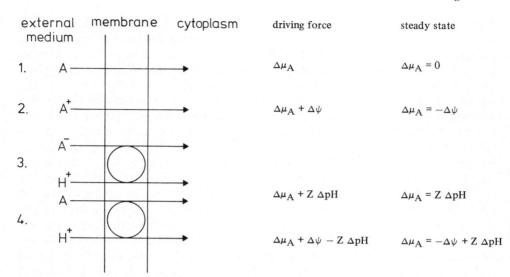

Fig. 9. Schematic presentation of four secondary transport processes. (1) Passive transport of a neutral solute; (2) passive transport of a cation; (3) facilitated transport of an anion in symport with one proton; (4) facilitated transport of a neutral solute in symport with one proton. For each process the driving force and the steady state accumulation level of the solute are indicated.

schematically four different translocation processes. In example 1 transport of a neutral solute via passive secondary transport is shown. The driving force of this transport will be supplied only by the solute gradient and will be equal to $\Delta\mu_A = A \log A_{in}/A_{out}$. A steady state will be reached when the solute gradient is dissipated ($\Delta\mu_A = 0$) and the internal solute concentration equals the external concentration. Example 2 shows transport of a monovalent positively charged solute by passive secondary transport. This transport will not only depend on the chemical concentration gradient of solute but also on the electrical potential. The driving force therefore will be $\Delta\mu_A = Z \log A_{in}/A_{out} + \Delta\psi$ and a steady state will be reached when $Z \log A_{in}/A_{out} = -\Delta\psi$. Unless the electrical potential is dissipated by movement of other ions across the membrane in the steady state, the internal concentration of solute will not be equal to the external concentration. In the absence of active primary transport systems net transport will already stop when the internal solute concentration is lower than the external concentration. However, when a $\Delta\psi$ (interior negative) is generated by other systems accumulation of solute can occur.

In a way similar to that shown above the driving forces and steady state levels of transport can be derived for other transport processes. This is done in example 3 for a symport system by which a negatively charged solute is translocated together with one proton. The overall charge during this translocation process is zero and due to the translocation of one proton only the ΔpH will contribute to the driving force of this translocation process. In example 4 the driving force is determined for a neutral solute symported with one proton. In this process charge and protons are translocated and consequently the total proton-motive force ($\Delta\psi$ and ΔpH) contributes to the driving force. In the transport processes 2, 3 and 4 the level of accumulation of solute will vary with the driving force supplied by the proton-motive force. In order to maintain the proton-motive force (interior negative and alkaline) during transport of solute continuous proton extrusion by primary transport systems has to occur. In the absence of such activity the translocation of solute will lead to a dissipation of the proton-motive force.

When the proton-motive force is maintained at a certain value by primary transport systems the level of accumulation of a solute will depend on the mechanism involved in the translocation. For instance, when the total proton-motive force is -180 mV, composed of a $\Delta\psi$ of -120 mV and a ΔpH of -60 mV the steady-state level of accumulation of solute A in example 2 will be 100-fold ($-60 \log A_{in}/A_{out} = -120$); in 3 10-fold ($-60 \log A_{in}/A_{out} = -60$) and in 4 1000-fold ($-60 \log A_{in}/A_{out} = -180$). Variations of the $\Delta\tilde{\mu}_{H^+}$ therefore affect the accumulation of solutes differently.

In 2, 3 and 4 a decrease of the driving force supplied by the proton-motive force will lead to a decreased steady-state solute accumulation and a decreased rate of accumulation. Because essentially all solutes are to a certain extent membrane permeable, in the absence of a proton-motive force leakage of internal metabolites will occur until the (electro)chemical potential gradients have all been dissipated. Ultimately, this leakage of internal metabolites will lead to the death of the organism.

5. Carriers for facilitated secondary transport

Facilitated secondary transport of solutes across the cytoplasmic membranes of bacteria is mediated by specific proteins, the carriers (previously termed permeases). Cytoplasmic membranes contain usually many of these carrier proteins, each having affinity for only one solute or a group of structurally related solutes.

Two general classes of molecular mechanisms have been proposed to account for the carrier-mediated transport processes: (i) "Mobile carrier model" in which the solute binds to a specific site in the carrier protein on one side of the membrane and is transferred across the membrane by the oscillation of the transport component, and (ii) the "Transport-channel model" in which the solute is translocated across the membrane via transport channels formed possibly by oligomeric subunit aggregates of the transport components (Fig. 10). In this latter model the binding of solute to a specific site on the channel protein on one side of the membrane results in some conformational change of the transport components which subsequently lead to the transfer of the solute (together with the solute recognition site) to the inner surface of the membrane. It is impossible to distinguish between these two transport models by kinetic studies of the transport systems. The biochemical information currently available [25] about some transport systems favors the transport channel model.

Carriers usually have a high affinity for their solutes. K_s-values range between

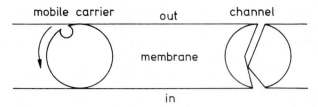

Fig. 10. Simplified mechanisms for a transport protein of the "mobile-carrier"- and the "transport-channel"-type, respectively.

10^{-5} and 10^{-6} M, but affinity constants as low as 10^{-8} M have also been reported.

Secondary facilitated systems have been demonstrated for a wide variety of solutes (for review see [4]): (i) Amino acids: in a number of bacteria transport of amino acids is mediated by at least 9 distinct carrier proteins. Transport systems for amino acids have also been demonstrated in bacteria which hardly utilize amino acids, like *Thiobacillus neapolitanus* [26]. This suggests that the presence of amino acid transport systems is a common feature of bacteria. (ii) Carboxylic acids: specific transport systems have been demonstrated in a number of bacteria for monocarboxylic acids (D-lactate, L-lactate, pyruvate), C_4-dicarboxylic acids (L-malate, fumarate, succinate, oxalate), tricarboxylic acids (citrate). (iii) Sugars: transport systems for the following sugars have been found: lactose, arabinose, glucuronate, gluconate, D-galactose, D-fructose, L-rhamnose, D-glucose. A number of bacteria possesses, besides these secondary transport systems, a group translocation system for other sugars and in some cases for the same sugars. (iv) Phosphorylated intermediates such as glucose-6-phosphate and glycerol-3-phosphate. (v) Inorganic cations: specific transport systems have been found for Na^+, K^+, Mn^{2+}, Ca^{2+}, Mg^{2+}. (vi) Inorganic anions: specific transport systems for phosphate and sulphate have been demonstrated.

The number of different carriers present in the cytoplasmic membrane will differ from organism to organism and will depend in a given organism on the growth conditions. For *E. coli* it has been estimated that about 60 different carriers are present in the cytoplasmic membrane. It seems reasonable to state that specific transport systems are present for all solutes which have to be accumulated (or excreted) at a high rate. Most likely this is also the case for membrane permeable solutes like acetate (permeable in undissociated form), ethanol or glycerol.

A surprisingly high number of carriers are present constitutively in the cytoplasmic membrane. One may speculate about the advantages for bacteria to possess always carriers for several solutes. Possibly the possession of constitute carriers would enable an organism to scavenge intermediates leaked passively out of the cell and/or allow the organism to react rapidly to changes in the external medium. Besides constitutive transport systems inducible transport systems are also found in bacteria such as those for lactose transport in *E. coli* or for citrate transport in *B. subtilis*.

It is important to realize that the carriers are embedded in the cytoplasmic membrane and that the available space is limited. When this is occupied by carriers and other membrane proteins (as is most likely the situation) an increased number of a specific carrier has to result in a decreased number of other membrane proteins, unless the cell is able to increase the surface area of the cytoplasmic membrane. Such changes of surface area have been observed in bacteria.

The surface to volume ratio is a function of specific growth rate. With decreasing growth rates, the surface to volume ratio increases. This holds for all bacteria. However, apart from this general phenomenon, the surface to volume ratio is relatively large in bacteria that grow relatively fast at extremely low concentrations of growth-limiting substrates. This is shown in Table 2 for a *Spirillum* spec. and a *Pseudomonas* spec., grown in a lactate-limited chemostat, which show crossing μ-s curves, the *Spirillum* growing faster at the lower concentration extreme.

Regulation of transport can occur at the level of carrier synthesis or at the level of carrier activity. Evidence has been presented that regulation of activity is exerted by the catalytic activity of protein components possibly by a regulation protein in the case of the PTS [29]. Regulation by intracellular sugar phosphates has also been suggested.

The action of the carriers is completely reversible and symmetrical. During solute transport the energy of the proton-motive force is used to generate a solute gradient and to concentrate solute internally. The reversed process can occur under conditions of an outwardly directed solute gradient and a low proton-motive force. The solute gradient will then drive solute efflux and a proton-motive force is generated because this efflux is accompanied in many systems (see Fig. 2) with protons and/or charge. In other words, the energy of the solute gradient is then used to generate a proton-motive force.

In bacteria such situations are very common during fermentation. The end products of fermentation are continuously produced internally and excreted into the external medium (Fig. 11). The energy yield by efflux of fermentation products can be quite considerable. Recently, Michels et al. [30] calculated on theoretical grounds that the additional energy yield during homolactic fermentation is in the order of

TABLE 2

Effect of dilution rate on surface to volume ratio of a *Spirillum* sp. (Sp.) and a *Pseudomonas* sp. (Ps.) grown under L-lactate limitation [26,27]

Dilution rate (h^{-1})	Steady-state L-lactate concentration (μM)		Surface/volume (μm^{-1})	
	Sp.	Ps.	Sp.	Ps.
0.06	4	10	8.05	6.24
0.16	20	30	7.82	5.66
0.35 (batch)	>200	110	6.27	–
0.64 (batch)	–	>200	–	4.59

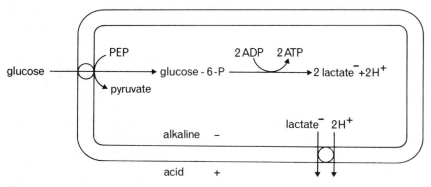

Fig. 11. Synthesis of ATP by substrate level phosphorylation and the generation of $\Delta\tilde{\mu}_{H^+}$ by lactate-proton efflux during homolactic fermentation.

30% of the energy produced by substrate level phosphorylation. Otto et al. [31] demonstrated that lactate efflux resulted in the generation of a $\Delta\tilde{\mu}_{H^+}$ in *Streptococcus cremoris*.

The proton-motive force generated by efflux of fermentation products can be used to drive other energy-consuming processes, such as solute transport and ATP-synthesis. The cell can therefore save its ATP produced by substrate level phosphorylation and use this predominantly for biosynthetic purposes. The proton-motive force generated by end product efflux will reach its maximal value when it is in equilibrium with the end-product gradient. The driving force for end-product efflux will then be zero, and consequently the internal concentration of end product will increase. The generation of a proton-motive force by other proton pumps, such as ATP-hydrolysis, will establish more rapidly this equilibrium. Such proton-pump activity therefore would lead to high internal end-product concentrations.

When, in the environment, the concentration of end products gradually increases, the end-product gradient decreases. In order to maintain a proton-motive force for a prolonged period of time, removal of end products from the environment is required. The presence of organisms which consume these will then be of benefit to the product-producing organism. Thus, this relation between a product-producer and -consumer represents a case of mutualism; for both organisms the interrelation is profitable.

6. *The phosphoenolpyruvate-dependent sugar phosphotransferase system*

(a) Group translocation

The types of transport discussed up to this point are all processes by which a compound X, originally located outside the membrane, is transferred across the membrane and appears as X, unmodified, inside the cell. The distinctive feature of group translocation types of transport is that, concomitant with transport, a chemical modification also occurs resulting in the appearance of X* inside the aqueous compartment enclosed by the membrane. By definition, the modification is an integral part of the transport process; without it transport by group translocation would not occur. Demonstration that transport is of the group-translocation type will be hindered if independent enzymatic reactions convert X to X* or X* to X at rates similar to the transport process itself. Doubtless, such difficulties are at least partially the reason why so few group translocation-type transport systems have been uncovered. At present the only thoroughly studied system of this type is the bacterial phosphoenolpyruvate-dependent sugar phosphotransferase system (PTS) [32,33] which catalyzes the transport of sugars according to the reaction

$$\text{Sugar}_{out} + \text{PEP} \underset{\text{HPr, Mg}^{2+}}{\overset{E_I E_{II}}{\rightleftharpoons}} \text{Sugar-P}_{in} + \text{Pyruvate}$$

Relative to the number of components involved, it is one of the simplest energy-dependent transport systems known. The overall transport which is catalyzed can be broken down into several independent phosphoryl-group transfer steps

$$\text{PEP} + E_I \overset{\text{Mg}^{2+}}{\rightleftharpoons} \text{P-}E_I + \text{Pyruvate}$$

$$\text{P-}E_I + \text{HPr} \rightleftharpoons \text{P-HPr} + E_I$$

$$\text{P-HPr} + E_{III}^X \rightleftharpoons \text{HPr} + \text{P-}E_{III}^X$$

$$\text{P-}E_{III}^X + \text{Sugar}_{out}^X \underset{\text{Mg}^{2+}}{\overset{E_{II}}{\rightleftharpoons}} \text{Sugar}^X\text{-P}_{in} + E_{III}^X$$

the last of which is coupled to the transport of the sugar across the cell membrane. E_I and HPr are usually designated the general proteins of the PTS since they are found in almost every known PTS. They are also considered cytoplasmic in origin because they are isolated from the soluble portion of broken-cell extracts. Each PTS sugar is transported by a specific E_{II} and E_{III}; therefore these two proteins are considered to be sugar specific proteins of the PTS. The nature of the enzymes, their cellular location and their sugar specificities vary from one microorganism to the next. *E. coli* contains a completely membrane bound E_{II} complex composed of three constitutive $E_{II}A$, one each for glucose, fructose and mannose and an $E_{II}B$. In the case of $E_{II}B$ no sugar specificity has been demonstrated even though its presence is certainly required for phosphorylation activity [34]. Another set of *E. coli* glucose specific PTS enzymes has also been identified and partially characterized [35]. It consists of a soluble E_{III}^{Glc} and a membrane-bound E_{II} distinct from the $E_{II}B$ above. The inducible PTS mannitol transport system in *E. coli* also works with a membrane-bound mannitol-specific E_{II}, but there is apparently no E_{III} required in this system [36]. The inducible PTS lactose transport system in *S. aureus*, on the other hand, requires a soluble E_{III}^{Lac} and a membrane-bound E_{II} [37] similar to the second of the glucose specific systems in *E. coli*.

(b) Purification and general properties

There is no single bacterial species from which all the PTS components have been purified, but HPr, E_I, E_{II} and E_{III} can now be obtained pure from a number of different bacteria [34,38–40].

HPr, 8000 to 10 000 daltons, is the smallest of the PTS proteins. It carries the phosphoryl group in P-HPr on a histidine residue.

E_I, classified as a cytoplasmic protein, possesses a very hydrophobic surface region, and has recently been purified by hydrophobic interaction chromatography [40]. The enzymatically active form is a 140 000-dalton dimer composed of 2 identical subunits [41]. The dimer is required both for the phosphorylation of E_I itself and the transfer of the phosphoryl group to HPr.

E_{III}^{Lac}. The only soluble sugar specific PTS component purified to homogeneity is E_{III}^{Lac} (Factor III^{Lac}) from the inducible lactose PTS in *S. aureus* [37]. It appears to be a 36 000-dalton trimer composed of 3 identical 12 000 dalton subunits. Each subunit can carry a phosphoryl group (in P-E_{III}^{Lac}) on a histidine residue.

E_{II}. The first membrane bound E_{II} was isolated from the *E. coli* constitutive PTS by Roseman's group [34] and separated into two protein components, a sugar-specific component designated $E_{II}A$ which became water-soluble during purification, and an integral membrane component, $E_{II}B$, which required the presence of detergents for solubility. Sugar phosphorylation activity could only be demonstrated when the two proteins were recombined with phosphatidyl glycerol and divalent cations.

More recently E_{II} from the *E. coli* inducible mannitol PTS has been purified [42]. It appears to be a single membrane-bound protein not requiring a second membrane-bound component for activity.

(c) Cellular localization of the PTS components

E_I, HPr and E_{III} are found in the soluble fraction of cell extracts and are considered cytoplasmic in origin. Nevertheless, the possibility that they are membrane-associated has never been ruled out. The extreme hydrophobic nature of at least some areas of E_I's surface and the apparent lack of specificity of these areas (they bind octyl chains and phenyl groups both very tightly) suggest that these areas may form a region for interaction with apolar portions of the membrane [40]. The appearance of E_I in solution after rupturing the cells can result from the mechanical force of the rupturing process or simply the effect of the dilution which occurs upon breakage.

(d) The complexity of the PTS

The multiple phosphorylated enzyme intermediates occurring in many bacterial PTS appear to be redundant if their only function is to phosphorylate E_{II}. Indeed, the *R*.

Transport across bacterial membranes 275

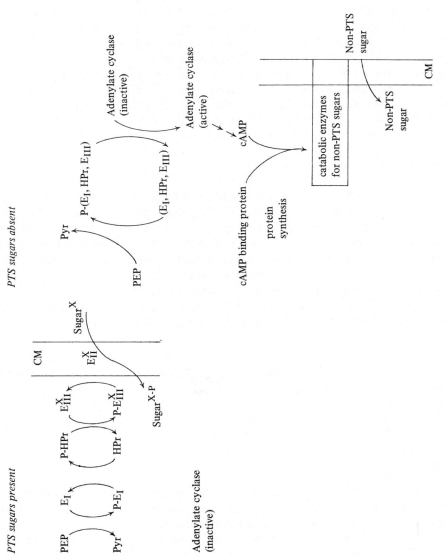

Scheme I.

sphaeroides and *R. rubrum* PTS are much simpler, consisting only of a membrane-bound E_I type enzyme which catalyzes the transfer of the phosphoryl group directly from PEP to the membrane bound E_{II}. The most likely reason for the existence of the additional proteins is that they, or their phosphoryl derivatives, are linked to other cellular processes such as the regulation of the activity of other transport systems. A general picture which has emerged from a variety of studies (for detailed reviews see [43–45]) is that one of the soluble PTS components modulates the activity of adenylate cyclase, the enzyme which synthesizes cAMP. As shown in Scheme I, when no PTS sugars are available, one of the PTS components in the phosphorylated form activates adenylate cyclase either by forming a complex with it or by phosphorylating it. The activated adenylate cyclase produces increased cAMP levels. The cAMP binds to the receptor protein and activates the initiation of transcription for synthesis of catabolic enzyme systems for non-PTS sugars [46]. If PTS sugars are available and can be transported, the PTS-phosphoprotein intermediates are shunted off into this sugar transport and phosphorylation cycle. They are not available to activate adenylate cyclase. Consequently cAMP levels remain low and synthesis of other catabolic enzyme systems is repressed.

(e) The specificity for phosphoenolpyruvate

The PTS is the only transport system which uses PEP instead of ATP or a $\Delta\tilde{\mu}_{H^+}$ as its energy source. Recently Dooyewaard and Robillard [39,40] proposed that through this specificity for PEP the expenditure of energy for glucose transport would be coupled to the metabolic state of the cell. When the microorganism finds itself in surroundings low in oxygen and carbon sources, fermentation-coupled growth predominates. ATP levels are low, pyruvate kinase is active and the carbohydrates are processed by the glycolytic pathway to lactic acid resulting in the production of up to 2 mol ATP/mol glucose. The overall result of the coordinate action of the PTS, glucokinase and pyruvate kinase is shown in Eqn. 4 of Scheme II.

Catalyst	Reaction	$\Delta G^{0\prime}$ (kcal/mol)
(1) PTS	$Glucose_{out} + PEP \rightleftharpoons Pyruvate + G\text{-}6\text{-}P_{in}$	−11.5
(2) Glucokinase	$ADP + G\text{-}6\text{-}P_{in} \rightleftharpoons ATP + Glucose_{in}$	+4.0
(3) Pyruvate kinase	$Pyr + ATP \rightleftharpoons PEP + ADP$	+7.5
(4)	$Glucose_{out} \rightleftharpoons Glucose_{in}$	0

Scheme II.

Equilibrium is established between internal and external glucose without *net* expenditure of metabolic energy. However, when oxygen and carbon sources are plentiful, respiration-coupled growth predominates in which glucose is processed through the Entner–Doudoroff pathway leading to as many as 36 mol ATP/mol glucose. These are conditions in which gluconeogenesis is favored. The glycolytic pathway is shut down by the high ATP levels which inhibit pyruvate kinase. However, the enzymes PEP synthase [47] and nucleoside monophosphate kinase are now active. They catalyze the attainment of the equilibria in reactions 5 and 6, respectively, in Scheme III. As can be seen in reaction 7 the coordinate action of the PTS, glucokinase, PEP synthase and nucleoside monophosphate kinase results in the transport of glucose into the cell at a *net* cost of 1 ATP/glucose. To summarize, if the microorganism has sufficient energy stores available, it will make a *net* expenditure of metabolic energy for the purpose of transporting glucose and storing it as glycogen for future use. However, if the energy stores are depleated no *net* expenditure of metabolic energy will be required. Glucose$_{out}$ will tend towards a direct equilibrium with glucose$_{in}$ and the organism will be free to devote its limited energy supplies to primary biosynthetic processes instead.

This versatility is a direct consequence of the specificity of the PTS for PEP; sugar transport is thereby harnessed to the stoicheiometric control mechanism for glycolysis vs. gluconeogenesis (i.e. control of the activity of pyruvate kinase and PEP synthase by the ATP concentration in the cell) and, as a consequence it will respond to the metabolic state within the cell.

Catalyst	Reaction	$\Delta G^{0\prime}$ (kcal/mol)
(1)+(2) PTS and GK	PEP+ADP+Glucose$_{out}$ ⇌ Pyr+ATP+Glucose$_{in}$	−7.5
(5) PEP synthase	Pyr+ATP ⇌ PEP+AMP+Pi	+0.2
(6) Nucleoside monophosphate kinase	AMP+ATP ⇌ 2 ADP	0
(7)	Glucose$_{out}$+ATP ⇌ Glucose$_{in}$+ADP+Pi	−7.5

Scheme III.

7. Interaction between energy-transducing processes

The information about energy-transducing processes presented above shows that these systems do not operate independently, but that they are closely coupled. This interaction between primary and secondary transport systems, substrate level phosphorylation and biosynthetic processes is given in the following scheme:

electron transfer ↔ $\Delta\tilde{\mu}_{H^+}$ ↔ ATP ← substrate level phosphorylation
    ↗ ↓   ↓
  secondary   flagellar   biosynthesis
  transport   movement

This scheme shows that the proton-motive force plays a central role in the coupling between energy-transducing processes. In a cell these processes continuously attempt to reach thermodynamic equilibrium. The systems behave therefore like connected vessels.

The scheme given above allows us to draw some general conclusions: (i) Changes in any one of the processes will lead to changes in the other processes. (ii) A simple relationship between electron transfer and ATP-synthesis does not exist and it is not justified to translate electron-transfer activity in ATP-synthesis activity. (iii) In cells which are not energy-limited the flow of energy will be in the direction of ATP-synthesis and the rate of ATP-consumption by biosynthesis will determine to a large extent the rate of ATP-synthesis. (iv) Cells which cannot generate a sufficiently high proton-motive force by electron transfer or secondary transport have to use ATP for the generation of a $\Delta\tilde{\mu}_{H^+}$. Consequently less ATP becomes available for biosynthetic purposes.

8. Methods for the determination of transmembrane gradients

Since the chemiosmotic concept emphasizes the importance of transmembrane gradients new techniques have been developed for their measurement. In these determinations it is essential to have procedures for the separation of the internal and external compartments of cells or closed membrane structures. This separation should be rapid enough to prevent leakage of the ion or solute under study from the internal compartment. This separation can be effected either through filtration on selective filters or by centrifugation through silicon oil. Quantitation of the transmembrane gradients can be inferred from the distribution of radioactively labeled solutes (e.g. amino acids) over the two compartments. However, for the determination of the components of the proton-motive force, the ΔpH and the $\Delta\psi$, indicator molecules have to be used. A large number of such indicator molecules have been described. For $\Delta\psi$ determinations, depending on the direction of the electrical potential gradient, cations or anions have to be used. In response to a $\Delta\psi$, internally negative, cations such as triphenylmethylphosphonium, tetraphenylphosphonium, dibenzyldimethylammonia and N-methyldeptropine accumulate in the membrane enclosed aqueous compartment. For a $\Delta\psi$, internally positive, the anions thiocyanate, phenyldicarbo-undecaborane and iodide can be used for quantitation of the gradient. The transmembrane pH gradient, inside alkaline, can be measured with the weak acids benzoic acid, acetate or 5,5-dimethyl-2,4-oxazolidinedione and pH gradients, inside acid, with the weak base methylamine.

Another method, which is routinely used for gradient determinations is flow dialysis [48]. In this method a compartment containing the membrane preparation and the radioactively labeled solute or indicator molecules is separated from a second compartment via a dialysis membrane. The solution in the second compartment is replaced with a constant rate and the radioactivity in the outflowing solution is monitored. The radioactivity monitored reflects the concentration of the radioactive probe in the external medium in the first compartment. Recently, an automatization of this technique has been described which considerably increases the applicability of flow dialysis [49]. Provided that suitable conditions are chosen, this technique can also be applied for kinetic studies. Ion-specific electrodes can also be used for determinations of changes in the concentration of a particular molecule in the external medium from which the transmembrane gradient can be calculated. A prerequisite is, of course, that a specific electrode is available for the ion of interest. Such electrodes have been described for many inorganic cations or anions and for compounds like tetraphenylphosphonium and dibenzyldimethylammonia. If a proton-specific electrode is used the internal buffer capacity of the membranes must be determined as well.

In all measurements an independent determination of the internal volume of the cells or membranes is required in order to allow calculation of the transmembrane gradients. The solute gradients can be expressed in mV by the Nernst-equation as shown in Fig. 9.

Various optical methods have also been used to estimate the two components of the proton-motive force. For $\Delta\psi$ measurements absorbance changes of carotenoids, chlorophyll and merocyanins have been determined, whereas for evaluation of changes of pH-gradients the fluorescence of compounds like 9-aminoacridine have been measured. These optical methods, however, are not of more than diagnostic value since their quantitative reliability is questionable.

Ionophores are important tools in the study of chemiosmotic systems. These are compounds which specifically increase the rate of one or more ion transfer processes across the membrane. Uncouplers (protonophores) which increase the membrane permeability for protons can be used to dissipate, more or less completely, the total $\Delta\tilde{\mu}_{H^+}$.

Many different protonophores, each with a different effectiveness, are available like 2,4-dinitrophenol (DNP) or 5-chloro-3-tert-butyl-2'-chloro-4'-nitro-salicyl-anilide (CCCP). Manipulation of the two components of the $\Delta\tilde{\mu}_{H^+}$ can be effected with two potassium ionophores. Valinomycin increases the electrogenic permeability of a membrane for potassium and leads to the dissipation of the $\Delta\psi$ if a high concentration of potassium is present (more than 10 mM). Nigericin catalyzes an electroneutral potassium proton exchange and thus dissipates the ΔpH under the same conditions. In combination these two ionophores function as an uncoupler.

9. Model systems for transport studies

Initially, the knowledge of transport systems in bacteria was obtained from studies with whole cells. The existence of primary and secondary transport systems was demonstrated and some information about the stoichiometry, specificity and kinetic constants of the transport proteins was obtained. In particular, the isolation of transport-deficient mutants established the functional role of specific transport systems. However, the information about molecular aspects of the transport processes which can be obtained from studies with intact cells is limited. The translocation process of the solute under study may be affected by the two additional cell-envelope layers which, apart from the cytoplasmic membrane, enclose the cell. Or the solute may be bound to or react with components which are present inside the cells. This restricts a clear-cut interpretation of the experimental data about the properties at the cytoplasmic membrane level.

These considerations urged the development of well-defined biochemical model systems. Ideally, such a system should consist of a phospholipid bilayer containing the transport systems and possibly the energy-generating systems. Furthermore, in order to allow measurements of transport activities an additional requirement is that the phospholipid bilayer forms a closed compartment. The isolation of cytoplasmic membrane vesicles, first described by Kaback [1], came very close to these requirements. The vesicles consist of cytoplasmic membranes of bacterial cells which form closed spherical structures and retain the physiologically active integrated membrane functions (Fig. 12). Secondary transport activities in these membrane vesicles can be studied, just as in whole cells, by measuring concentration changes of solutes in the external and/or internal compartment, but without interference of cytoplasmic components. A two-step procedure can be used for the isolation of membrane vesicles; the organism is first converted into an osmotically sensitive form and subsequently lysed under controlled conditions in the presence of nucleases and a chelating agent. The membrane vesicles retain a large number of membrane-

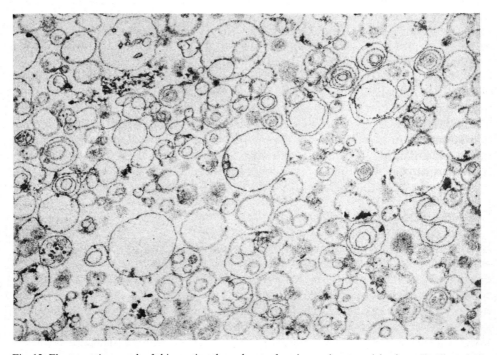

Fig. 12. Electron micrograph of thin section through cytoplasmic membrane vesicles from *Bacillus subtilis*.

associated enzymes and perform several integrated membrane functions. Several techniques have been used to demonstrate that a large majority of the vesicles, obtained by osmotic lysis, are oriented in the same way as the plasma membrane of intact cells.

Also procedures for the isolation of inside-out membranes, by French Press treatment of intact bacteria, have been described. In phototrophic organisms, these membranes are derived from the invaginations of the plasma membrane and are called chromatophores. These preparations have been extensively used for studies on light-dependent cyclic electron transfer and photophosphorylation. In non-phototrophic bacteria the resulting structures are often called inverted membranes or membrane particles, in analogy with sub-mitochondrial particles. Amongst others, these preparations have been isolated from *Azotobacter vinelandii* and *E. coli*. These inverted membranes can be used for the study of oxidative phosphorylation and the determination of H^+/e stoicheiometries since the enzymatic machinery for these processes is located on the external surface of these membranes. Also excretion of ions (like Ca^{2+}) from intact cells can be studied conveniently in these preparations because these ions are accumulated in inverted membranes.

For certain studies, however, even the isolated biological membranes are too complex to be analysed. Therefore, simpler model systems, like artificial vesicles prepared from purified lipids and one or more primary or secondary transport proteins, have been developed. The use of these reconstituted systems is the method of choice if one wants to prove the function of a specific transport protein or if one wants to develop mathematical relations to describe ion translocation initiated by a primary ion pump. During the past decade important progress has been made in our knowledge of these artificial membranes. Their physicochemical properties have been extensively characterized and several techniques are available for the incorporation of (transport) proteins [19,50]. Reconstitution experiments with primary proton (ion) pumps are facilitated by the fact that many of these primary transport systems can be assayed biochemically (ATP-hydrolysis, oxidoreduction etc.). Most promising for the preparation of homogeneous vesicles are those techniques in which the transport proteins are incorporated in preformed lipid vesicles. Reconstitution of primary transport systems from eukaryotic origin (Na^+/K^+-; Ca^{2+}- and H^+-ATPases; H^+-oxidoreductases and H^+-transhydrogenases) as well as from prokaryotic origin (H^+-ATPase and a light-driven H^+-pump) have been described. In addition, the structural and functional characteristics of vesicles containing bacteriorhodopsin have been extensively described. Reconstitution of secondary facilitated transport systems has been hampered by the lack of biochemical assay procedures for these systems. Despite this difficulty, the reconstitution of alanine carriers, purified from a thermophilic bacterium and from *B. subtilis,* has been reported. Also reconstitution experiments with secondary facilitated transport systems from eukaryotic cells have been described.

The co-reconstitution of two transport systems in one liposome opens possibilities for studies on transport system interactions. Reconstitution of oxidative and photophosphorylation has been demonstrated and the results of these studies contributed significantly to the appreciation of the chemiosmotic theory. Analogous experiments in which a secondary transport system is reconstituted with a primary transport system have also been reported. These experiments will allow the study of energy coupling between primary and secondary transport systems in the absence of intervening ion translocation systems.

Acknowledgement

The authors wish to thank Mr. M. Veenhuis for supplying us with the electron micrograph of *Bacillus subtilis* and Mrs. M.Th. Broens-Erenstein and Mrs. M. Pras for their help in the preparation of this manuscript.

References

1. Kaback, H.R. (1971) in W.B. Jakoby (Ed.), Methods in Enzymology, Vol. 22, pp. 99–120, Academic Press, New York.
2. Mitchell, P. (1977) in B.A. Haddock and W.A. Hamilton (Eds.), Microbial Energetics, pp. 383–423, Cambridge University Press, Cambridge.
3. Konings, W.N. and Michels, P.A.M. (1980) in C.J. Knowles (Ed.), Diversity of Bacterial Respiratory Systems, CRC Press, West Palm Beach, pp. 33–86.
4. Kaback, H.R. (1979) in E. Quagliariello, F. Palmieri, S. Papa and M. Klingenberg (Eds.), Function and Molecular Aspects of Biomembrane Transport, pp. 229–238, Elsevier/North-Holland, Amsterdam.
5. Harold, F.M. (1977) in D. Rao Sanadi (Ed.), Current Topics in Bioenergetics, Vol. 6, pp. 84–149, Academic Press, New York.
6. Mitchell, P. (1961) Nature 191, 144–148.
7. Mitchell, P. (1968) Chemiosmotic Coupling and Energy Transduction, Glynn Research Ltd., Bodmin, England.
8. Mitchell, P. (1962) J. Gen. Microbiol. 29, 25–37.
9. Mitchell, P. (1966) Biol. Rev. 41, 445–502.
10. Slater, E.C. (1953) Nature 172, 975–978.
11. Mitchell, P. (1975) FEBS Lett. 56, 1–6.
12. Papa, S. (1976) Biochim. Biophys. Acta 456, 39–84.
13. Williams, R.J.P. (1978) FEBS Lett. 85, 9–19.
14. Lehninger, A.L. (1979) Abstr. XIth Intn. Congr. Biochemistry, Toronto, pp. 414.
15. Kagawa, Y. (1978) Biochim. Biophys. Acta 505, 45–93.
16. Simoni, R.D. and Postma, P.W. (1975) Annu. Rev. Biochem. 44, 523–554.
17. Boyer, P.D., Chance, B., Ernster, L., Mitchell, P., Racker, E. and Slater, E.C. (1977) Annu. Rev. Biochem. 46, 995–1026.
18. Oesterhelt, D. and Stoeckenius, W. (1971) Nature New Biol. 233, 149–152.
19. Hellingwerf, K.J. (1979) Structural and functional studies on lipid vesicles containing bacteriorhodopsin, Ph.D. Thesis, Univ. of Amsterdam.
20. Stoeckenius, W., Lozier, R.H. and Bogomolni, R.A. (1978) Biochim. Biophys. Acta 505, 215–278.
21. Henderson, R. and Unwin, P.N.T. (1975) Nature 257, 28–32.
22. Khorana, H.G., Gerber, G.E., Herlihy, W.C., Gray, C.P., Anderegg, R.J., Nihei, K. and Biemann, K. (1979) Proc. Natl. Acad. Sci. USA 76, 5046–5050.
23. Stoeckenius, W. and Lozier, R.H. (1974) J. Supramol. Struct. 2, 769–774.

24 Hwang, S.-B., Korenbrot, J.I. and Stoeckenius, W. (1978) Biochim. Biophys. Acta 509, 300–317.
25 Lo, T.C.Y. (1977) J. Supramol. Struct. 7, 463–480.
26 Matin, A., Konings, W.N., Kuenen, J.G. and Emmens, M. (1974) J. Gen. Microbiol. 83, 311–318.
27 Matin, A. (1979) in M. Shilo (Ed.), Strategies of Microbial Life in Extreme Environments, pp. 323–339, Verlag Chemie, D-6940 Weinheim, Germany.
28 Matin, A. and Veldkamp, H. (1978) J. Gen. Microbiol. 105, 187–197.
29 Saier, M.H. and Moczydlowski, E.G. (1978) in B. Rosen (Ed.), Bacterial Transport, Vol. 4, pp. 103–127, Marcel Dekker, New York.
30 Michels, P.A.M., Michels, J.P.J., Boonstra, J. and Konings, W.N. (1979) FEMS Microbiol. Lett. 5, 357–364.
31 Otto, R., Sonnenberg, A.S.M., Veldkamp, H. and Konings, W.N. (1980) Proc. Natl. Acad. Sci. USA 77, 5502–5506.
32 Roseman, S. (1969) J. Gen. Physiol. 54, 138s–180s.
33 Kundig, W. and Roseman, S. (1971) J. Biol. Chem. 246, 1393–1406.
34 Kundig, W. and Roseman, S. (1971) J. Biol. Chem. 246, 1407–1418.
35 Kundig, W. (1974) J. Supramol. Struct. 2, 695–714.
36 Jacobson, G.R., Lee, C.A. and Saier, M.H. (1979) J. Biol. Chem. 254, 249–252.
37 Hays, J., Simoni, R.D. and Roseman, S. (1973) J. Biol. Chem. 248, 941–956.
38 Anderson, B., Weigel, N., Kundig, W. and Roseman, S. (1971) J. Biol. Chem. 246, 7023–7033.
39 Dooyewaard, G., Roossien, F.F. and Robillard, G.T. (1979) Biochemistry 18, 2990–2996.
40 Robillard, G.T., Lolkema, J. and Dooyewaard, G. (1979) Biochemistry 18, 2984–2989.
41 Misset, O., Brouwer, M. and Robillard, G.T. (1980) Biochemistry 9, 883–890.
42 Jacobson, G.R., Lee, C.A. and Saier, M.H. (1979) J. Biol. Chem. 254, 249–252.
43 Saier, M.H. (1977) Bacteriol. Rev. 41, 856–871.
44 Peterkofsky, A. and Gazdar, C. (1978) J. Supramol. Struct. 9, 219–230.
45 Postma, P.W. and Roseman, S. (1976) Biochim. Biophys. Acta 457, 213–257.
46 Bolshakova, T.N., Gabrielyon, T.R., Bourd, G.I. and Gershanovitch, V.N. (1978) Eur. J. Biochem. 89, 483–490.
47 Cooper, R.A. and Kornberg, H.L. (1965) Biochim. Biophys. Acta 104, 620–623.
48 Ramos, S., Schuldiner, S. and Kaback, H.R. (1976) Proc. Natl. Acad. Sci. USA 73, 1892–1896.
49 Hellingwerf, K.J. and Konings, W.N. (1980) Eur. J. Biochem. 106, 431–437.
50 Kagawa, Y. (1978) Biochim. Biophys. Acta 505, 45–93.

CHAPTER 11

Coupled transport of metabolites

P. GECK and E. HEINZ *

*Gustav Embden-Zentrum der Biologischen Chemie,
J.W. Goethe-Universität, Frankfurt/Main, F.R.G., and
* Department of Physiology, Cornell University
Medical College, New York, NY, U.S.A.*

1. Introduction

The present chapter will deal mainly with coupled transport of hydrophilic nutrients such as amino acids and sugars, and their derivatives by animal cells, as too little is known whether the uptake of lipophilic solutes, such as fatty acids, is also coupled.

There are several reviews on this topic written from different points of view [1–16], which the reader interested in more details should consult.

The amount of nutrients in the ambient media from which living organisms take them up is often very low, but powerful mechanisms have been developed to transport these substances actively into the organism, up to the concentration suitable for further utilization. In single cells these uptake mechanisms are located in the cytoplasmic membrane, which has therefore to be considered a cellular organelle with highly specialized functions. In multicellular organisms, such as higher animals, the uptake of metabolites occurs through epithelial cell layers into body fluids. One used to distinguish this transcellular transport from the homeocellular transport of monocellular organisms, but the difference between these two kinds of uptake mechanisms is probably not fundamental. It appears that also in transcellular transport the crucial process is located in the cytoplasmic membrane, whereas the passage through the cell interior is merely by diffusion, which because of the small dimensions is rapid enough not to limit the overall uptake rate.

Active transport requires the expenditure of energy, which is supplied by a simultaneous exergonic process through a "coupling" device in such a way that the energy liberated by the latter can be utilized to drive the transport [17–19].

(a) Source of energy

The ultimate source of energy for all transport processes is metabolism, i.e. the energy derived from respiration or fermentation. As a rule, the driving forces for transport processes are not directly coupled to the ultimate oxygen consuming or glycolytic reactions, but the transfer of the energy is rather mediated through the chain of intermediate reactions. The reaction to which the transport process is coupled directly, which for convenience we may call "driver" process, can be either a chemical reaction, such as the hydrolysis of energy-rich phosphates, or an osmotic process, e.g. the downhill flow of another solute, predominantly Na^+ in animal cells and H^+ in microorganisms. In the first case, the transport is called "primary active" and in the second case, "secondary active", as the gradient of the driver species must be continuously restored, presumably by a primary ion-transport mechanism. In primary active transport, the driving energy derives from the "affinity" of the chemical reaction, defined as the change in free energy per "degree of advancement" of the chemical reaction. In secondary active transport, on the other hand, it derives from the "electrochemical potential difference" of the driving solute, Na^+ or H^+, respectively, defined as the change of free energy by the movement of one mole of solute across the barrier.

(b) Principles of coupling

As to the mechanical aspect of coupling the general principle underlying this is that of the common intermediate, which means here that the coupled processes must have at least one step in common, which ties them together so that the one cannot proceed without involving the other, except for uncoupled "leakage" pathways. Most chemical reactions and transport processes can also proceed by such parallel leakage pathways which naturally reduce the "degree of coupling" (q) [20] and as a consequence the efficiency of energy transfer as will be discussed further below.

The coupling in primary active transport poses some conceptual problems in that the (driving) chemical reaction is a scalar, i.e. without a distinct direction in space, whereas the transport process, on the other hand, is clearly a vector, i.e. oriented in space in the direction perpendicular to the membrane area. How can a chemical reaction, in which all the intermediate steps are at random with respect to direction in space, share a step with the specially directed process of transport? There are the following possibilities to overcome this theoretical obstacle which may be illustrated by a conventional carrier model [21]: In most general terms this describes mediated transport as being effected by a "translocator" (carrier) in a cyclic process which includes four sequential steps (Fig. 1): (1) Recognition and binding of the transportable solute to the translocator (carrier) on the trans side. (2) Translocation of the solute across barrier. (3) Release of transported solute on the *trans* side. (4) Restitution of the translocator to its initial state. Within this cycle, only Step 2, and presumably Step 4, are vectorial whereas Steps 1 and 3 can be scalar. To effect active transport, at least one of these four steps has to be coupled to the (driver) reaction. This could be Step 1 and/or Step 3 in that the translocator was either activated on the *cis* side to be able to forcibly bind the solute ("push" effect), or inactivated on the *trans* side to release the solute, respectively ("pull" effect). In either case the translocation itself would occur passively (downhill). This principle of coupling underlies most models of primary active transport proposed and might be characterized as the "source and sink" principle.

An alternative model in conceivable in which the (vectorial) Step 2 is directly

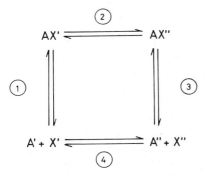

Fig. 1. General model for mediated transport. Explanations see text.

coupled. Some enzymes can be envisaged to be oriented within a barrier so that it can accept its substrate(s) from one side of the barrier, but release the product(s) of the reaction on the opposite side. Clearly there is a vectorial step built in the sequence of steps within such a reaction. Such a vectorial reaction may be suitable to effect translocation through that membrane. In contrast to the previous principle, we would call the present one "conveyor" principle, since here it is the translocation itself (Step 2 in the general model) that is directly linked to the chemical (driver) reaction. Active transport based on this principle has so far been ascertained only for the phosphotransferase system transporting some sugars in microorganisms [22]. The "glutamyl transferase" system, postulated for amino acid transport in animal cells [23], is also based on the conveyor principle, but has not been verified yet experimentally. In contrast, there is some good experimental evidence that it cannot be involved in amino acid uptake [24]. We therefore shall not discuss this principle in more detail.

The distinction between the two kinds of direct coupling between transport and chemical reaction does not apply to secondary active transport, as the coupled processes are both vectorial. This transport is assumed to involve a "ternary" complex, i.e. between the translocator (carrier), the transportable solute and the co-ion. The latter is assumed to influence the translocation of the solute in two ways [25]: (1) it may favor the formation of the ternary complex, in that its binding to the translocator increases the affinity of the latter to the transportable solute and vice versa (affinity effect) or, (2) it may increase the velocity at which the solute is translocated through the barrier, e.g. in that the ternary complex moves faster across the barrier than do the two binary complexes (velocity effect). In natural systems both effects appear to occur, separately as well as mixed. A crude distinction is often attempted on the basis of the two Michaelis–Menten parameters: the maximum velocity (V_{max}) and the half saturation constant (K_m) in the assumption that the former is altered in the velocity effect (V-type) and the latter, in the affinity effect (K-type). The relationship is often more complicated especially if electric potentials are involved [26].

(c) Materials for transport studies

Information on accumulative uptake of different organic solutes was obtained from experimental studies on materials of very different complexity. Clearance or balance studies can be performed with the whole body of higher animals. The next step in reduction are experiments with organs in situ, e.g. by (micro-)perfusion studies [27]. In investigations on transepithelial transport, often isolated epithelia were used to separate two experimentally well controlable and accessable aqueous phases, e.g. in Ussing-chamber [28] or everted sac technic [29]. Such experimental steps enable relatively simple chemical, physicochemical, or electric measurements in these phases. A new direction in experimental work on transepithelial transport was opened by studies with tissue cultures derived from epithelial cells that under suitable conditions in vitro form epithelium-like structures with polarized cells and typical tight junctions [30]. Such arrangements seem to be very appropriate for studies on specific aspects of epithelial transport. For many questions it is necessary to study single cells in suspension, e.g. red and white blood cells, several tumor cell lines in their ascites form, or cells cultivated in submersed culture. From different organs, epithelia as well as others, single cells can be obtained by incubations with different hydrolytic enzymes, especially with hyaluronidase, collagenase, or trypsin. All these procedures lead to simplifications of the geometry of the extracellular compartment. The complex structure of the intracellular compartment, however, is not affected by these procedures. Therefore, determinations of intracellular electrochemical activities of the organic solutes and ions remain difficult; the chemical as well as the physicochemical methods used show problems.

In order to reduce such interferences, successful efforts have been made to isolate the cell membranes, or even their transport-active constituents. One way to achieve this is by preparation of isolated membranes which have a natural tendency to form closed and homogenous vesicles [31,32]. Another approach is by "reconstituted systems", i.e. to isolate membrane components involved in specific transport processes, and to incorporate them in artificial lipid membranes, usually liposomes [33,34]. Vesicles have been successfully prepared from various cells and tissues and tested for transport activities. Whenever membranous material has been isolated from other cellular components it tends to form vesicles spontaneously, sometimes with an uniform orientation, right-side-out or inside-out vesicles, respectively. For mixtures of vesicles of the two orientations, methods were developed to separate the two polarities. Furthermore, one can separate vesicles from different cell types or even from different regions of the cell, e.g. brush-border membranes form basal lateral ones [35,36].

Similar behavior was observed with reconstituted systems. Sometimes there seems to be a tendency for incorporation of the protein into the liposomes in an uniform orientation; for random incorporation, separation technics were worked out.

Investigations at all the above-mentioned degrees of complexity are valuable to elucidate the processes going on in nature, since each technic has its specific advantage and disadvantage.

(d) Electrochemical potential difference

The electrochemical potential differences of Na^+ and H^+, supposed to drive the (secondary active) transport of various metabolites into and across cells, is composed of an electrical and an chemical portion. Energetically the two portions are equivalent and hence at a given total it should not make any difference which of them predominates at any given moment. This may be of special significance as they often do not change according to the same pattern, e.g. if an electrogenic pump of the ion concerned is involved. Whenever such a pump is started the electric potential difference may rise much faster than the chemical potential difference, as the generation of the former requires a much smaller amount of ions to be translocated than does the generation of a chemical potential difference. Hence under suitable conditions most of the driving force derivable from the electrogenic pump may become available with little delay after starting the pump, presumably in less than 100 ms, whereas the steady state of the pump may take a few minutes to be established in full [37].

It should be mentioned at this point that the energetic equivalence between electrical and chemical driving force discussed above does not necessarily apply to the kinetics [26]. There are considerable methodical difficulties in determining both components of the electrical driving force, that seem to be overcome incompletely only. Calculations of the chemical activity of cytosolic Na^+ from chemical determinations of Na^+ content of a tissue have to make assumption on the cytoplasmic activity coefficient as well as on the degree of compartmentalization inside the cell. These problems do not play any role in direct electrochemical determinations of cytosolic Na^+ using ion-specific microelectrodes [38] that, however, show a limited specificity. Direct determinations of the membrane potential using microelectrodes [39] were not possible for all relevant situations. Calculations using the Nernst equation for the distribution of the lipid soluble ions [40] or indirect optical methods [41] are often difficult to interpret.

2. Types of energization

(a) Primary active transport

In animal cells and tissues, primary active transport of nutrients and metabolites seems to be rare or may not even exist at all. There is no doubt that the active transport of certain ions, e.g. Na^+, K^+, H^+, Ca^{2+} and others, is primary active, being directly driven either by the hydrolysis of ATP or, in the case of mitochondrial proton transport, by a redox reaction. However, no system as to the transport of nutrients or metabolites has been proved beyond doubt to be linked directly to ATP hydrolysis or another chemical reaction [1], even though in some cases such linkage cannot be completely excluded either. A typical system of primary transport of amino acids that has been postulated for animal cells is the glutamyl system [23].

But, as has been mentioned before, evidence of its existence is still not complete. Moreover, it has been shown that certain cells deficient of some crucial component of this system have unimpaired ability to transport and accumulate amino acids [42,43]. Hence in the present context little remains to be said for primary active transport of nutrients and metabolites, at least as far as animal cells and tissues are concerned.

The situation may be different with microorganisms, although here, too, secondary active transport of nutrients and metabolites appears to be firmly established for many transport systems. Especially the "membrane bound" (shock-resistant) systems which appear to be firmly incorporated in the cytoplasmic membrane are rather clearly identified as secondary active transport systems, as they require an "energized state" of the membrane, which means that they are driven by a protonmotive force, i.e. an electrochemical potential gradient of H^+-ions [3].

On the other hand, there are two groups or systems for various nutrients and metabolites in which transport of certain metabolites postulated to be driven directly by the hydrolysis of ATP or other energy-rich compounds without the requirement of a protonmotive force: these are the "binder"-requiring systems and the phosphotransferase systems [22]. "Binders" are periplasmic proteins, which occur between the cytoplasmic and the outer membrane of microorganisms. They appear to be loosely attached to the outer face of the cytoplasmic membrane, from which they can be removed, for instance, by mild osmotic shock. However, the mentioned hypothesis that the (shock-sensitive) binder systems of active transport are primary active does not appear to be fully established mainly because a distinction between primary and secondary active transport is not possible in the presence of a functioning H^+-transporting ATPase. The main evidence in favor of primary active transport in shock-sensitive transport systems has been derived from a Ca^{2+}, Mg^{2+}, ATPase-less mutant of *E. coli*. There are, however, some findings which do not seem to support this hypothesis.

Therefore, the questions cannot be answered whether there is any type of primary active transport of organic solutes besides that by the phosphotransferase system, discussed below.

(i) Phosphotransferase systems

In the phosphotransferase systems the transport of a sugar is coupled directly to the hydrolysis of the energy-rich phosphate bound of phosphoenolpyruvate (PEP) [22], and is therefore primary active. The linkage between the translocation of the sugar into the cell, and the phosphorylation of the sugar by a kinase reaction, both result from the same step. The catalyzing kinase is built into the barrier in such a way, that it can accept the sugar only from the outside but can deliver the product, the glucose-6-phosphate, only to the inside of the cell. Such reactions have been called "vectorial" or "heterophasic". It looks as if the catalyzing enzyme, or at least its reaction site are encased by two barriers, one towards the outside and another towards the inside of the cell. Hence the former transmits glucose but excludes glucose phosphate, whereas for the latter the opposite is true. As a consequence the enzyme is not accessible to sugar coming from the inside of the cell, nor to glucose-6-phosphate coming from the outside. Obviously the principle underlying this coupling is fundamentally different from the "source-and-sink" principle assumed for other transport models and has therefore been called the "conveyor" principle [6]. This system also differs from the others in that it does not primarily translocate the solute glucose, but the glycosyl group as the immediate result of the translocation process in the appearance of glucose phosphate inside the cell. In order to make this system a real transport system, a second reaction is required inside the cell, namely a splitting of the glucose-6-phosphate by an appropriate phosphatase, which, however, need not be built into the membrane, so that finally free glucose accumulates inside the cell as the result of the overall transferase process. The detailed thermodynamic analysis of this model, which has been given elsewhere [6], shows quite clearly that the system is active, in that at least part of the free energy of the hydrolysis of the energy rich phosphate appears in the final accumulation of the sugar. A detailed description of this transport system is given elsewhere in this volume [45].

(b) Secondary active transport

For secondary active transport of organic solutes two different types of transport systems are necessary, as shown schematically in Fig. 2. A primary active transport system is responsible for generating an electrochemical potential gradient of a cation, H^+ in microorganisms and Na^+ in animal cells. These transport systems are discussed elsewhere in this volume [44]. The energy source generated by the ion pump can be used by cotransport systems to accumulate several organic solutes. There may be several cotransport systems specific for different groups of solutes. The kinetics of these cotransport systems is discussed as follows.

(i) Carrier model of cotransport
The stoichiometric coupling between the transport of an organic substrate (A) and a cation can be fairly easily represented by the carrier model of cotransport (Fig. 3). It is assumed that the transport is mediated by a membrane component (X) which can alternate between two states: X' and X", respectively. X' is assumed to communicate with the '-phase (e.g. extracellular phase) from which it can bind both ligands (A) and (B) to form the two binary complexes AX' and BX' and also a ternary complex ABX'. On the other hand, the state X" communicates with the "-phase (e.g. intracellular phase) from which it may bind A and/or B to form the complexes AX", BX" and ABX". The translocation of A and B between the two phases, ' and ", is assumed to be effected by the interconversion between the two states of each species of X, as has been described in more detail elsewhere [6,10,25,46,47]. The transfer of energy between the transfers of A and B via cotransport requires that they move predominantly by ternary complex (ABX), and this implies either that the two binary complexes AX and BX are less ready to interconvert between the '-state and "-state, respectively, than is the ternary complex, or that the two binary

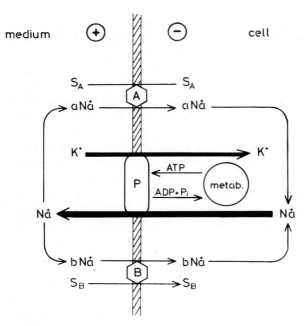

Fig. 2. Secondary active transport. Explanations see text.

complexes form much less abundantly than does the ternary complex, owing to cooperativity between the binding of the two ligands. A full cycle of the transfer from the '- to the ''-compartment therefore involves four distinct steps: (1) formation of ABX' (A' + B' + X' $\rightleftharpoons$ ABX'), (2) the conversion of the complex from the '-state to the ''-state (ABX' $\rightleftharpoons$ ABX''), (3) release of the two ligands on the ''-side (ABX'' $\rightleftharpoons$ A'' + B'' + X''), and (4) restoration of the '-state of the free carrier (X'' $\rightleftharpoons$ X'). For reasons of simplicity, it is usually assumed that the two transitional steps (2 and 4) are slow as compared to the binding steps (1 and 3) so that the ternary complex on

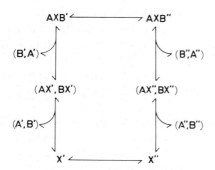

Fig. 3. Carrier-mediated cotransport. Explanations see text.

Coupled transport of metabolites

each side can be treated as being in quasi equilibrium with the freely dissolved ligand in the adjacent phases. No assumption is necessary concerning the nature of the transition between the '- and "-states of the species involved which could occur through change in conformation through diffusion or rotation, each of which would even lead to identical kinetic equations, as long as no transfer of an electric charge is involved in any of these steps.

The kinetic treatment becomes more complicated if electrical effects are introduced [26] which are likely to exist for many cotransport systems in which the transport of a cation is linked with that of an electrically neutral organic substrate. In the following, various possibilities of electrical effects on the transport kinetics will be discussed separately.

(ii) Kinetics of influx

To the extent that the binary complexes can be neglected the following equations hold for influx of A and B and for binding of an inhibitor (K), which competes with A but is itself not transportable:

$$J_a = J_b = x_t \cdot \tilde{p}_0 \cdot \alpha' \cdot \tilde{\beta}'/N$$

$$(K)_{bound} = x_t \cdot (\tilde{p}'_0/\tilde{p}'_c) \cdot \kappa \cdot \tilde{\beta}'/N$$

$$N = 1 + (\tilde{p}'_0/\tilde{p}'_c) \cdot (\alpha' + \kappa') \cdot \tilde{\beta}' + (\tilde{p}'_0/\tilde{p}''_0) \cdot (1 + \alpha' \cdot \tilde{\beta}')$$

This gives the following transport kinetic constants:

$$(J_a)_{max} = x_t/(1/\tilde{p}'_0 + 1/\tilde{p}''_0)$$

$$K_{t,a} = (K_a/\tilde{\beta}') \cdot \frac{1/\tilde{p}'_0 + 1/\tilde{p}''_0}{1/\tilde{p}'_c + 1/\tilde{p}''_0}$$

$$K_i = (K_\kappa/\tilde{\beta}') \cdot (1 + \tilde{p}'_0/\tilde{p}''_0) \cdot (\tilde{p}'_c/\tilde{p}'_0)$$

$(J_a)_{max}$	maximal velocity for influx of A
K_a, K_b, K_k	apparent dissociation constants for the different carrier substrate complexes
$K_{t,a}$	transport constant for A
K_i	inhibitor constant
a', b', k'	concentrations of A, B, K at the '-side
$\tilde{p}'_0$, $\tilde{p}'_c$, $\tilde{p}''_0$	permeation probablities
p'_0, p'_c, p''_0	permeation probablities at zero membrane potential
m	symmetry factor ($|m| \leq 1$)
x_t	total amount of carrier
α', β', κ'	relative concentrations of A, B and K (a'/K_a; b'/K_b; k'/K_k)
$\tilde{\beta}'$	relative electrochemical activity of B ($\tilde{\beta}' = \beta' \cdot \xi^{(m+1)/2}$)

ξ electrochemical activity coefficient ($\xi = \exp(-F\psi/RT)$)
ψ membrane potential

(iii) Effect of membrane potential

Information about detailed features of the transport system should be available from the influence of changes of membrane potential on the kinetic parameters of a transport system [26]. According to certain assumptions concerning details of the given transport mechanism, one may expect differential effects on the transfer constants ($\tilde{p}_i$). Let us compare three possible models:

(z = 0). In this case, the unloaded carrier is electrically neutral but accepts the positive charge upon formation of the ternary complex. Hence the transfer of the ternary complex involves the translocation of a positive electric charge. Thus we would expect that p_0' and p_0'', the transfer constants of the unloaded carrier, are independent of the potential, whereas the corresponding transfer constant of the ternary complex depends on the membrane potential according to the following expression:

$$\tilde{p}_c' = p_c' \xi^{(1+m)/2} \qquad \tilde{p}_c'' = p_c'' \xi^{-(1-m)/2}$$

Binding and release of the cation to the transport system should not be affected by an electrical potential.

(z = −1): As in this case the unloaded carrier has a negative charge, the ternary complex should be electrically neutral. Accordingly, the transfer constants of the ternary complex $\tilde{p}_c'$ and $\tilde{p}_c''$ should not depend on the membrane potential whereas the corresponding parameters for the unloaded carrier do according to the following equations:

$$\tilde{p}_0' = p_0' \cdot \xi^{-(1+m)/2} \qquad \tilde{p}_0'' = p_0'' \cdot \xi^{+(1-m)/2}$$

as in case 1, binding release of the cation to the translocator should not be affected by the electrical potential.

A gated channel: If one assumes that mobile carriers are not involved, but some type of gate mechanisms, as the one devised by Patlak many years ago [48], neither the rate coefficients of the loaded nor those of the unloaded carrier can be directly affected by an electrical potential. Then binding of the cation B to the transport site from the '-phase and/or its release to the "-phase must be affected by the electrical potential. Hence in contrast to the model involving mobile carriers we would not expect any effect of the electrical potential on the translocation parameters, $\tilde{p}_i$, but on the binding constants only. Such a mechanism seems to be distinguishable from the mobile carrier according to its response to changes in membrane potential. However, to design unequivocal kinetic tests to distinguish between mobile carrier and gated channel more theoretical work has to be done.

(iv) Predictions of kinetic feature from a model

It can be shown that Na^+ enhances the transport of A (amino acid or sugar) by reducing K_m while the maximum rate is little affected. Only if the contributions of the binary complexes are not neglected, contrary to what we suggested before, one may under certain conditions also obtain a change in the maximal transport rate by B. Inhibition constant (K_i) for a competitor of A (e.g. phlorizine) is also reduced by B.

(v) Effects of electrical potential

The effects of the membrane potential on the kinetic functions of cotransport are shown in Fig. 4. For simplicity it is assumed that the system is symmetric and the transfer coefficients are the same for free carrier and ternary complex. Considering the above three models, one may note these differences with respect to dependence

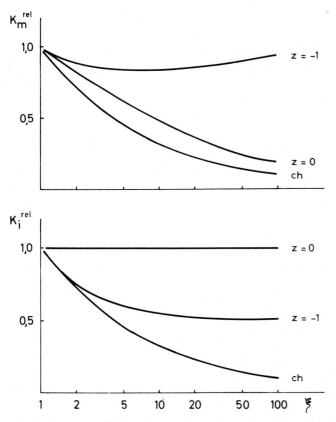

Fig. 4. Influence of membrane potential on K_t and K_i. K_t and K_i (in relative units) are plotted vs. the electrochemical activity coefficient $\xi = \exp(-F\psi/RT)$. For details see text. $z=0$, unloaded carrier uncharged; $z=-1$, unloaded carrier negatively charged; ch, gated channel.

of K_m and K_i on the electrical potential. Both in the gate model and in the first carrier model ($z=0$), K_m is strongly reduced by an electrical potential but is barely changed in the second carrier model ($z=-1$). By contrast, K_i is not affected by the membrane potential at $z=0$ but is reduced in the gate model, and to a smaller extent, in the carrier model at $z=-1$. Hence studying the dependence of electrical PD of each K_m and K_i one might be able to distinguish between the three possibilities as follows:

(a) A decrease in both K_m and K_i after applying a membrane PD is best compatible with a gated channel (model 3).
(b) If under the same conditions K_m is reduced whereas K_i is not affected, the first carrier model ($z=0$) would best explain the result.
(c) If under the mentioned conditions K_i is reduced whereas K_m is not appreciably affected, the second carrier model ($z=-1$) would best explain the results.
(d) If neither constant is significantly affected by the membrane PD, none of the three models would explain the results.

In addition, the ratio between substrate flux and inhibitor binding is related to the electrical potential in a simple manner which is characteristic for each of the three above-mentioned models; for instance, hyperpolarization of the membrane potential should not change this ratio with the gate type model nor with the carrier model 2 ($z=-1$) whereas under the same conditions that ratio should increase in carrier model 1 ($z=0$). As this relationship does not depend on the composition of the two adjacent fluid phases it should be suitable for further testing the model underlying a given transport mechanism.

(vi) Effect of cotransport on the membrane potential

Cotransport between a cation and an electroneutral solute, which involves the translocation of an electrical charge across the barrier, should not only depend on the electrical membrane potential but also by itself affect the latter [49–51]. The extent of such changes in membrane potential relative to the translocation of unit charge by cotransport does no longer depend on special properties of the cotransport system but only on those ion flows which will accompany cotransport in order to maintain electroneutrality. This is depicted for Na^+ cotransport system of animal cells in Fig. 2. As usual, we assume the plasma membrane has an electrogenic Na^+,K^+-pump which moves three Na^+ out of the cell and two K^+ into the cell per hydrolyzed ATP. Such secondary flows may consist of free ion flows as well as of other rheogenic cotransport systems. For steady-state in respect to ion distributions following Mullins and Noda's procedure [52] a relationship between membrane potential change and cotransport rate was derived [50]. The effect of cotransport on the electric potential is independent of the nature of the cotransported organic solute as long as it is neutral but strongly depends on the stoichiometry between Na^+ ions and organic solute molecule during cotransport. Hence all cotransport systems with the same stoichiometry should show the same dependence between membrane potential change and cotransport under otherwise equal conditions. This relationship should enable us to draw some information about the stoichiometric ratio from such dependence of the changes in membrane potential on cotransport rate.

(vii) Pseudocompetition [50]

All rheogenic cotransport processes, as they depend on the electrical potential, should be affected by all measures which change the membrane potential. In some cases it is difficult to tell whether the inhibitor affects the transport system directly or by abolishing or eliminating the driving potential. Such inhibitors are those which inhibit the electrogenic ion pump, e.g. ouabain [53] or ionophores, which change the membrane diffusion potential through changing the permeability of the membrane for such ions. As rheogenic cotransport processes, according to what has been discussed before, also influence membrane potential they should by the same token also influence other rheogenic cotransport systems. As a consequence, fluxes through different cotransport systems which depend both on the electrochemical potential of Na^+ should affect each other through this change in electrochemical potential difference of Na^+ and the kinetic features of this kind of mutual inhibition is very similar to that of direct competition between the cotransport systems, e.g. in that it increases K_m of the secondarily inhibited transport system. This "pseudocompetition" can be distinguished from true direct competition only if also the membrane potential is measured simultaneously. The relationship is represented in Fig. 5, in which the transport rate of the test substance is plotted vs. the membrane potential. All points which are on the inhibitory line drawn in this figure are exclusively affected through the electrical membrane-potential change. Points which are situated above this line appear to indicate that simultaneously with the inhibition, there is another stimulation of the same transport system. For points which are below this line direct inhibition has to be postulated. The true competitive inhibition can be obtained by correcting the inhibition for the change in potential.

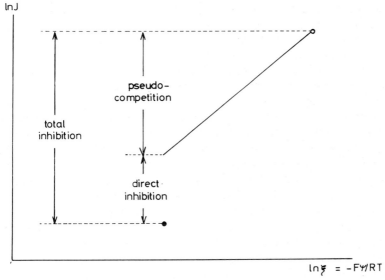

Fig. 5. Pseudocompetition. Explanation see text.

3. Special systems

(a) Nonpolarized cells

All animal cells studied up to now show the ability to accumulate certain amino acids against their concentration gradients, while for sugars such accumulation was only observed in polarized cells, e.g. cells from small intestine or proximal tubule. Ehrlich cells and other tumor cell lines [53–57], tissue culture cells, avian erythrocytes [58], isolated cells and slices from different tissues [59] were studied for their ability to accumulate amino acids. Apart from investigations with epithelia, studies, on Ehrlich cells, of the energetics and mechanism of the accumulative uptake of amino acids play a fundamental role in working out the concept of cotransport between Na^+ and amino acids by animal cells.

In the early fifties, Christensen's group tested several different cells and tissues for their ability to accumulate amino acids [54]. These investigations have shown Ehrlich cells to be most suitable for studying the mechanism and energetics of accumulative amino acid transport, a phenomenon first described by van Slyke and Meyer in 1913. Christensen's investigations have shown that substitution of extracellular Na^+ by K^+ reduces amino acid accumulation drastically. These results were interpreted as a consequence of a countertransport of K^+ and amino acids [60]. In 1963, however, Kromphardt et al. [61] have shown that not the reduction in electrochemical potential gradient of K^+, but that of extracellular Na^+ explained this phenomenon, and that amino acid influx is activated by extracellular Na^+ (Fig. 6). For this observation two different explanations were discussed: (1) the electrochemical potential difference of Na^+ might activate a primary active transport system and (2) the electrochemical potential difference of Na^+ might act as driving force in cotransport between Na^+ and amino acids as was postulated by Crane's gradient hypothesis for sugar resorption by the intestine [62].

Experiments were presented seemingly supporting the gradient hypothesis and/or disproving primary active transport and others giving just the opposite evidence. In favor of primary active transport and against cotransport as the only process for amino acid accumulation, experiments were presented showing that the electrochemical driving force for Na^+ is not always sufficient to explain the accumulation ratios for amino acids observed [63–66] and others, showing that amino acid transport and the conjugated driving force for Na^+ amino acid cotransport [5] were not always congruent [67]. Arguments drawn from these two types of experiments depend critically on an accurate determination of the electrochemical potential gradient of Na^+. Yet, both components of this quantity, the osmotic as well as the electrical one, are difficult to determine. To calculate the chemical potential difference (osmotic component) one has to know the activity coefficient of cytosolic Na^+ as well as the fact whether or not Na^+ is compartmentalized inside the cell so that the cytosolic Na^+ concentration differs from the mean intracellular concentration. On both questions reliable information is not available. Determinations of the mem-

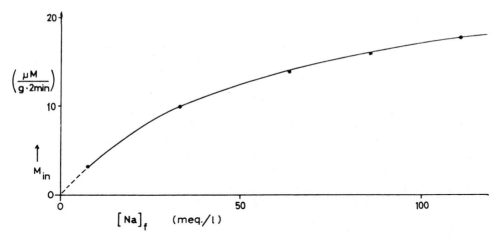

Fig. 6. Na^+-dependence of glycine-influx in Ehrlich cells. Glycine uptake was plotted against the Na^+-concentration in the incubation medium. (From [61].)

brane potential (electrical component), too, give experimental problems. Direct determinations of membrane potential of Ehrlich cells using microelectrodes do not show stable recording and were not performed for all situations relevant for testing the gradient hypothesis. Calculations using the Cl^--distribution are, at least under some conditions, leading to wrong values since Cl^- does not distribute passively, according to the Nernst equation, but is involved in cotransport [68] and countertransport [69] processes. Calculating an upper limit for the membrane potential from the K^+ distribution is not justified since if there are electrogenic pumps operating its value can be significantly higher than the K^+ equilibrium potential [70]. There is good evidence that also in Ehrlich cells the Na^+,K^+-pump is electrogenic and that under experimental conditions which most strongly seem to contradict the gradient hypothesis the pump is activated. It was shown that under this condition the membrane potential is considerably higher than the K^+ equilibrium potential [53,71]. For this purpose the potential determinations can be made by help of the distribution ratio of the lipophilic cation tetraphenylphosphonium (TPP) [50] or fluorescence quenching of cyanine dyes [72,73]. Both methods show a strong dependence of the membrane potential on the rate of the electrogenic Na^+,K^+-pump. Therefore, these experiments cannot disprove the gradient hypothesis but are very well compatible with it. Evidence for the gradient hypothesis and against primary active transport of amino acids results mainly from the following observations: In ATP-depleted cells an accumulative amino acid uptake is observed as long as a sufficient Na^+-gradient can be maintained [74,75] (Fig. 7). As the gradient breaks down, the accumulated amino acid leaves the cell [76]. While coupling between amino acid transport and ATP hydrolysis could not be demonstrated, coupling to Na^+ transport could be shown in metabolically inhibited cells [57,77,78] as well as in respiring ones. Further support for the gradient hypothesis results from electrical effects predicted by the gradient hypothesis as described before and verified experimentally. So it was shown that amino acid transport is rheogenic since it depolarizes partially the membrane potential in a concentration-dependent manner [49,50,71–73]. Variations in membrane potential influence the amino acid transport [26,50,70,75]. This influence is more pronounced at a lower amino acid concentration than at a higher one, indicating that mainly K_t is affected [26,50] (Fig. 8). As shown before, this behavior is best explained assuming as transport mechanism an uncharged carrier or a gated pore. As a consequence of these two effects, transport processes through different rheogenic transport mechanisms must influence each other indirectly, competing for the same energy source, mainly its electrical component (the membrane potential). Since this inhibition influences mainly K_t it might be misinterpreted as a competition for the same transport site. Tests do distinguish this indirect inhibition (pseudocompetition) from true competition were worked out [50]. All these electrical effects are only observed if the extracellular medium contains Na^+. The partial inhibition of amino acid transport by ouabain, which shows no time lag and which is much faster than the brake down of the chemical component of the Na^+ gradient, is explained by electrical effects since reduction in membrane potential as a consequence of blocking the electrogenic pump is very fast. Therefore,

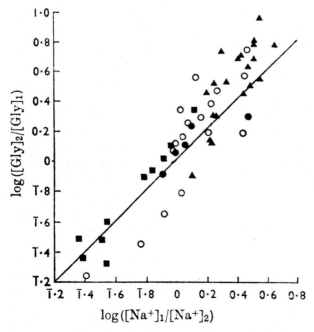

Fig. 7. Relationship between glycine and Na^+ distribution ratio. Transient accumulation ratio for glycine plotted against that of Na^+ for metabolically inhibited Ehrlich cells. (From [64].)

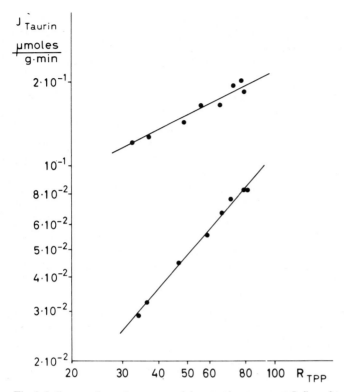

Fig. 8. Influence of membrane potential on taurine transport. Influx of taurine into Ehrlich cells is plotted against the distribution ratio of the lipophilic cation tetraphenylphosphonium (TPP) as a measure of the electrochemical activity coefficient $\xi = \exp(-F\psi/RT)$. (From [50].)

postulating a direct action of ouabain on amino acid transport is not justified.

Summarizing these results, one has good evidence that at least in Ehrlich cells the accumulative amino acid transport is coupled to an Na^+ flux in the same direction (cotransport) as postulated by the gradient hypothesis, while no hard evidence can be presented for primary active transport of amino acids.

(b) Epithelia

The energetics of transepithelial active transport was mainly studied for amino acids and sugars by small intestine and proximal tubule of the kidney. The transport systems in both organs seem to be very similar. The experimental material for other solutes and other epithelia is less numerous. As for nonpolarized cells, two hypotheses on the energetics were discussed: primary and secondary active transport. There is general agreement that transport through the paracellular pathway is exclusively passive, therefore the accumulative transepithelial transport must proceed through the cells of the epithelium (Fig. 9). Special aspects as compared to nonpolarized cells result from the fact that in transcellular transport the solutes have to cross two membranes. For accumulative transport it is necessary that these two membranes, the luminal and the contraluminal one, have different features. The "active" step responsible for feeding in the energy necessary for the transepithelial accumulation of organic solutes may in principle be located at the luminal or at the contraluminal membrane. To distinguish between these two possibilities one has to determine the intracellular electrochemical potential of the solute transported. If it is lower as in the two extracellular solutions the active step must be located at the contraluminal membrane actively pumping the solute out of the cell, while its entry through the luminal membrane is passive following the electrochemical potential gradient of the solute. If, on the other hand, the intracellular electrochemical potential of the solute is higher than in the two extracellular solutions there must be an active transport of the solute into the cell at the luminal membrane while the exit at the contraluminal side is passive. For sugars as well as for amino acids it was clearly shown that they were accumulated by the epithelial cells and therefore the "active" step must be located at the luminal membrane. Furthermore, it was shown that this accumulation needs Na^+ in the incubation medium.

The earliest observation on a role of Na^+Cl^- on resorption of sugars was reported by Reid 80 years ago. This information, however, was forgotten when 20 years ago the role of Na^+ in sugar resorption was rediscovered. These investigations led to the formulation of Crane's gradient hypothesis [62] for sugar resorption by

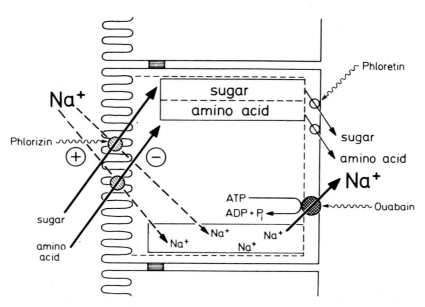

Fig. 9. Model for transepithelial transport of sugars and amino acids. Explanations see text. (From [100].)

small intesti For amino acids, too, a secondary active transport mechanism was postulated [10]. All predictions from the model could be verified experimentally: e.g. ouabain inhibition of the Na^+,K^+-pump, which is located at the contraluminal side of the epithelial cell, prevents transepithelial accumulation of sugars as well as of amino acids [79]. Influx [80] (Fig. 10) and steady-state distribution depend on the Na^+ concentration in the lumen. Furthermore, coupling between transport of Na^+ and of the solutes across the luminal membrane could be shown [80]. Arguments against explaining the transepithelial active uptake of the organic solutes by secondary active transport arose mainly from calculations that the energy available from the Na^+ gradient across the luminal membrane seemed to be insufficient to explain the accumulation inside the epithelial cell. These results, however, were not confirmed, since direct measurements using microelectrodes to determine the membrane potential of the luminal membrane [81–83] as well as the intracellular activity of Na^+ [38] have shown that the Na^+ gradient is sufficient to explain the intracellular accumulation of the solutes. The direct measurement of the membrane potential across the luminal membrane in the presence of actively accumulated sugars or amino acids shows a concentration-dependent depolarization that requires Na^+ in the lumen [14,51,84] (Fig. 11). This clearly indicates that the cotransport of Na^+ with these solutes is rheogenic. The depolarization reduces the Na^+-gradient and therefore leads to a reduction in transport by other rheogenic cotransport systems

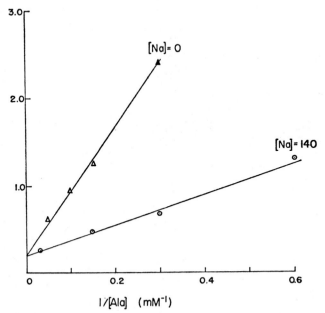

Fig. 10. Na^+-dependence of alanine influx in small intestine. Double reciprocal plot of alanine influx against alanine concentration in mucosal solution; effect of two different Na^+-concentrations. (From [80].)

and complicates the analysis of the specificity of different cotransport systems present in the same cell. Presumably this pseudocompetition is the reason for the mutual inhibition between sugar and amino acid transport [85], which had been interpreted by some authors postulating a polyfunctional carrier, responsible for the transport of sugars as well as amino acids and by others as a competition for the same energy source [85–88]. The specificity of different transport systems for sugars and amino acids was studied very intensively for small intestine [89] and kidney [90,91]. Further information on specificity of transport systems was obtained from studies on the transport defects in patients with well defined inborn errors of transport [92,93].

The transport of sugars is specifically inhibited by phlorizin in a competitive manner. Studies with this inhibitor led to valuable information on the cotransport systems, so by binding studies the numbers of carriers per cell and from these the turnover numbers for cotransport were estimated. Potential dependence of the binding constant for phlorizin and of K_t for glucose transport were discussed to get information on the charge transferring step of the carrier model (see Section 2b) and seems to be best explained assuming some type of gated channel as cotransport system for glucose [94,95].

Besides investigations with whole tissue or isolated cells, studies with vesicles isolated from luminal or basal-lateral membranes of small intestine or kidney gave

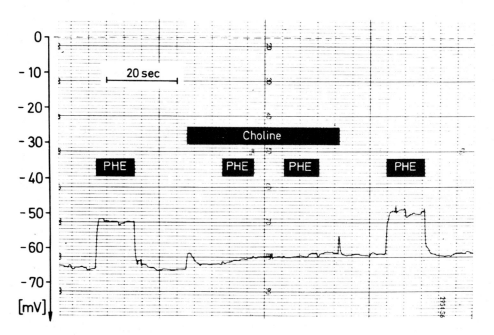

Fig. 11. Influence of cotransport on membrane potential. Na^+-dependent depolarisation of intracellular electrical potential at proximal tubule of rat kidney perfused with 5 mM L-phenylalanine. (From [14].)

valuable information for understanding the mechanism of transepithelial transport of sugars and amino acids. Special aspects of these studies are described in the next section.

In recent years some progress was obtained in isolating the cotransport system characterizing it chemically and incorporating it into vesicles to demonstrate its transport ability.

Intensive studies to prove the γ-glutamyl cycle for active amino acid transport [23] led to no results supporting this hypothesis but increasing evidence against the cycle was obtained [24,42,43]. Probably γ-glutamyl transferase, the key enzyme of the cycle, is involved in the degradation of certain γ-glutamyl peptides, as ophthalmic acid, glutathione, and its S-substituted derivatives [96,97].

(c) Cell-free systems (vesicles, liposomes)

To evade the interferences due to metabolism or intracellular compartmentalization and sequestration, isolated membrane preparations in the form of vesicles have proved useful, and can be obtained either directly from isolated natural membranes [32,36] or, going one step further, by extracting "transport-related" proteins and reconstituting them into artificial phospholipid vesicles (liposomes) [34,98]. The preparation of the natural membrane vesicles is aided by the natural tendency of membrane fragments to form closed vesicles spontaneously under suitable conditions. Reconstituted vesicles are more difficult to obtain, for even rather pure preparations of transport components show little tendency to spontaneously integrate themselves in an artificial lipid membrane. Nonetheless, some successful attempts have been described in the literature of such an incorporation.

Either kind of vesicles has been prepared from both animal cells and microorganisms. Besides being devoid of metabolism and subcompartments, these vesicles have the advantage that they can be separated according to their origin from different regions of the cell. So in studies of epithelial cells vesicles of the brush border can be separated from those of the basal lateral membrane to study the transport behavior of the entry step at the luminal membrane and the exit step at the basal lateral membrane separately. The absence of metabolism, although advantageous in some respect, precludes studying active transport under experimentally convenient steady-state conditions. Instead, the active nature of a given transport system may manifest itself only through highly transient phenomena, such as "overshoot" phenomenon (Fig. 12) which is of central significance in transport studies with cell-free systems such as vesicles and reconstituted liposome preparations. It is a transient accumulation of the transported solute in the vesicles which will last as long as a gradient of driver ion species, Na^+ or H^+, respectively, is maintained. Usually this gradient is produced by the sudden addition of the driver ion (e.g. Na^+) to a suspension of Na^+-free vesicles.

The electrical potential can be modified by special ionophores or by varying the anion of the added Na^+ salt; a more mobile ion should tend to increase the electrical potential whereas the opposite should be expected by a slow anion. In

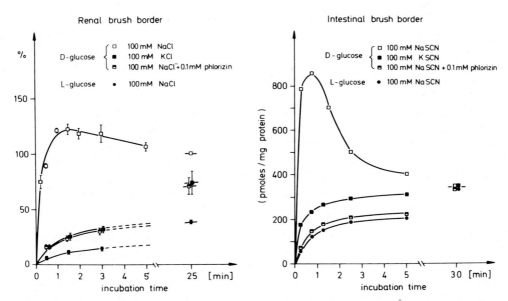

Fig. 12. Overshoot in glucose-uptake by vesicles. D-Glucose uptake by brush-border vesicles from kidney and intestine under different ion gradients. (From [100].)

vesicles the membrane potential cannot be measured directly using microelectrodes. Usually it is monitored indirectly using fluorescence-quenching technics [99].

The ion gradient will subsequently dissipate at a rate which depends on the permeabilities of both Na^+ and the accompanying anion (or countertransport cation).

Any solute that is driven by this ion gradient tends to equilibrate with the electrochemical potential difference of the ion species. If its entrance by Na^+-linked cotransport is fast enough as compared to the Na^+ leak pathway, then its distribution will approach rapidly equilibrium with the electrochemical activity ratio of Na^+ long before the latter has dissipated to a major extent. At this point the distribution of the solute (e.g. sugar) is at its peak, as here the uptake curve of the organic solute and decay curve of the ion meet. Subsequently, as the Na^+ gradient continues to decay, the sugar distribution will go down concomitantly.

As the mathematical representation of the complete overshoot curve is very complicated, only the amplitude of the peak is usually taken as an indicator of secondary active transport, provided that other causes of an overshoot phenomenon, such as osmotic swelling or temporary changes in membrane potential, can be excluded.

With the above described membrane preparations, natural or reconstituted vesicles, evidence has been strengthened that the active transport of metabolites in animal cells is almost exclusively secondary active. In particular the active transport of certain sugars, of certain amino acids, of anionic metabolites, such as acidic amino acids, lactate, citrate, and bile acids across the epithelial layer of certain regions of the intestinal tract or of the kidney tubule, is carried out by cotransport with Na^+ through the brush-border membrane, while its exit across the basal lateral membranes is by facilitated diffusion. With vesicles it was also confirmed that certain such metabolites are not transported actively but translocated by facilitated diffusion, such as the amino acids transported by the L system or glucose into most cells other than the intestinal and renal epithelial cells and nucleosides.

As to microorganisms, studies with natural and of reconstituted vesicles have also revealed or confirmed that many of the common nutrients and metabolites are transported actively by cotransport with H^+. The generation of a "protonmotive force", in analogy to the electrical potential gradient of Na^+, has previously often been called "energization of the membrane" at a time when the ion gradient had not clearly been identified.

Investigations on primary active transport are also performed with vesicles. For inside-out vesicles no additional experimental problems arise, since the enzymatic site of the transport system located at the cytoplasmic site of the membrane is freely accessible to the substrate of the exergonic driving reaction from the incubation medium. Right-side-out vesicles, however, have to be prepared in the presence of the driver substrate, e.g. ATP, or a system for its regeneration, e.g. ATP/ADP + phosphoenolpyruvate + pyruvate kinase.

4. Conclusion

With the exception of the phosphotransferase system that is responsible for uptake of several sugars by bacteria, the active uptake of organic solutes is secondary active and coupled via cotransport to the downhill transport of a cation, Na^+ in animal cells and H^+ ions in microorganisms. In transcellular transport in epithelia, such as small intestine and proximal tubule of the kidney, the solutes are accumulated inside the cell via a cotransport mechanism at the luminal membrane, and leave the cell passively presumably by facilitated diffusion at the contraluminal side.

References

1 Crane, R.K. (1977) Rev. Physiol. Pharmacol. 78, 99–159.
2 Curran, P.F. (1973) in H.H. Ussing and N.A. Thorn (Eds.), Transport Mechanisms in Epithelia, pp. 298–312, Academic Press, New York.
3 Harold, F.M. (1972) Bacteriol. Rev. 36, 172–230.
4 Heinz, E. (1972) in L.E. Hokin (Ed.), Metabolic Pathways, Vol. 6, pp. 455–501, Academic Press, New York.
5 Heinz, E. (1974) in F. Bronner and A. Kleinzeller (Eds.), Current Topics in Membranes and Transport, Vol. 4, pp. 137–159, Academic Press, New York.
6 Heinz, E. (1978) Mechanisms and Energetics of Biological Transport, Springer, Berlin.
7 Lerner, J. (1978) A Review at Amino Acid Transport Processes in Animal Cells and Tissues, Orono University Press, Orono.

8 Mitchell, P. (1973) Bioenergetics 3, 63–91.
9 Schultz, S.G. (1978) in T.E. Andreoli, J.F. Hoffman and D.D. Fanestil (Eds.), Physiology of Membrane Disorders, pp. 273–286, Plenum, New York.
10 Schultz, S.G. and Curran, P.F. (1970) Physiol. Rev. 50, 637–718.
11 Silbernagl, L.S., Foulkes, E.C. and Deetgen, P. (1975) Rev. Physiol. Biochem. Pharmacol. 74, 105–167.
12 Silverman, M. (1976) Biochim. Biophys. Acta 457, 303–351.
13 Ullrich, K.J. (1976) Kidney Int. 9, 134–148.
14 Ullrich, K.J. (1979) in G. Giebisch, D.C. Tosteson and H.H. Ussing (Eds.), Membrane Transport in Biology, Vol. IV, pp. 413–448, Springer, Berlin.
15 Schultz, S.G. (1979) in G. Giebisch, D.C. Tosteson and H.H. Ussing (Eds.), Membrane Transport in Biology, Vol. IV, pp. 749–780, Springer, Berlin.
16 West, I.C. (1980) Biochim. Biophys. Acta 604, 91–126.
17 Kedem, O. (1961) in A. Kleinzeller and A. Kotyk (Eds.), Membrane Transport and Metabolism, pp. 87–93, Academic Press, New York.
18 Ussing, H.H. (1952) Adv. Enzymol. 13, 21–65.
19 Katchalsky, A. and Curran, P.F. (1965) Nonequilibrium Thermodynamics in Biophysics, Harvard University Press, Cambridge, MA.
20 Kedem, O. and Caplan, S.R. (1965) Trans. Faraday Soc. 61, 1897–1911.
21 Jacquez, J.A. (1961) Proc. Natl. Acad. Sci. USA 47, 153–163.
22 Postma, P.W. and Roseman, S. (1976) Biochim. Biophys. Acta 457, 213–257.
23 Meister, A. (1973) Science 180, 33–39.
24 Silbernagl, S., Wendel, A., Pfaller, W. and Heinle, H. (1978) in H. Sies and A. Wendel (Eds.), Functions of Glutathione in Liver and Kidney, pp. 60–72, Springer, Berlin.
25 Heinz, E., Geck, P. and Wilbrandt, W. (1972) Biochim. Biophys. Acta 255, 442–461.
26 Heinz, E. and Geck, P. (1977) in J.F. Hoffman (Ed.), Coupled Transport Phenomena in Cells and Tissues, The Peter F. Curran Memorial Symposium, pp. 13–30, Raven Press, New York.
27 Ullrich, K.J., Frömter, E. and Baumann, K. (1969) in H. Passow and R. Stämpfli (Eds.), Laboratory Techniques in Membrane Biophysics, pp. 106–129, Springer, Berlin.
28 Ussing, H.H. and Zerahn, K. (1951) Acta Physiol. Scand. 23, 110–127.
29 Wilson, T.H. (1962) Intestinal Absorption, Saunders, Philadelphia.
30 Handler, J.S., Perkins, F.M. and Johnson, J.P. (1980) Am. J. Physiol. 238, F1–F9.
31 Kaback, H.R. and Stadtman, E.R. (1966) Proc. Natl. Acad. Sci. USA 55, 920–927.
32 Hopfer, U., Nelson, K., Perrotto, J. and Isselbacher, K.J. (1973) J. Biol. Chem. 248, 25–32.
33 Racker, E., Knowles, A.F. and Eyton, E. (1975) Ann. N.Y. Acad. Sci. 264, 17–31.
34 Crane, R.K., Malathi, P. and Preiser, H. (1976) Biophys. Biochem. Res. Commun. 67, 214–216.
35 Sachs, G. and Kinne, R. (1978) in T.E. Andreoli, J.F. Hoffman and D.D. Fanestil (Eds.), Physiology of Membrane Disorders, pp. 95–105, Plenum, New York.
36 Murer, H. and Kinne, R. (1980) J. Membr. Biol. 55, 81–95.
37 Heinz, E. (1981) in C. Slayman (Ed.), Electrogenic Pumps, Yale University Press, New Haven, in press.
38 Khuri, R.N. (1979) in G. Giebisch, D.C. Tosteson and H.H. Ussing (Eds.), Membrane Transport in Biology, Vol. IV, pp. 47–95, Springer, Berlin.
39 Lassen, U.V. and Rasmussen, B.E. (1979) in G. Giebisch, D.C. Tosteson and H.H. Ussing (Eds.), Membrane Transport in Biology, Vol. 1, pp. 169–203, Springer, Berlin.
40 Skulachev, V.P. (1972) J. Bioenerg. 3, 25–38.
41 Hoffman, J.F. and Laris, P.C. (1974) J. Physiol. 239, 519–552.
42 Schulman, J.D., Goodman, S.I., Mace, J.W., Patrick, A.D., Tietze, E. and Butler, E.J. (1975) Biophys. Biochem. Res. Commun. 65, 68–74.
43 Pellefigue, F., DeBrohun Butler, J., Spielberg, S.P., Hollenberg, M.D., Goodman, S.I. and Schulman, J.D. (1976) Biophys. Biochem. Res. Commun. 73, 997–1002.
44 Schuurmans Stekhoven, F.M.A.H. and Bonting, S.L. (1981) This volume, Chapter 6.
45 Konings, W.N., Hellingwerf, K. and Robbilard, G.T. (1981) This volume, Chapter 10.

46 Jacquez, J.A. (1972) Biochim. Biophys. Acta 318, 411–425.
47 Semenza, G. (1967) J. Theor. Biol. 15, 145–152.
48 Patlak, C.S. (1957) Bull. Math. Biol. 19, 209–217.
49 Eddy, A.A., Philo, R. (1976) in S. Silbernagl, F. Lang and R. Greger (Eds.), Amino Acid Transport and Uric Acid Transport, pp. 27–33, Thieme, Stuttgart.
50 Geck, P. (1978) Habilitationsschrift, Univ. Frankfurt.
51 Okada, Y., Tsuchiya, W., Irimjiri, A. and Inoyue, A. (1976) J. Membr. Biol. 31, 221–232.
52 Mullins, L.J. and Noda, K. (1963) J. Gen. Physiol. 47, 117–132.
53 Pietrzyk, C., Geck, P. and Heinz, E. (1978) Biochim. Biophys. Acta 513, 89–98.
54 Christensen, H.N. (1969) Adv. Enzymol. 36, 1–20.
55 Heinz, E. and Walsh, P.M. (1959) J. Biol. Chem. 233, 1488–1493.
56 Johnstone, R.M. and Scholefield, P.G. (1961) J. Biol. Chem. 236, 1419–1424.
57 Eddy, A.A. (1968) Biochem. J. 108, 195–206.
58 Vidaver, G.A. (1964) Biochemistry 3, 662–667.
59 Bégin, N. and Scholefield, P.G. (1965) J. Biol. Chem. 240, 332–337.
60 Riggs, T.R., Walker, L.M. and Christensen, H.N. (1958) J. Biol. Chem. 233, 1479–1484.
61 Kromphardt, H., Grobecker, H., Ring, K. and Heinz, E. (1963) Biochim. Biophys. Acta 74, 549–551.
62 Crane, R.K. (1962) Fed. Proc. 21, 891–895.
63 Jacquez, J.A. (1972) in E. Heinz (Ed.), Na^+-linked Transport of Organic Solutes, pp. 4–14, Springer, Berlin.
64 Eddy, A.A. (1968) Biochem. J. 108, 489–498.
65 Johnstone, R.M. (1974) Biochim. Biophys. Acta 356, 319–330.
66 Potashner, S.J. and Johnstone, R.M. (1970) Biochim. Biophys. Acta 233, 91–103.
67 Schafer, J.A. and Heinz, E. (1971) Biochim. Biophys. Acta 249, 15–33.
68 Geck, P., Pietrzyk, C., Burckhardt, B.-C., Pfeiffer, B. and Heinz, E. (1980) Biochim. Biophys. Acta 600, 432–447.
69 Villereal, M.L. and Levinson, C. (1977) J. Cell Physiol. 90, 553–564.
70 Gibb, L.E. and Eddy, A.A. (1972) Biochem. J. 129, 979–981.
71 Heinz, E., Geck, P. and Pietrzyk, C. (1975) Ann. N.Y. Acad. Sci. 264, 428–441.
72 Laris, P.C., Pershadsingh, H.A. and Johnstone, R.M. (1976) Biochim. Biophys. Acta 436, 475–488.
73 Burckhardt, G. (1977) Biochim. Biophys. Acta 468, 227–237.
74 Eddy, A.A. and Hogg, M.C. (1969) Biochem. J. 114, 807–814.
75 Reid, M., Gibb, L.E. and Eddy, A.A. (1974) Biochem. J. 140, 383–393.
76 Eddy, A.A. (1972) in E. Heinz (Ed.), Na^+-linked Transport of Organic Solutes, pp. 28–38, Springer, Berlin.
77 Schafer, J.A. and Jacquez, J.A. (1967) Biochim. Biophys. Acta 135, 1081–1083.
78 Vidaver, G.A. (1964) Biochemistry 3, 803–808.
79 Schultz, S.G., Fuisz, R.E. and Curran, P.F. (1966) J. Gen. Physiol. 49, 849–866.
80 Curran, P.F., Schultz, S.G., Chez, R.A. and Fuisz, R.E. (1967) J. Gen. Physiol. 50, 1261–1286.
81 Boulpaep, E.L. (1979) in G. Giebisch, D.C. Tosteson and H.H. Ussing (Eds.), Membrane Transport in Biology, Vol. IV, pp. 97–144, Springer, Berlin.
82 Windhager, E.E. and Giebisch, G. (1965) Physiol. Rev. 45, 214–244.
83 Frömter, E. (1974) Fortschr. Zool. 23, 248–263.
84 Armstrong, W.McD. (1975) in T.Z. Czaky (Ed.), Intestinal Absorption and Malabsorption, pp. 45–65, Raven, New York.
85 Newey, H. and Smyth, D.H. (1964) Nature 202, 400–401.
86 Munck, B.G. (1980) Biochim. Biophys. Acta 597, 411–417.
87 Murer, H., Sigrist-Nelson, K. and Hopfer, U. (1975) J. Biol. Chem. 250, 7392–7396.
88 Kimmich, G.A. and Carter-Su, C. (1978) Am. J. Physiol. 235, C73–C81.
89 Crane, R.K. (1967) Physiol. Rev. 40, 789–825.
90 Ullrich, K.J., Frömter, E., Hinton, B.T., Rumrich, G. and Kleinzeller, A. (1976) in U. Schmidt and U.C. Dubach (Eds.), Biochemical Aspects of Kidney Function, pp. 356–368, Huber, Bern.

91 Silbernagl, S. and Völkl, H. (1977) Pflügers Arch. 367, 221–227.
92 Scriver, C.R. and Bergeron, M. (1974) in W.L. Nyhan (Ed.), Heritable Disorders of Amino Acid Metabolism, pp. 515–530, Wiley, New York.
93 Silbernagl, S., Foulkes, E.C. and Deetgen, P. (1975) Rev. Physiol. Biochem. Pharmacol. 74, 105–167.
94 Toggenburger, G., Kessler, M., Rothstein, A., Semenza, G. and Tannenbaum, C. (1978) J. Membr. Biol. 40, 269–290.
95 Aronson, P.S. (1978) J. Membr. Biol. 42, 81–98.
96 Orlowski, M., Wilk, S. (1977) in W.G. Guder and U. Schmidt (Eds.), Current Problems in Clinical Biochemistry, Vol. 8, Biochemical Nephrology, pp. 66–72, Huber, Bern.
97 Wendel, A., Heinle, H. and Silbernagl, S. (1977) in W.G. Guder and U. Schmidt (Eds.), Current Problems in Clinical Biochemistry, Vol. 8, Biochemical Nephrology, pp. 73–80, Huber, Bern.
98 Kinne, R. and Faust, R.G. (1978) Biochem. J. 168, 311–314.
99 Beck, J.C. and Sacktor, B. (1978) J. Biol. Chem. 253, 7158–7162.
100 Ullrich, K.J., Frömter, E. and Murer, H. (1979) Klin. Wochenschr. 57, 977–991.

CHAPTER 12

The coupled transport of water

ALAN M. WEINSTEIN, JOHN L. STEPHENSON
and KENNETH R. SPRING *

*Section on Theoretical Biophysics, National Heart, Lung and Blood Institute and Mathematical Research Branch, National Institute of Arthritis, Metabolism, and Digestive Diseases, National Institutes of Health, Bethesda, MD 20205, and * Laboratory of Kidney and Electrolyte Metabolism, National Heart, Lung and Blood Institute, National Institutes of Health, Bethesda, MD 20205, U.S.A.*

1. Introduction

The transport of salt solutions by epithelia has been the focus of active physiological investigation for over a century. Part of the impetus for this activity was the early recognition that with respect to water transport, epithelia differed from passive membranes. The transepithelial movement of water is a process that is coupled to the vital activity of the living tissue. It has been the task of physiologists to make precise our understanding of the nature of this coupling.

Many of the early observations on water transport were made on experimental preparations of mammalian intestine, a tissue whose natural function includes the transfer of large volumes of salt solution from the intestinal lumen to the blood. In 1902, Waymouth Reid [1] presented his perspective on epithelial water transport and emphasized the differences between tissue and simple membrane. Citing his own work, Reid stressed that volume transport occurred spontaneously across an intestine placed between bathing media of identical composition. A necessary condition for this transport was an intact layer of healthy epithelial cells lining the luminal surface. Indeed, transfer of water from lumen to blood could occur against an adverse osmotic gradient. Reid also knew from work of Gummilewsky and Heidenhain that when the intestinal lumen contained a salt solution at roughly equal concentration to that in blood, intestinal transport of salt and water would proceed leaving the luminal salt concentration essentially unaltered.

In the modern era, this basic phenomenology of coupled water transport has been confirmed in a variety of tissues. For example, Diamond [2–4] studied the transport properties of the fish gallbladder. In his experiments the gallbladder was excised from the fish and suspended in a relatively large bath so that the composition of both luminal and (serosal) bath solutions could be controlled. Volume transport out of the lumen was manifest as weight loss of the gallbladder lifted out of the bath. Fig. 1 shows the results of a study of the effect of osmotic gradients on transepi-

Bonting/de Pont (eds.) Membrane transport
© *Elsevier/North-Holland Biomedical Press, 1981*

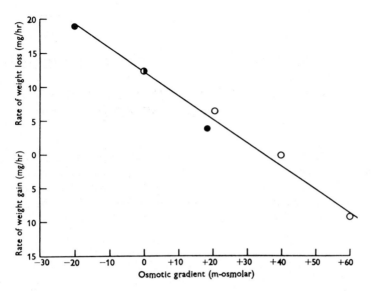

Fig. 1. Effect of osmotic gradients on water absorption by a normal gallbladder. Abcissa, osmolarity of luminal solution minus osmolarity of outer solution. Ordinate, rate of change of weight of gallbladder preparation (weight loss means that fluid is being absorbed from the lumen). Open circles denote experiments in which the luminal solution is different from the reference and the bath solution is at reference; closed circles, experiments in which the bath is different from reference and the lumen at reference; the half-filled circle, both solutions at reference. (From [4].)

thelial volume flow (from [4]). In particular it demonstrates the capability of the tissue to transport water against an adverse osmotic gradient of up to 40 mosM. By keeping track of both the weight and composition of the luminal contents, Diamond could calculate the composition of the transported fluid. Fig. 2 shows the results of experiments in which the gallbladder contents and bath were initially identical solutions [2]. Even after gallbladders had transported up to 75% of their luminal contents into the surrounding bath, no significant osmotic gradients between the two media were established. This latter phenomenon has come to be termed "isotonic transport".

The effort to understand the interrelationship of physical forces and cellular activity in the transport of salt solutions was advanced considerably by the experiments of Curran and Solomon [5] on the rat small intestine. These workers perfused the intestinal lumen of the rat with solutions all isotonic to plasma but with varying NaCl concentration. They measured both the solute flux and volume flow out of the lumen and their results are plotted in Fig. 3. These data are taken as strong evidence for the primacy of solute flux as the determinant of transepithelial water flow. This view was confirmed in experiments by Windhager et al. [6] in which the epithelium under investigation was the renal proximal tubule of the amphibian, *Necturus*.

Coupled transport of water

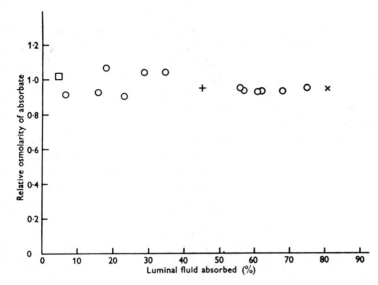

Fig. 2. Osmolarity of the absorbate, relative to Ringer's solution, at various stages in the absorption process. Abscissa, percentage of the initial luminal volume reabsorbed by the end of the experiment; ordinate, ratio of the osmolarity of the absorbate to the osmolarity of Ringer's solution. Gallbladders were allowed to absorb varying fractions of the luminal fluid, then the absorbate osmolarity was computed from the amount of water and ions absorbed. (Identical solutions bathed both sides of the gallbladder except in the experiment indicated by x, in which the outer solution was Ringer's and the luminal solution K-free. The circles denote experiments performed with Ringer's solution; the square, that with Na_2SO_4 solution; and the +, that with a K-rich solution.) (From [2].)

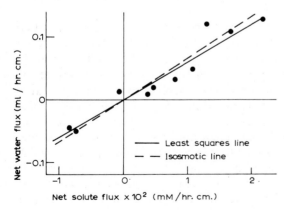

Fig. 3. Relation between net solute flux (net Na flux + 1/2 net mannitol flux) and net water flux. The intestinal lumen of the rat was perfused in vivo with solutions all isotonic to plasma but with varying NaCl concentration. The solid line has been drawn through the points by the method of least squares; the broken line is the expected relation for transport of solute in isotonic concentration. (From [5].)

Again, the tubule lumen was perfused with several isotonic solutions of varying salt content. A graph of water flow against solute flux revealed linear dependence, with zero volume flow in the absence of solute flux. In each of these investigations, the transepithelial solute flux was largely that of NaCl. Thus it was established that metabolically driven transport of salt by the epithelial cells was responsible for producing volume flow. At this point the task of relating water flow to the vital activity of the tissue was reduced to deciphering the details of the coupling of transepithelial salt and water fluxes.

Of inestimable importance in the establishment of lines of experimental inquiry into coupled water transport has been the formalism of non-equilibrium thermodynamics. Non-equilibrium thermodynamics provides a concise mathematical representation of the passive permeability properties of simple membranes. Furthermore, the properties of composite membrane systems can be predicted in terms of the parameters associated with the component membranes [7–9]. the modern effort to understand intraepithelial solute-solvent coupling may be summarized as the attempt to represent the epithelium as an array of compartments bounded by simple membranes. The epithelial cells are polarized with a highly infolded apical (luminal) surface separated by tight junctions from the basolateral (serosal) membrane. The tight junctions form belt-like structures around the epithelial cells and prevent the paracellular movement of large molecules across the tissue. Small molecules pass readily through these junctions [10]. Beneath the junctions the cells are separated by lateral intercellular spaces, and across the basolateral cell membrane there is metabolically driven transport of solute into these intercellular spaces. The aim of such model systems is the accurate simulation of the whole tissue phenomenology. This type of representation has in turn stimulated experimental effort to determine the morphology and dimensions of the intraepithelial structures. Further, the passive permeability properties of the component membranes has been an area of intensive investigation.

At this time, much of our intuition about the coupled transport of water is still far from being considered experimentally secure. Indeed, areas of frank controversy remain [11] and the modern debate seems no less animated than that reported 60 years ago [12]. the framework of this chapter will be, therefore, a survey of the mathematical models used to describe solute-solvent coupling. Within the context of the appropriate model, pertinent experimental observations will be discussed and points of uncertainty will be indicated. The material of the chapter divides naturally into two sections that parallel the experimental division of macroscopic observations on tissues and microscopic physiology. In the first section, phenomenological models of transport will be considered. We present a general mathematical framework with which the basic notions of coupled water transport may be defined and which may be applied to more detailed models. The linear phenomenology of non-equilibrium thermodynamics is developed within this general framework. In this section, the effect of unstirred layers on solute-linked water transport is discussed. In the second section, deterministic models of intraepithelial solute-solvent coupling are taken up, and these are essentially models of the lateral intercellular space. We begin with an

Coupled transport of water

elementary compartment model of the paracellular channel and indicate the function of the several model parameters in the coupling of transepithelial water flow. Next, a model that permits intraepithelial concentration gradients is displayed and coupled water transport by this "standing gradient" model is described according to our general framework. Finally, the effect of introducing a permeable tight junction is discussed for the compartment model of the interspace and some of the efforts to elaborate more comprehensive simulations are indicated.

2. Phenomenological models of transport

(a) General formulation

Any mathematical model of an epithelium provides a means of computing the flows of volume and solutes across the tissue in terms of the pressures and concentrations in the bathing media separated by the tissue. If we restrict our attention to the case of a single neutral solute we may write formally

$$J_s = J_s(C_M, C_S, P_M, P_S; C_0, P_0)$$
$$J_v = J_v(C_M, C_S, P_M, P_S; C_0, P_0) \tag{1}$$

where J_s and J_v are transepithelial solute and volume flux from mucosal to serosal baths, C_0 and P_0 are reference concentrations and pressures, C_M and P_M are the deviations from reference of concentration and pressure within the mucosal bath, and C_S and P_S are the deviations from reference of concentration and pressure within the serosal bath. Denote by $(J_s)_0$ and $(J_v)_0$, the solute and volume flux in the case that both bathing media are identical and at the reference conditions. The concentration of the solute within the transported solution is naturally defined by the relation

$$C_R = \frac{J_s}{J_v} \tag{2}$$

and for the case of equal bathing media

$$(C_R)_0 = \frac{(J_s)_0}{(J_v)_0} \tag{3}$$

For the remainder of this analysis we shall understand the concentrations to be the osmotic activity of NaCl in the respective solutions. For the epithelial models under consideration we shall suppose that the volume flow from mucosa to serosa is positive when the baths are identical and that this flow declines with increases in mucosal bath salt concentration. Denote by $\hat{C}$ that increase in mucosal bath concentration that just nullifies transepithelial volume flow. Thus

$$J_v(\hat{C},0,0,0;C_0,P_0)=0 \tag{4}$$

and term $\hat{C}$ the strength of transport of the epithelium. Given a model relation represented in Eqn. 1, Eqn. 4 specifies an equation that may, in theory, be solved for $\hat{C}$.

There are two model equations that correspond to the two types of experimental observation of isotonic transport. In the first case, the mucosal bath volume is small relative to a large serosal bath, initially at the reference concentration. With time, a substantial fraction of the mucosal contents is transferred to the serosa and the mucosal bath concentration tends to a limiting value. Because of its large size, the serosal bath concentration is little changed. The mucosal bath, however, comes to transport equilibrium, with the concentration of the transported solution essentially equal to the mucosal bath concentration. This may be written

$$C_R(C_M^*,0,0,0;C_0,P_0)=C_0+C_M^* \tag{5}$$

The condition of isotonic transport means that the mucosal deviation C_M^*, that satisfies Eqn. 5 is at most a few percent of the reference, C_0. The second type of experiment is that of a small serosal bath relative to a mucosal bath at the reference concentration. This includes the type of experiment in which the organ lumen is filled with the reference solution, the organ suspended, and the serosal drainage collected. (See, for example, Diamond [13] using rabbit gallbladder or Lee [14] using rat small intestine.) In this case the serosa is at transport equilibrium so that

$$C_R(0,C_S^*,0,0;C_0,P_0)=C_0+C_S^* \tag{6}$$

Again, isotonic transport implies that the serosal deviation, C_S^*, that satisfies Eqn. 6 is at most a few percent of C_0.

It is worth emphasis that the verification of isotonic transport for a particular model necessitates the solution of Eqns. 5 and 6. In general, the simple calculation of $(C_R)_0$, the reabsorbate concentration with exactly equal bathing media, is not sufficient. Nevertheless, $(C_R)_0$ is an important theoretical aspect of a model and has been used to define the coupling coefficient of osmotic transport [15] by

$$\gamma=\frac{C_0}{(C_R)_0}=\frac{(J_v)_0}{(J_s)_0/C_0} \tag{7}$$

Thus, the osmotic coupling coefficient is the ratio of the actual volume flow observed with equal bathing media relative to the virtual volume flow that would be necessary for reabsorption to be isotonic. In the models we shall consider, $0 \leq \gamma \leq 1$. For tight coupling, $\gamma \approx 1$; in an uncoupled system $\gamma = 0$. γ has also been termed the efficiency of osmotic transport [16].

It will be useful for relating the general framework outlined above to the experimentally oriented models to consider the linear model that best approximates Eqn. 1. To do this, represent J_s and J_v by their respective Taylor series

$$J_s = (J_s)_0 + \left(\frac{\partial J_s}{\partial C_M}\right)_0 C_M + \left(\frac{\partial J_s}{\partial C_S}\right)_0 C_S + \left(\frac{\partial J_s}{\partial P_M}\right)_0 P_M + \left(\frac{\partial J_s}{\partial P_S}\right)_0 P_S$$

$$J_v = (J_v)_0 + \left(\frac{\partial J_v}{\partial C_M}\right)_0 C_M + \left(\frac{\partial J_v}{\partial C_S}\right)_0 C_S + \left(\frac{\partial J_v}{\partial P_M}\right)_0 P_M + \left(\frac{\partial J_v}{\partial P_S}\right)_0 P_S \quad (8)$$

The applicability of such approximate linear models to the experimental range of values of the independent variables is certainly suggested by Fig. 1. In what follows we shall assume both bathing media at the reference pressure. The reabsorbate tonicity may then be written

$$C_R = \frac{(J_s)_0 + \left(\frac{\partial J_s}{\partial C_M}\right)_0 C_M + \left(\frac{\partial J_s}{\partial C_S}\right)_0 C_S}{(J_v)_0 + \left(\frac{\partial J_v}{\partial C_M}\right)_0 C_M + \left(\frac{\partial J_v}{\partial C_S}\right)_0 C_S} \quad (9)$$

The strength of transport for this model may be estimated, following Eqn. 4, by

$$\hat{C} = -(J_v)_0 / \left(\frac{\partial J_v}{\partial C_M}\right)_0 \quad (10)$$

The derivative $(\partial J_v/\partial C_M)_0$ is the change in transepithelial volume flow with respect to a change in mucosal bath osmolality, or essentially, epithelial water permeability. Intuitively, Eqn. 10 states that the osmotic gradient across the epithelium, $\hat{C}$, produces a serosal to mucosal water flow of magnitude $\hat{C}(\partial J_v/\partial C_M)_0$, which just nullifies the coupled water flow from mucosa to serosa, $(J_v)_0$.

The achievement of isotonic transport may be assessed by substituting Eqn. 9 into Eqns. 5 and 6 to obtain

$$C_0 + C_M^* = \frac{(J_s)_0 + \left(\frac{\partial J_s}{\partial C_M}\right)_0 C_M^*}{(J_v)_0 + \left(\frac{\partial J_v}{\partial C_M}\right)_0 C_M^*} \quad (11)$$

and

$$C_0 + C_S^* = \frac{(J_s)_0 + \left(\frac{\partial J_s}{\partial C_S}\right)_0 C_S^*}{(J_v)_0 + \left(\frac{\partial J_v}{\partial C_S}\right)_0 C_S^*} \tag{12}$$

Eqns. 11 and 12 are quadratic in C_M^* and C_S^* but in the case that transport is close to isotonic one may estimate the solutions by

$$C_M^* = -\frac{(J_s)_0 - C_0(J_v)_0}{-C_0\left(\frac{\partial J_v}{\partial C_M}\right)_0 + \left(\frac{\partial J_s}{\partial C_M}\right)_0 - (J_v)_0} \tag{13}$$

$$C_S^* = \frac{(J_s)_0 - C_0(J_v)_0}{C_0\left(\frac{\partial J_v}{\partial C_S}\right)_0 - \left(\frac{\partial J_s}{\partial C_S}\right)_0 + (J_v)_0} \tag{14}$$

In the case that osmotic forces act equally from either side of the epithelium the transport equations depend only on the difference $(C_M - C_S)$, and

$$\left(\frac{\partial J_s}{\partial C_M}\right)_0 = -\left(\frac{\partial J_s}{\partial C_S}\right)_0 \text{ and } \left(\frac{\partial J_v}{\partial C_M}\right)_0 = -\left(\frac{\partial J_v}{\partial C_S}\right)_0 \tag{15}$$

so that

$$C_S^* = \frac{(J_s)_0 - C_0(J_v)_0}{-C_0\left(\frac{\partial J_v}{\partial C_M}\right)_0 + \left(\frac{\partial J_s}{\partial C_M}\right)_0 + (J_v)_0} \tag{16}$$

Eqns. 13 and 14 or 16 will provide the basis for our understanding of the factors responsible for transport isotonicity. It will be helpful, therefore, to cast them in terms of the osmotic coupling coefficient and strength of transport. For the sake of simplicity we shall suppose that $(\partial J_s/\partial C_M)_0$ is dwarfed by the other terms in the denominator. This means that near the reference condition, the transepithelial salt flux is relatively insensitive to variation in the salt concentrations of the bathing media. The experimental justification for this simplification has been given for rabbit gallbladder by Whitlock and Wheeler [17] and for rat small intestine by Parsons and Wingate [18]. Substituting Eqns. 7 and 10 into 13 and 16 and setting $(\partial J_s/\partial C_M)_0 = 0$, we obtain

$$-C_M^* = \frac{\hat{C}}{1 - \hat{C}/C_0}(1/\gamma - 1) \tag{17}$$

$$C_S^* = \frac{\hat{C}}{1+\hat{C}/C_0}(1/\gamma - 1) \tag{18}$$

These relations, Eqns. 17 and 18, show that for an epithelium to transport isotonically and to be capable of transport against a substantial gradient, it must be the case that coupling is tight ($\gamma \approx 1$). If, for example, isotonicity were verified to be within 2%, ($C_M^*/C_0 \leq 0.02$), in a rabbit gallbladder epithelium that could transport water against a 75 mosM osmotic gradient, ($\hat{C}/C_0 = 0.25$), then according to Eqn. 17, $0.02 \geq 1/3(1\gamma - 1)$, or $\gamma \geq 0.94$.

(b) Thermodynamic formulation

The non-equilibrium thermodynamic approach to models of epithelial transport begins with the identification of a system of relevant fluxes, J_i, and conjugate driving forces, X_i [19]. With this system the energy dissipation of the model, ϕ, may be expressed as a sum of the products of conjugate flows and forces

$$\phi = \sum_i J_i X_i \tag{19}$$

Linear phenomenological equations, accurate in the region near equilibrium, may be written relating the flows and forces either in a conductive form

$$J_i = \sum_j L_{ij} X_j \tag{20}$$

or a resistive form

$$X_j = \sum_i R_{ji} J_i \tag{21}$$

Due to the work of Onsager, the matrices of phenomenological coefficients, (L_{ij}) and (R_{ij}), are known to be symmetric ($L_{ij} = L_{ji}$ and $R_{ij} = R_{ji}$). Essig and Caplan [20] have used the resistive form of the phenomenological equations to define coupling coefficients between the ith and jth flows by

$$q_{ij} = \frac{R_{ij}}{\sqrt{R_{ii} R_{jj}}} \tag{22}$$

These might be used, for example, to define the coupling of salt flux or water flux to a metabolic reaction. Our aim, however, in the introduction of the osmotic coupling coefficient, was to define a parameter that would reflect the relation between metabolically coupled salt flux and volume flow.

A phenomenological model of epithelial transport in the conductive form Eqn. 20

can be considered systematically within the general framework of Section 2a. For simplicity we suppose a single metabolic reaction driving transepithelial salt and water flux. A system of conjugate flows and forces that could be employed for this model [19] is reaction rate, J_r, with affinity, A; volume flow, J_v, with hydrostatic pressure difference, $\Delta P = P_M - P_S$; and solute velocity, J_D, with osmotic pressure $\Delta \pi$. Here

$$\Delta \pi = RT(C_S - C_M) \tag{23}$$

and

$$J_D = \frac{J_s}{\bar{C}_s} - J_v \tag{24}$$

where $\bar{C}_s$ is the logarithmic mean concentration given by

$$\bar{C}_s = \frac{C_M - C_S}{\ln(C_0 + C_M) - \ln(C_0 + C_S)} \tag{25}$$

or $\bar{C}_s = C_0$ if $C_M = C_S = 0$. The phenomenological equations are written in conductive form

$$J_v = L_P \Delta P + L_{PD} \Delta \pi + L_{Pr} A$$
$$J_D = L_{DP} \Delta P + L_D \Delta \pi + L_{Dr} A$$
$$J_r = L_{rP} \Delta P + L_{rD} \Delta \pi + L_r A \tag{26}$$

so that

$$J_s = \bar{C}_s(L_P + L_{DP})\Delta P + \bar{C}_s(L_{PD} + L_D)\Delta \pi + \bar{C}_s(L_{Pr} + L_{Dr})A \tag{27}$$

In particular

$$(C_R)_0 = C_0 \left(1 + \frac{L_{Dr}}{L_{Pr}}\right) \tag{28}$$

and

$$\frac{1}{\gamma} = 1 + \frac{L_{Dr}}{L_{Pr}} \tag{29}$$

which means that the osmotic coupling coefficient depends only on the phenomenological coefficient and not explicitly on the reaction affinity or rate.

It remains to cast the phenomenological equations for volume and solute flux in terms of more familiar coefficients. Following Katchalsky and Curran [19]

$$L_{PD} = -L_p\sigma$$

$$L_D = \frac{h_s}{RT\bar{C}_s} + L_p\sigma^2 \tag{30}$$

where σ is the reflection coefficient for salt and h_s the salt permeability at zero volume flow. The reflection coefficient, σ, is a measure of the effectiveness of salt concentration differences across the epithelium in driving transepithelial water flow. It should be recalled [19] that for $\sigma < 1$ any transepithelial water flow will induce a convective component of transepithelial salt flux, while for $\sigma = 1$ all the salt is "reflected" from the epithelial surface (see Eqn. 35). Denote by N the metabolically driven solute flow

$$N = \bar{C}_s(L_{Pr} + L_{Dr})A = \frac{\bar{C}_s}{\gamma}L_{Pr}A \tag{31}$$

and set

$$\hat{C} = \frac{\gamma N}{RTL_p\sigma\bar{C}_s} = \frac{L_{Pr}A}{RTL_p\sigma} \tag{32}$$

Then the equation for volume flow becomes

$$J_v = L_P\Delta P + RTL_p\sigma\left[(C_S - C_M) + \hat{C}\right] \tag{33}$$

and as in Section 2a $\hat{C}$ appears as the strength of transport. The equation for solute flux is

$$J_s = \bar{C}_s L_p(1-\sigma)\Delta P$$
$$+ \left[RTL_p\sigma\bar{C}_s(1-\sigma)\right](C_S - C_M) + h_s(C_M - C_S) + N \tag{34}$$

or

$$J_s = \bar{C}_s(1-\sigma)J_v + h_s(C_M - C_S) + N\left[1 - \gamma(1-\sigma)\right] \tag{35}$$

The Eqns. 33 and 34 or 35 constitute a model of the form Eqn. 1, provided we know how to compute N and $\hat{C}$, as well as the parameters of passive transport L_P, σ and h_s, as functions of the bath conditions. For the interpretation of experimental data the parameters of passive transport are usually assumed to be constant. (An intriguing exception to this is the study by Diamond [21] suggesting the dependence of L_P on the mean concentration of an impermeant species.) For active solute transport across a simple membrane $\hat{C} = \gamma = 0$ and N might be specified as fixed or by Michaelis–Menten kinetics. In the second section we shall consider some epithelial models in which both N and $\hat{C}$ are constant.

In the case that the passive membrane properties are fixed, one may use the Eqns. 33 and 34 to assess the isotonicity of transport. For this system, the fluxes between equal bathing media are

$$(J_v)_0 = RTL_P\sigma\hat{C}$$
$$(J_s)_0 = N \tag{36}$$

and at the reference state the derivatives are

$$\left(\frac{\partial J_v}{\partial C_M}\right)_0 = -RTL_P\sigma + RTL_P\sigma\left(\frac{\partial \hat{C}}{\partial C_M}\right)_0$$

$$\left(\frac{\partial J_v}{\partial C_S}\right)_0 = RTL_P\sigma + RTL_P\sigma\left(\frac{\partial \hat{C}}{\partial C_S}\right)_0$$

$$\left(\frac{\partial J_s}{\partial C_M}\right)_0 = -RTL_P\sigma C_0(1-\sigma) + h_s + \left(\frac{\partial N}{\partial C_M}\right)_0$$

$$\left(\frac{\partial J_s}{\partial C_S}\right)_0 = RTL_P\sigma C_0(1-\sigma) - h_s + \left(\frac{\partial N}{\partial C_S}\right)_0 \tag{37}$$

For the case that $\hat{C}$ and N are constant we substitute Eqns. 36 and 37 into 13 and 14 to obtain

$$C_M^* = -\frac{(N/C_0)(1-\gamma)}{RTL_P\sigma(\sigma - \hat{C}/C_0) + h_s/C_0} \tag{38}$$

and

$$C_S^* = \frac{(N/C_0)(1-\gamma)}{RTL_P\sigma(\sigma + \hat{C}/C_0) + h_s/C_0} \tag{39}$$

These expressions for the osmotic deviations reveal the isotonicity of transport to be dependent upon (1) the solute transport rate, N, (2) the tightness of coupling, γ, and (3) the whole epithelial permeabilities L_P and h_s. Thus, for a tightly coupled epithelium a substantial osmotic gradient does not develop between the bathing media because salt and water are transferred at an isotonic ratio even in the absence of external gradients. For a very permeable tissue the large water flows in response to osmotic gradients tend to keep these gradients small. The Eqns. 38 and 39 show these effects to be synergistic.

We may consider in this light some of the epithelial parameters that have been measured. It is important to note here that the experimental determination of the strength of transport as well as the water permeability, L_P, is often made using osmotic gradients established with an impermeant species such as sucrose. To include the effects of such species Eqn. 33 is commonly modified [19] to

$$J_v = L_P \Delta P + RTL_P(C_{iS} - C_{iM}) + RTL_P \sigma \left[(C_S - C_M) + \hat{C} \right] \qquad (40)$$

where C_{iM} and C_{iS} represent the concentrations of the impermeant solute in the mucosal and serosal baths. Thus

$$RTL_P = -\frac{\partial J_v}{\partial C_{iM}} \qquad (41)$$

gives a direct determination of the epithelial water permeability, and the nulling of volume flow by mucosal sucrose yields the value of $\sigma \hat{C}$. (The extension of Eqn. 34 to include multiple species poses more serious theoretical problems. See the discussion of Sauer [22].) Data from five epithelia are displayed in Table 1. The first column is labeled N/C_0 to indicate the volume flow required for isotonic transport. For the tightly coupled gallbladder and intestinal epithelia this is virtually identical to $(J_v)_0$. For the proximal tubules the situation is more ambiguous. Since transport against an osmotic gradient is not observed in proximal tubule, it is likely that $\hat{C}/C_0 \ll \sigma$. Thus, the denominators of Eqns. 38 and 39 are roughly equal and we may write the approximate osmotic deviations

$$-C_M^* \approx C_S^* \approx C^* = \frac{(N/C_0)(1-\gamma)}{RTL_P \sigma^2 + h_s/C_0} \qquad (42)$$

For the rat proximal tubule data in Table 1, the magnitude of the transepithelial osmotic gradient required for isotonic transport is $C^* = (1-\gamma)(16.)$ mosM. The actual magnitude of the gradient in vivo, or conversely, the significance of intraepithelial coupling, remains uncertain. (Andreoli and Schafer [23–25] discuss in detail the several driving forces for water across rabbit proximal tubule.)

(c) The effect of unstirred layers

In their discussion of water transport across the rat colon, Curran and Schwartz [26] speculated on the mechanism of coupled volume flow. They suggested that the active transport of solute might create a region of hypertonicity on the serosal side of the epithelium and that water would move in response to favorable transepithelial osmotic gradients. For the experiment in vivo, the intestinal capillary osmolality reflects the fluxes of salt and water out of the intestinal lumen as well as arterial

TABLE 1

Tissue	N/C_0 $\cdot 10^6$ ml/s.cm²	$\sigma \hat{C}^n$ mosM	RTL_P $\cdot 10^4$ cm/s.osm	σ_{NaCl}	h_{NaCl} $\cdot 10^6$ cm/s
Fish gallbladder	4.2 [a]	40 [b]	1.0 [b]	0.93 [b]	1.9 [b]
Rabbit gallbladder	14.7 [c]	80 [c]	0.44 [d] – 1.7 [e]	0.85 [c]	
Rat small intestine	33.7 [f]	200 [f]	1.7 [f]		
Necturus proximal tubule	0.9 [g] – 1.64 [i]		0.37 [h] – 0.97 [j]	0.7 [k]	3.0 [i]
Rat proximal tubule	48 [m]		45–54 [l]	0.7 [l]	150 [l]

[a] [2]; [b] [4]; [c] [68]; [d] [13]; [e] [39]; [f] [69]; [g] [6]; [h] [70]; [i] [61]; [j] [37]; [k] [71]; [l] [72]; [m] [73].
[n] Strength of transport determined against an adverse osmotic gradient established with an impermeant species.

plasma osmolality. With sufficiently slow intestinal blood flow a favorable gradient for water flow could be established. In the case of an in vitro experiment, however, in which water transport proceeds against an adverse osmotic gradient between two large bathing media a more subtle argument is required. As outlined by Dainty [27], despite stirring of the bulk solutions bathing the tissue in vitro, viscous flow requires that there be regions adjacent to the tissue in which the velocity of the solution is small in directions parallel to the tissue. In these regions, or "unstirred layers", the flow of water is essentially perpendicular to the epithelium and is just the transepithelial water flow. The flow of solute is governed by convection and diffusion. Solute transport across the epithelium in this model is viewed as the removal of salt from the mucosal bath just adjacent to the epithelial surface and its deposition at the serosal surface of the epithelium. Thus, when the bathing media are of identical composition, the region of the mucosal bath adjacent to the epithelium should be slightly hypotonic relative to the bulk solution; the region adjacent to the serosal surface should be hypertonic (Fig. 4c). This difference in osmolality at the epithelial surfaces produces transepithelial water flow. Thus, because of the presence of unstirred layers, one observes transepithelial water flow coupled to solute transport even in the absence of intra-epithelial coupling. This extra-epithelial coupling is amenable to an analysis along the lines of Sections 2a and 2b.

We shall suppose the two bathing media at bulk osmolalities $C_0 + C_M$ and $C_0 + C_S$, separated by an idealized epithelium positioned at $x = 0$. Distance, x, is positive measured from the serosal surface of the membrane, and negative into the mucosal bath. Denote by δ_M and δ_S the thickness of the unstirred layers and by D_M and D_S the diffusion coefficient for salt within these layers. Denote by $C_0 + C_{M,int}$ and $C_0 + C_{S,int}$ the osmolalities just adjacent to the epithelium. Suppose that the transepithelial flows are as described by Eqns. 33 and 34 with constant metabolically driven solute transport, N, and coupling only in the unstirred layers ($\hat{C} = 0$). Then in the absence of pressure gradients

$$J_v = RTL_P \sigma (C_{S,int} - C_{M,int}) \tag{43}$$

and

$$J_s = [RTL_p\sigma(1-\sigma)\bar{C}_s - h_s](C_{S,int} - C_{M,int}) + N \tag{44}$$

It will be a convenience to combine Eqns. 43 and 44 to obtain

$$J_s = \left[(1-\sigma)\bar{C}_s - \frac{h_s}{RTL_p\sigma}\right]J_v + N \tag{45}$$

Within an unstirred layer the diffusion equation takes the form

$$J_v C - D\frac{dC}{dx} = J_s \tag{46}$$

where $C = C(x)$ is the osmolality at x. Since we are concerned with only steady-state phenomena, J_s and J_v are constant, independent of x. Eqn. 46 has the solution

$$C = \frac{J_s}{J_v} - K\exp\left(\frac{J_v x}{D}\right) = C_R - K\exp\left(\frac{J_v x}{D}\right) \tag{47}$$

where $C_R = J_s/J_v$ is the osmolality of the transported solution, and K is a constant of integration. On the serosal side, K is determined so that

$$C(\delta_S) = C_0 + C_S \tag{48}$$

or

$$C(x) = C_R + (C_0 + C_S - C_R)\exp\left(\frac{J_v}{D_S}(x - \delta_S)\right) \tag{49}$$

In particular

$$C(0) = C_0 + C_{S,int} = C_R + (C_0 + C_S - C_R)\exp\left(\frac{-J_v\delta_S}{D_S}\right) \tag{50}$$

Within the mucosal bath, K is determined so that

$$C(-\delta_M) = C_0 + C_M \tag{51}$$

or

$$C(x) = C_R + (C_0 + C_M - C_R)\exp\left(\frac{J_v}{D_M}(x + \delta_M)\right) \tag{52}$$

and in particular

$$C(0) = C_0 + C_{M,int} = C_R + (C_0 + C_M - C_R)\exp\left(\frac{J_v\delta_M}{D_M}\right) \qquad (53)$$

Thus, subtracting Eqn. 53 from Eqn. 50 we obtain

$$C_{S,int} - C_{M,int} = (C_0 + C_S - C_R)\exp\left(\frac{-J_v\delta_S}{D_S}\right)$$

$$- (C_0 + C_M - C_R)\exp\left(\frac{J_v\delta_M}{D_M}\right) \qquad (54)$$

Following the analysis of Dainty and House [28] we may simplify Eqn. 54 by approximating the exponentials by the first terms of their Taylor expansions

$$\exp\left(\frac{-J_v\delta_S}{D_S}\right) \approx 1 - \frac{J_v\delta_S}{D_S}$$

$$\exp\left(\frac{J_v\delta_M}{D_M}\right) \approx 1 + \frac{J_v\delta_M}{D_M} \qquad (55)$$

The justification of this approximation depends upon showing the exponents $J_v\delta_M/D_M$ and $J_v\delta_S/D_S$ to be small. Table 1 shows the relevant transepithelial volume flow, J_v, to be of the order 10^{-5} ml/s.cm². Tormey and Diamond [29] have indicated that the diffusion coefficient of salt in connective tissue is roughly three-fourths of that in free solution so that we may suppose $D_M \approx D_S \approx 10^{-5}$ cm²/s. Finally, Dainty and House [28] have shown that the mucosal unstirred layer thickness, δ_M, is about 0.005 cm; the observations of Tormey and Diamond suggest the serosal unstirred layer, δ_S, might be as large as 0.05 cm. With these parameters, Eqns. 55 are sufficiently accurate for our concerns (within 1%) and may be substituted into Eqn. 54 to yield

$$C_{S,int} - C_{M,int} = C_S - C_M - \left[\frac{(C_0 + C_S)\delta_S}{D_S} + \frac{(C_0 + C_M)\delta_M}{D_M}\right]J_v$$

$$+ \left[\frac{\delta_S}{D_S} + \frac{\delta_M}{D_M}\right]J_s \qquad (56)$$

Using Eqn. 43

$$J_v = \left[\frac{1}{RTL_p\sigma} + \frac{(C_0+C_S)\delta_S}{D_S} + \frac{(C_0+C_M)\delta_M}{D_M}\right]^{-1}(C_S - C_M)$$

$$+ \left[\frac{1}{RTL_p\sigma} + \frac{(C_0+C_S)\delta_S}{D_S} + \frac{(C_0+C_M)\delta_M}{D_M}\right]^{-1}\left[\frac{\delta_S}{D_S} + \frac{\delta_M}{D_M}\right] J_s \quad (57)$$

The Eqns. 45 and 57 constitute a system of the form Eqn. 1 and may be analyzed according to the methods of Section 2a.

The strength of transport may be determined exactly. Setting $J_v = 0$ in Eqn. 57 we find

$$\frac{\hat{C}}{C_0} = \left[\frac{\delta_S}{D_S} + \frac{\delta_M}{D_M}\right]\frac{N}{C_0} \quad (58)$$

It is of interest to note that the strength of transport depends only on the metabolically driven solute transport rate and the unstirred layer properties. It is independent of the passive properties, and in particular the water permeability, of the epithelium. Using our parameter estimates in Eqn. 58 we find $\hat{C}/C_0 = 0.055$ or $\hat{C} = 16.5$ mosM for $C_0 = 300$ mosM. Fig. 4b shows the concentration profiles within the unstirred layers when the bulk mucosal osmolality has been raised sufficiently to just null transepithelial volume flow; in this case $C_{S,int} - C_{M,int} = 0$. For rabbit gallbladder, then, with a serosal unstirred layer between 0.025 and 0.05 cm one would need to retard the mobility of solute from the membrane by a factor of from 5 to 10 to obtain the observed strength of transport.

To compute the osmotic coupling coefficient for this model one evaluates Eqn. 57 when $C_M = C_S = 0$, so that

$$(J_v)_0 = \left[\frac{1}{RTL_p\sigma} + \frac{C_0\delta_S}{D_S} + \frac{C_0\delta_M}{D_M}\right]^{-1}\left[\frac{\delta_S}{D_S} + \frac{\delta_M}{D_M}\right](J_s)_0 \quad (59)$$

Thus

$$\gamma = \left[\frac{1}{RTL_p\sigma C_0} + \frac{\delta_S}{D_S} + \frac{\delta_M}{D_M}\right]^{-1}\left[\frac{\delta_S}{D_S} + \frac{\delta_M}{D_M}\right] \quad (60)$$

and

$$\frac{1}{\gamma} - 1 = \frac{N/C_0}{RTL_p\sigma\hat{C}} = \frac{1}{RTL_p\sigma C_0}\left[\frac{\delta_S}{D_S} + \frac{\delta_M}{D_M}\right]^{-1} \quad (61)$$

In particular we observe that γ depends only on the passive permeability properties of the epithelium and the thickness of the unstirred layers. The degree of osmotic coupling within the unstirred layers is independent of the solute transport rate, N.

Concentration profiles for the conditions of transport equilibrium are shown schematically in Figs. 4d and 4e. For serosal equilibrium, Eqn. 6 applies so that $C_R = C_0 + C_S^*$, and substitution in Eqn. 49 shows $C(x) = C_0 + C_S^*$ at all points on the serosal side of the epithelium. This means that at serosal transport equilibrium there is no concentration gradient at all within the serosal unstirred layer. This condition has been recognized by Hill and Hill [16] in their analysis of the unilateral sac preparation of the gallbladder. In the case of mucosal equilibrium (Fig. 4e) Eqns. 5 and 52 show a uniform concentration, $C_0 + C_M^*$, in the mucosal bath and throughout the mucosal unstirred layer. In the case that $\sigma \approx 1$ and $h_s \approx 0$, we may derive simple and intuitively useful expressions for the osmotic deviations relevant to the assessment of transport isotonicity. For this situation the Eqns. 17 and 18 apply and we may write, using Eqn. 61

$$-C_M^* = \frac{N/C_0}{RTL_P(1 - \hat{C}/C_0)} \tag{62}$$

$$C_S^* = \frac{N/C_0}{RTL_P(1 + \hat{C}/C_0)} \tag{63}$$

We have already seen above that the term $\hat{C}/C_0$ is small for physiologically relevant parameters so that C_M^* and C_S^* are quite close in magnitude. The actual value of these deviations is essentially just the solute transport rate relative to the epithelial water permeability. This is precisely what one would estimate even in the absence of coupled water transport.

The question remains, however, as to the value of L_P to use in Eqn. 60 to evaluate γ or in Eqns. 62 and 63 for the calculation of the osmotic deviations. Put conversely, we seek to know how the presence of unstirred layers distorts the evaluation of epithelial water permeability. When one applies an osmotic gradient across the epithelium, solute may accumulate at the interface of lower osmolality and be swept away from the interface at higher osmolality (Figs. 4a and 4f). This "solute polarization effect" may make $C_{S,\text{int}} - C_{M,\text{int}}$ smaller in magnitude than $C_S - C_M$. When one calculates epithelial water permeability as the observed volume flow relative to the driving force $C_S - C_M$, one is thus overestimating the true driving force and, hence, underestimating permeability.

To gain a sense of the magnitude of this solute polarization effect we will consider again the case when $\sigma \approx 1$ and $h_s \approx 0$. The observed water permeability of the tissue, $L_{P,\text{obs}}$, may be identified with the change in volume flow produced by a change in osmolality of one of the bathing media. Thus, differentiating Eqn. 57 with respect to C_S

Coupled transport of water

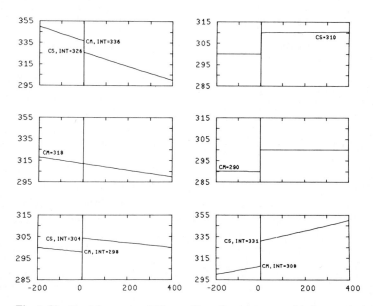

Fig. 4. Unstirred layer osmolality profile adjacent to an epithelium actively transporting solute. Each figure corresponds to a different set of osmolalities in the bulk bathing solutions. Distance from the epithelium is measured in μm and shown on the abscissa; distance into the serosal bath is positive and into the mucosal bath, negative. Osmolality is indicated in the ordinate of each figure; within the serosal layer it is determined according to Eqn. 49 and in the mucosal layer according to Eqn. 52. Transepithelial volume flow is given by Eqn. 57. For each figure the diffusion coefficients are $D_M = D_S = 10^{-5}$ cm^2/s; the layer thicknesses are $\delta_M = 0.02$ cm and $\delta_S = 0.04$ cm; and the transepithelial solute flux is 3 nmol/s.cm^2. The epithelial L_P was 10^{-3} cm/s.osm, taken high to accentuate the layer profiles. (a) Bulk mucosal concentration is 350 mosM and serosal concentration is 300 mosM. Volume flow is from serosa to mucosa, but the effective driving force is only 10.6 mosM. (b) Transepithelial volume flow has been nullified by bringing the mucosal bath to 318 mosM. Thus, 18 mosM is the strength of transport ($\hat{C}$) for this model. (c) Transport between bathing solutions of equal composition. Mucosa to serosa volume flow is a consequence of the unstirred layer gradients. (d) Serosal transport equilibrium: The serosal bath osmolality is equal to the reabsorbate osmolality, 310.4 mosM (mucosa at 300 mosM). (e) Mucosal transport equilibrium: The mucosal bath osmolality is equal to the reabsorbate osmolality, 290.2 mosM (serosa at 300 mosM). (f) Bulk mucosal concentration is 300 mosM and serosal concentration 350 mosM. The effective driving force across the epithelium is 23.3 mosM.

$$RTL_{P,obs} = \left(\frac{\partial J_v}{\partial C_S}\right)_0$$

$$= \frac{1}{C_0}\left[\frac{1}{RTL_P C_0} + \frac{\delta_S}{D_S} + \frac{\delta_M}{D_M}\right]^{-1}$$

$$- \frac{\delta_S}{D_S C_0^2}\left[\frac{1}{RTL_P C_0} + \frac{\delta_S}{D_S} + \frac{\delta_M}{D_M}\right]^{-2}\left[\frac{\delta_S}{D_S} + \frac{\delta_M}{D_M}\right] N \qquad (64)$$

and by applying Eqns. 60 and 58 we obtain

$$RTL_{P,obs} = RTL_P(1-\gamma) - \frac{\delta_S \hat{C}}{D_S}\left[RTL_P(1-\gamma)\right]^2 \quad (65)$$

Similarly, differentiating with respect to C_M

$$RTL_{P,obs} = -\left(\frac{\partial J_v}{\partial C_M}\right)_0 = RTL_P(1-\gamma) + \frac{\delta_M \hat{C}}{D_M}\left[RTL_P(1-\gamma)\right]^2 \quad (66)$$

It is important to note that the expressions shown on the right hand sides of Eqns. 65 and 66 are different. This difference suggests that the observed water permeability of the epithelium will depend upon the direction of the imposed osmotic gradient. Such directional dependence of osmotic water permeability is commonly referred to as rectification, although we shall see that with the unstirred layer parameters as estimated above, such rectification is quantitatively negligible. Eqns. 65 and 66 are quadratic in the unknown $RTL_P(1-\gamma)$. When

$$RTL_{P,obs} \ll \frac{D_S}{4\delta_S \hat{C}} = 3.1 \cdot 10^{-3}$$

and

$$RTL_{P,obs} \ll \frac{D_M}{4\delta_M \hat{C}} = 3.1 \cdot 10^{-2} \quad (67)$$

as is the case with observed water permeabilities $RTL_{P,obs} = 10^{-4}$ cm/s.osm (Table 1) we have the approximation

$$L_P = \frac{1}{1-\gamma} L_{P,obs} \quad (68)$$

equivalent to the formula stated by Diamond [11]. (Note that in Figs. 4a and 4f rectification is significant due to the high value of L_P used in this calculation in order to accentuate unstirred layer profiles.)

We may now compute the osmotic coupling coefficient of this model by rewriting Eqn. 61 as

$$\gamma = RTL_{P,obs} C_0 \left[\frac{\delta_S}{D_S} + \frac{\delta_M}{D_M}\right] \quad (69)$$

so that $\gamma = 0.17$. Thus, these unstirred layers fail to cause substantial underestimation of the epithelial water permeability. The osmotic deviations may also be

Coupled transport of water

evaluated according to Eqns. 62 and 63.

$$-C_M^* \approx C_S^* \approx \frac{(N/C_0)(1-\gamma)}{RTL_{P,\text{obs}}} \tag{70}$$

For this example the right hand side of Eqn. 70 equals 83 mosM. Thus in addition to being an inadequate explanation for the strength of transport of rabbit gallbladder, the external unstirred layers also fail to account for transport isotonicity. The understanding of these properties of transporting tissue must therefore be pursued via models of intraepithelial coupling.

3. Solute-solvent coupling in the lateral intercellular space

(a) Elementary compartment model

Curran pursued the problem of uphill water transport with the proposal of an intraepithelial series membrane system [30]. Depicted in Fig. 5, the essentials of this scheme are three compartments separated by two membranes. The "thin" membrane between compartments A and B is assumed to be such that salt concentration differences between the compartments are manifest as substantial osmotic forces across the membrane. Across the "thick" membrane between compartments B and C, salt concentration differences result in negligible osmotic forces. In terms of the phenomenological parameters, the reflection coefficient of the thin membrane is close to one, that of the thick membrane close to zero. For this system, direct salt input into the middle compartment, B, would generate strong osmotic forces favoring flow of water from A to B but not from C to B. Any resulting rise in hydrostatic pressure within the middle compartment would produce volume flow

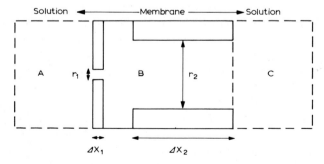

Fig. 5. Schematic model system for water transport. A and C represent external solutions. B is the "middle compartment". The membrane between A and B has salt reflection coefficient close to one; that between B and C has salt reflection coefficient close to zero. r and Δx represent pore radius and membrane thickness respectively. (From [30].)

from B to C and possibly diminish by a fraction the water flow from A to B. In sum, solute input into the middle compartment could produce net water flow from compartments A to C *.

The intuitive appeal of this scheme was strengthened by the actual construction of such a composite system and the observation of its functioning as predicted [31–32]. The analysis of the model behavior in terms of the component membrane properties was undertaken by Patlak et al. [33]. These workers derived expressions for the net solute and volume flows in terms of the rate of solute transport into the middle compartment and the solute concentrations in the outer solutions. The essential message of this mathematical effort was that the middle compartment model would be compatible, at least qualitatively, with the observed physiology of epithelial transport.

In his proposal of this scheme, Curran [30] also speculated on the realization of this membrane system within the epithelium. He suggested that the thin membrane might be one of the epithelial cell membranes while the thick membrane might be one of the submucosal layers. Whitlock and Wheeler [17], with reference to their observations on rabbit gallbladder, proposed that the middle compartment of the Curran scheme might well be the lateral intercellular space. This latter suggestion, it should be noted, contains implicitly the strong prediction that transepithelial water flow traverses the lateral intercellular spaces. Although no direct observation of water flow has been made to confirm this prediction an abundance of circumstantial evidence appears to favor it. Electron microscopic observations in rabbit gallbladder [29,35], rat intestine [36], and *Necturus* proximal tubule [37] demonstrate lateral interspace distention with enhanced volume flow and interspace collapse with inhibition of volume flow. Williams [36] also documented the transition over time from collapse to distention with the establishment of mucosal to serosal water flow. Spring and Hope [38] have shown that in the *Necturus* gallbladder epithelium, the interspace volume is a sensitive function of pressure. If, as is likely, the interspace distention with enhanced volume flow is due to increased pressure, this is another point of compatibility with the middle compartment scheme.

The quantitative predictions of the middle compartment model may be appreciated from the analysis of a compartment model of the lateral intercellular space. The purpose of this analysis is to display the behavior of the composite system in terms of the component membrane parameters. Conversely, the phenomenological observations of coupled water transport impose certain restrictions on models of the lateral intercellular space. With this analysis one may also see how these restrictions translate into parameter selection.

Consider the lateral intercellular space, or channel (Fig. 6) as a compartment bounded by apical (A), basal (B), and lateral (L) membranes. In this model, the apical membrane corresponds to the tight junction, the basal membrane to the combined basement membrane plus subepithelial tissue, and the lateral membrane

---

* In concept, the operation of this middle compartment model and that of the "double-membrane" system described by Durbin [34], are virtually identical.

Coupled transport of water

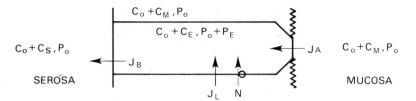

Fig. 6. Schematic representation of the lateral intercellular space. The cell and mucosal medium are assumed to be at the same osmolality and pressure.

to the lateral cell membrane. The channel is at osmolality $C_0 + C_E$ and pressure $P_0 + P_E$ and is separated by the apical membrane from a well stirred mucosal bath at osmolality $C_0 + C_M$ and by the basal membrane from a well stirred serosal bath at osmolality $C_0 + C_S$. For simplicity, we shall suppose that the lateral membrane separates the channel from a cellular compartment isotonic to the mucosal bath. To each membrane there is associated an area A_α and phenomenological coefficients $L_{P\alpha}$, σ_α, and h_α where α refers to A, B, or L. We denote $L_\alpha = A_\alpha L_{P\alpha}$ and $H_\alpha = A_\alpha h_\alpha$. Across each membrane there is volume flow $J_{v\alpha}$ and solute flux $J_{s\alpha}$. It is assumed that there is metabolically driven solute transport into the channel across the lateral membrane at constant rate N.

For what we term the "elementary compartment model", a particularly simple set of component membrane parameters is assumed. In this model the tight junction is totally impermeable ($L_A = H_A = 0$), the lateral membrane is impermeable to solute ($H_L = 0$, $\sigma_L = 1$), and the basal membrane has a reflection coefficient of zero ($\sigma_B = 0$). Thus the lateral membrane corresponds to the "thin" membrane and the basal membrane to the "thick" membrane in the Curran scheme. Across the component membranes one may write the relations for volume flow

$$J_{vL} = L_L\left[(P_M - P_E) + RT(C_E - C_M)\right] \tag{71}$$

$$J_{vB} = L_B[P_E - P_S] \tag{72}$$

and solute flux

$$J_{sL} = N \tag{73}$$

$$J_{sB} = J_{vB}(C_0 + \bar{C}_B) + H_B(C_E - C_S) \tag{74}$$

where

$$C_0 + \bar{C}_B = \frac{C_E - C_S}{\ln(C_0 + C_E) - \ln(C_0 + C_S)} \tag{75}$$

is the mean osmolality of the solution across the basal membrane. The equations expressing volume and solute conservation in the steady state are

$$J_v = J_{vL} = J_{vB}$$
$$J_s = J_{sL} = J_{sB} \tag{76}$$

and indicate that volume and solute entry into the channel across the lateral membrane must precisely balance that leaving the channel across the basal membrane. These two Eqns. 76, along with the defining relations, Eqns. 71–75, constitute the model that determines the two dependent variables P_E and C_E (and hence J_v and J_s) in terms of the independent variables C_M, C_S, P_M, P_S and N.

It may be observed that the only non-linearity in the model equations lies in the solvent drag term of the expression for solute flux across the basal membrane (Eqn. 74). Furthermore it is clear that if $C_M = C_S = P_M = P_S = N = 0$ then $C_E = P_E = J_v = J_s = 0$. It is natural therefore, to linearize the model equations about this reference point and consider the approximate solute flux relation

$$J_{sB} = J_{vB} C_0 + H_B (C_E - C_S) \tag{77}$$

With respect to the complete non-linear model equations, this approximation has been shown to be accurate provided the bathing media do not deviate too far from the reference concentration [15]. With this approximation the mass balance Eqns. 76 may be rewritten

$$J_v = L_L [(P_M - P_E) + RT(C_E - C_M)] = L_B (P_E - P_S) \tag{78}$$
$$J_s = N = J_v C_0 + H_B (C_E - C_S) \tag{79}$$

If one solves these two linear equations to eliminate P_E and C_E one finds

$$J_v = \frac{L_{LB} H_B}{H_B + RT L_{LB} C_0} \left[(P_M - P_S) + RT \left(C_S - C_M + \frac{N}{H_B} \right) \right] \tag{80}$$

$$J_s = N \tag{81}$$

where

$$L_{LB} = \frac{L_L L_B}{L_L + L_B} \tag{82}$$

is a composite parameter whose value reflects the water permeabilities of the bounding membranes.

Eqns. 80 and 81 have brought the elementary compartment model into the form considered in Section 2b (Eqns. 33 and 35). For this model the whole epithelial phenomenological coefficients are seen to be

$$L_P = \frac{L_{LB} H_B}{H_B + RTL_{LB} C_0} \qquad (83)$$

$$\sigma = 1 \qquad H_S = 0$$

and the strength of transport

$$\hat{C} = \frac{N}{H_B} \qquad (84)$$

It is a simple matter to verify that the coupling coefficient for this model is

$$\gamma = \frac{RTL_{LB} C_0}{H_B + RTL_{LB} C_0} = 1 - \frac{L_P}{L_{LB}} \qquad (85)$$

and substitution in Eqns. 38 and 39 yields the osmotic deviations from isotonicity

$$-C_M^* = \frac{(N/C_0)}{RTL_{LB}(1 - \hat{C}/C_0)} = \frac{\hat{C}\left(\frac{1}{\gamma} - 1\right)}{1 - \hat{C}/C_0} \qquad (86)$$

$$C_S^* = \frac{(N/C_0)}{RTL_{LB}(1 + \hat{C}/C_0)} = \frac{\hat{C}\left(\frac{1}{\gamma} - 1\right)}{1 + \hat{C}/C_0} \qquad (87)$$

The right-most equality in Eqns. 86 and 87 is just the verification of Eqns. 17 and 18 for this particular model.

The relations Eqns. 84 through 87 are formally identical to those derived in the unstirred layer analysis, Eqns. 58 through 63, using the correspondences

$$\frac{1}{H_B} \sim \left[\frac{\delta_S}{D_S} + \frac{\delta_M}{D_M}\right] \quad \text{and} \quad L_{LB} \sim L_P \qquad (88)$$

and much of the intuition gained from that analysis remains applicable. Thus, for the compartment model, the strength of transport is independent of the water permeabilities of the bounding membranes and is determined by the rate of solute transport relative to solute trapping within the interspace. Here again, the achievement of isotonic transport is only minimally dependent upon solute trapping effects, although it is crucially dependent upon the component membrane water permeabilities. We see from these considerations that although discussions of uphill water transport and isotonic transport are often subsumed under the category of solute-solvent coupling, these two phenomena depend on quite distinct epithelial properties.

Finally, Eqn. 85 shows that in the case of tight coupling the L_P that is determined experimentally is only a small fraction of the parameter, L_{LB}, relevant to transport isotonicity. Further, the Eqns. 86 and 87 recall the point made in Section 2a that when both the strength of transport is substantial and the osmotic deviations from isotonicity are small, coupling is tight and γ must be close to 1. This means that for a tissue like rabbit gallbladder or rat intestine intraepithelial solute polarization effects must be large and will vitiate any attempt to interpret L_P data in terms of cell-membrane water permeability. Van Os et al. [39] in their determination of rabbit gallbladder L_P tried to estimate the magnitude of the solute polarization effect from considerations of a diffusion potential set up across the tight junction. Their analysis shows that the change in salt concentration within the interspace reduces by at least a factor of five the driving force for water across the "thin" membrane. From the perspective of this model, the estimate of Van Os et al. means that $\gamma > 0.8$, and not that the "true" L_P is five times that actually measured.

In his early analysis of isotonic transport, Diamond [13] tried to use measured values of the whole epithelial water permeability, L_P, in place of the quantity L_{LB}. The unacceptably large osmotic deviations from isotonicity that he computed caused him to reject the elementary compartment model of the lateral intercellular space. We should reconsider, therefore, the requirements imposed upon the elementary compartment model by the experimental data on rabbit gallbladder (Table 1). For $N/C_0 = 1.47 \cdot 10^{-5}$ cm/s and $\hat{C}/C_0 = 0.27$, Eqn. 84 requires $H_B = 5.5 \cdot 10^{-5}$ cm/s. Eqn. 86 may be used to give a lower bound on the coupling coefficient. If mucosal equilibrium is within 2% of exact isotonicity then $-C_M^*/C_0 = 0.02$ so that $\gamma = 0.95$. Thus, if $L_P = 1.7 \cdot 10^{-4}$ cm/s.osm, Eqn. 85 implies L_{LB} is at least $34 \cdot 10^{-4}$ cm/s.osm. It remains to consider these model predictions for H_B and L_{LB} in relation to the pertinent experimental data.

Simple diffusion from a flat surface through an unstirred layer 0.025 cm thick would result in a solute permeability $20 \cdot 10^{-5}$ cm/s. Thus, the value of H_B computed for this model, $5.5 \cdot 10^{-5}$ cm/s, signifies that solute transport into an interspace bounded by a basement membrane retards its diffusion from the region adjacent to the cell membrane by a factor of four. The permeability H_B can also be translated into an electrical resistance, R_B, by use of the formula

$$R_B = \frac{RT}{H_B C_0 F^2} \tag{89}$$

where $RT = 2.5 \cdot 10^3$ J/mol and $F = 96\,500$ C/mol. For $H_B = 5.5 \cdot 10^{-5}$ cm/s, $R_B = 16$ Ωcm². This may be compared with the measured value of 11 Ωcm² for the subepithelial resistance of rabbit gallbladder obtained by Henin et al. [40], who pierced the epithelium from the mucosal side with microelectrodes.

The water permeability of the lateral membrane L_L may be obtained from L_{LB}

and the basement membrane permeability L_B by rewriting Eqn. 82 as

$$L_L = \frac{L_{LB}}{1 - L_{LB}/L_B} \approx L_{LB} \tag{90}$$

The water permeability for proximal tubule basement membrane has been measured by Welling and Grantham [41] at 2 cm/s.osm. Thus, even for channel mouth area 2% of epithelial area $L_{LB}/L_B < 0.1$ and the approximation of Eqn. 90 is justified. Sha'afi et al. [42] have measured the unit water permeability of the human red cell membrane at $2.2 \cdot 10^{-4}$ cm/s.osm. Thus, if the lateral cell membrane has the same unit permeability as the red cell and a total permeability $L_L = 34 \cdot 10^{-4}$ cm/s.osm then lateral membrane area must be 15 times greater than epithelial area. The lateral membrane area, enhanced by folding, has been measured in rabbit gallbladder by Blom and Helander [43] and found to be approx. 30 cm^2/cm^2 epithelial area.

In sum, this elementary compartment model of the lateral intercellular space appears capable of providing a realistic representation of coupled water transport across rabbit gallbladder. The solute transport rate, N, is set according to experimental observation. The observed strength of transport and the experimental estimate of the osmotic deviations required for transport equilibrium together determine the osmotic coupling coefficient, γ, for the model. (Eqns. 86 and 87). Given N, one solves for the basement membrane solute permeability, H_B, according to Eqn. 84 to obtain the desired strength of transport. Given γ and the observed epithelial osmotic water permeability, L_P, one solves Eqn. 85 for the lateral cell membrane water permeability, L_L. The correspondence of the estimated values of H_B and L_L with independent experimental determinations provides a measure of confirmation of the adequacy of this compartment model to represent the lateral intercellular space.

(b) Standing gradient interspace models

As we have seen, it was, perhaps, an unfair judgment that dismissed compartment models of the lateral intercellular space as unsatisfactory. Nevertheless, because it was supposed that the models, with a well mixed interspace could not realistically represent tight intraepithelial coupling, the standing gradient model was introduced (Diamond and Bossert [44]). In the standing gradient model (Fig. 7), the flow of solute and volume along the length of the paracellular channel is described by the diffusion equation. In this way, one can represent the osmotic equilibration of the channel contents with the cell as flow proceeds toward the channel mouth. In comparison with external unstirred layers, solute-solvent coupling within the channel is enhanced due to (1) small-channel cross-section relative to epithelial area (and hence retarded diffusion of solute from the cell surface) and (2) large-cell surface area across which there is osmotic flow of water. Thus, in contrast to compartment models, the standing gradient models are crucially dependent on the details of channel geometry and have, in part, motivated morphologic investigation of transporting epithelia.

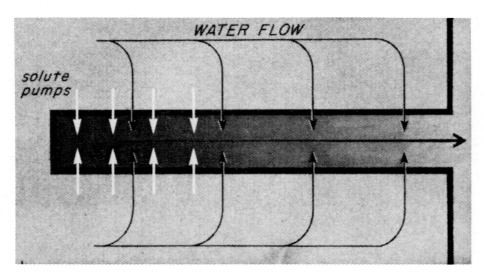

Fig. 7. Diagram of a standing gradient flow system, consisting of a long and narrow channel closed at one end. (From [44].)

The model equations of Diamond and Bossert [44], suppose a channel of length L, cross-section area A, and circumference S. An equivalent channel radius, r, may be defined by

$$r = \frac{2A}{S} \tag{91}$$

Distance along the channel, x, is measured from the tight junction ($x=0$) to the channel mouth ($x=L$). The cell is at the mucosal bath osmolality, $C_0 + C_M$, and the serosal bath is at $C_0 + C_S$. The lateral cell membrane has unit water permeability RTL_P and transports solute into the interspace at rate $N(x)$ (mosm/cm^2.s). The solute transport rate is written as a function of position because in their model, solute transport was assumed to occur only in an apical segment of the channel. In this apical segment transport, N, is constant. If λ is the fraction of channel length over which solute transport occurs, then

$$J_s(L) = \lambda SLN \tag{92}$$

is the total transepithelial solute flux. The model variables are the channel concentration, $C_0 + C_E(x)$, volume flux, $J_v(x)$, and solute flux, $J_s(x)$.

The solute flux in the channel is written as a sum of convective and diffusive terms in the equation

$$J_s = (C_0 + C_E)J_v - DA\frac{dC_E}{dx} \tag{93}$$

The model equations are completed by specifying the steady-state mass conservation relations for volume

$$\frac{dJ_v}{dx} = SRTL_P(C_E - C_M) \tag{94}$$

and solute

$$\frac{dJ_s}{dx} = SN \tag{95}$$

The system of three first-order equations, 93–95, requires three boundary conditions. In the original standing gradient model, the boundary data were that for a sealed tight junction, $J_v(0) = 0$ and $J_s(0) = 0$, and for the channel opening directly into the serosal bath, $C_E(L) = C_S$. For their model calculations, Diamond and Bossert considered only the case of exactly equal bathing media, namely $C_M = C_S = 0$. Thus, their calculations of C_R display, in essence, the parametric dependence of the tightness of coupling. They found that coupling became tighter for small diffusion coefficients or small channel radius or for large cell water permeability or long channel length. Coupling was quite independent of total solute transport rate but tightness of coupling decreased substantially if solute transport was permitted over the whole channel length ($\lambda = 1$).

Considerable insight was gained into the performance of this model from the analytical treatment of Segel [45] (also Lin and Segel [46]). Segel found that little accuracy was lost by approximating the diffusion Eqn. 93 by

$$J_s = C_0 J_v - DA\frac{dC_E}{dx} \tag{96}$$

This so-called "isotonic convection approximation" results in a system of model equations, 94–96, that are all linear and can be solved analytically. With the same boundary conditions as Diamond and Bossert, Segel obtained for the reabsorbate tonicity

$$\frac{(C_R)_0}{C_0} = \left[1 - \frac{\sinh(\lambda K)}{\lambda K \cosh(K)}\right]^{-1} \tag{97}$$

where

$$K^2 = \frac{SRTL_P C_0 L^2}{AD} = \frac{RTL_P C_0}{D}\frac{2L^2}{r} \tag{98}$$

In the case that solute transport is restricted to a very small fraction of the channel length then $\sinh(\lambda K) \approx \lambda K$ so that

$$\frac{(C_R)_0}{C_0} \approx \left[1 - \frac{1}{\cosh(K)}\right]^{-1} \tag{99}$$

Thus, the transport tonicity depends on only a single composite parameter, K. Segel added intuitive strength to this result by rewriting K as

$$\frac{1}{2}K^2 = \frac{[SLRTL_P\bar{C}_E]C_0}{[AD\bar{C}_E/2L]} \tag{100}$$

where $\bar{C}_E$ is the deviation from reference of the channel osmolality at $x = 1/2$. According to Eqn. 100, K is the ratio of the mean convective flow of solute through the channel to the mean diffusive flow. For K large, coupling tightens and $(C_R)_0$ approaches C_0.

To some degree, the analytical approach of Segel has been extended by the work of Weinbaum and Goldgraben [47] who admitted an additional variable for hydrostatic pressure along the channel, and by the work of King-Hele [48] who included a permeable tight junction. Nevertheless, the standing gradient model has received a certain measure of criticism. Hill's objection has simply been that the parameters necessary to achieve tight coupling are not physiologically realistic [16,49,50]. However, the validity of Hill's own parameter choices has been disputed [11,51]. The localization of solute pumps to the channel apex has also been criticized. Localization studies of the putative solute pump along the lateral cell membrane, Na^+-K^+ ATPase, have all shown a uniform pump distribution [52–54]. The tightness of coupling is substantially impaired when solute pumps are located along the whole channel length. Finally, and perhaps the most subtle of the "difficulties" of standing gradient theory, has been the observation that when channel models are solved for the case of equal bathing media and uniform solute pumping, no significant standing gradients are observed [55–57].

The significance of these numerical results may be appreciated by re-examining the standing gradient theory within the general framework of Section 2a. Here again the approximate Eqns. 94–96 are sufficient to reveal the essentials of the model. It is useful first to combine this system of equations into a single second-order equation for the channel concentration. Differentiating Eqn. 96 and substituting Eqns. 94 and 95 for the flux derivatives one obtains

$$\frac{d^2C_E}{dx^2} = \left(\frac{K}{L}\right)^2 (C_E - C_M) - \frac{SN(x)}{AD} \tag{101}$$

where K is as in Eqn. 98 and $N(x)$ is either a constant or zero and satisfies Eqn. 92. For simplicity, the junction will be sealed so that $dC_E/dx(0) = 0$ and at the channel mouth $C_E(L) = C_S$. Following Segel [45], Eqn. 101 is solved for the region $0 \leq x \leq \lambda L$

Coupled transport of water 341

and then for $\lambda L \leq x \leq L$. The two solutions are matched so that $C_E(x)$ and dC_E/dx are continuous across $x = \lambda L$. With $C_E(x)$ one obtains the transepithelial volume flow from

$$J_v(L) = \int_0^L SRTL_P(C_E - C_M)\,dx \qquad (102)$$

From the analysis outlined above, one finds

$$J_v = \frac{AD}{C_0 L} K \tanh(K)(C_S - C_M) + \frac{J_s}{C_0}\left[1 - \frac{\sinh(\lambda K)}{\lambda K \cosh(K)}\right] \qquad (103)$$

Eqns. 103 and 92 constitute a model of the form Eqn. 1. Inspection of Eqn. 103 confirms the coupling coefficient, γ, to be that found by Segel (Eqn. 97). Further, the strength of transport of the standing gradient model is seen to be

$$\frac{\hat{C}}{C_0} = \frac{L(J_s/C_0)\gamma}{ADK \tanh(K)} \qquad (104)$$

or

$$\frac{\hat{C}}{C_0} = \frac{1}{\sqrt{AD/L}} \frac{1}{\sqrt{SLRTL_P C_0}} \frac{(J_s/C_0)\gamma}{\tanh(K)} \qquad (105)$$

Eqn. 105 reveals that in a standing gradient model, the strength of transport is weakened either by a large diffusion coefficient, D, or a large-cell-membrane water permeability, L_P. This is in sharp contrast to tightness of coupling and, as indicated below, transport isotonicity, which are both enhanced by high lateral membrane water permeabilities.

Since J_s is constant for this model, Eqns. 17 and 18 apply for the assessment of the osmotic deviations. The important part of this calculation is to note

$$\frac{C^*}{C_0} = \frac{\hat{C}}{C_0}\left(\frac{1}{\gamma} - 1\right) = \frac{(J_s/C_0)}{SLRTL_P C_0} \frac{\sinh(\lambda K)}{\lambda \sinh(K)} \qquad (106)$$

The mucosal deviation, C_M^*, will be a bit larger than C^* and the serosal deviation, C_S^*, a bit smaller. When transport is uniform along the channel, $\lambda = 1$, and

$$\frac{C^*}{C_0} = \frac{(J_s/C_0)}{SLRTL_P C_0} \qquad (107)$$

This is identical to the result computed for the external unstirred layer problem (cf. Eqns. 62 and 63) and is essentially that computed for the compartment model (Eqns.

86 and 87). The osmotic deviation from isotonicity depends upon the water permeability of the lateral cell membrane and is largely independent of the effect of a finite diffusion coefficient. When solute transport is restricted to a region near the apex, $\lambda \approx 0$, and

$$\frac{C^*}{C_0} = \frac{(J_s/C_0)}{SLRTL_\mathrm{p}C_0} \frac{K}{\sinh(K)} \tag{108}$$

In this case, diffusion limitation within the interspace will act to decrease any deviation from isotonicity.

Intraepithelial solute polarization effect may also be estimated from the standing gradient model. From Eqn. 103 one sees

$$L_{\mathrm{P,obs}} = -\frac{\partial J_v}{\partial C_M} = SLRTL_\mathrm{p}\frac{\tanh(K)}{K}$$

$$= SLRTL_\mathrm{p}(1-\gamma)\frac{\lambda \sinh(K)}{\sinh(\lambda K)} \tag{109}$$

Thus, the measured epithelial water permeability is always less than the water permeability relevant to isotonic transport. For the case of uniform solute transport along the channel length ($\lambda = 1$), the right hand side of Eqn. 109 reduces to $SLRTL_\mathrm{p}(1-\gamma)$. The magnitude of the solute polarization effect is then just $1-\gamma$, as was found for external unstirred layers Eqn. 68 and the elementary compartment model Eqn. 85.

At this point a numerical evaluation of some of these parameters may be helpful in showing how isotonic transport is achieved by this standing gradient model. Blom and Helander [43] have indicated that for the rabbit gallbladder the effective channel radius, $r = 2A/S = 0.15 \cdot 10^{-4}$ cm. Further, they note that the channels occupy about 15% of epithelial volume, so we may say $A = 0.15$ cm^2/cm^2 epithelium. This means that $S = 2 \cdot 10^4$ cm/cm^2 epithelium. We suppose $D = 10^{-5}$ cm^2/s and estimate the unit membrane water permeability by that for the human red cell $RTL_\mathrm{P} = 2.2 \cdot 10^{-4}$ cm/s.osm [42]. Then, if channel length is $5.0 \cdot 10^{-3}$ cm, $\sqrt{AD/L} = 1.7 \cdot 10^{-2}$ and $\sqrt{SLRTL_\mathrm{p}C_0} = 8.1 \cdot 10^{-2}$ (so that $K = 4.8$). For the case of uniform solute input ($\lambda = 1$), Eqn. 97 yields $\gamma = 0.8$, so that when $J_s/C_0 = 10^{-5}$ cm^3/s (Table 1), $C^*/C_0 = 1.5 \cdot 10^{-3}$ (using Eqn. 107). For the case of apical solute input ($\lambda = 0$), Eqn. 99 yields $\gamma = 0.98$, so that when $J_s/C_0 = 10^{-5}$ cm^3/s, $C^*/C_0 = 0.12 \cdot 10^{-3}$. In either case, transport is isotonic to within 0.2%. It should be emphasized that although with uniform solute transport $(C_R)_0$ is 25% greater than C_0, no significant osmotic gradient will develop across this tissue if placed between initially equal bathing media.

The strength of transport for any pattern of salt transport is approx. $\hat{C}/C_0 = 0.007$ and is seriously smaller than the observed value $\hat{C}/C_0 = 0.270$. It is not clear that any simple parameter alteration can remedy this weakness. Although the interspace area tends to diminish with the imposition of a hypertonic mucosal solution, the area would have to collapse to 1% of its reference value to transport against a gradient of 20 mosM. This degree of collapse is simply not observed [58]. This is a serious limitation of a standing gradient model unaided by the inclusion of a basement membrane.

Finally, we consider the implications of the absence of osmotic gradients along the channel length. The concentration at the interspace apex, $C_E(0)$, may be determined in the process of solving Eqn. 101. In the case of equal bathing media

$$\frac{C_E(0)}{C_0} = \frac{(J_s/C_0)}{SLRTL_pC_0}\left[\frac{\tanh(K)\sinh(K\lambda)}{\lambda} + \frac{1-\cosh(K\lambda)}{\lambda}\right] \qquad (110)$$

so that for uniform pumps ($\lambda = 1$)

$$\frac{C_E(0)}{C_0} = \frac{(J_s/C_0)}{SLRTL_pC_0}\left[1 - \frac{1}{\cosh(K)}\right] = \frac{C^*}{C_0}\left[1 - \frac{1}{\cosh(K)}\right] \qquad (111)$$

and for apical pumps ($\lambda = 0$)

$$\frac{C_E(0)}{C_0} = \frac{(J_s/C_0)}{SLRTL_pC_0}[K\tanh(K)] = \frac{C^*}{C_0}\left[\cosh(K) - \frac{1}{\cosh(K)}\right] \qquad (112)$$

Thus, even for the worst case (apical transport), in our numerical example the apical concentration is only 2 mosM greater than at the channel mouth. Nevertheless, solute polarization is still substantial with the measured $L_{P,obs}$ only 20% of the lateral membrane L_P. This means that the absence of perceptible "standing gradients" does not guarantee negligible effects of solute-solvent coupling within the lateral intercellular space.

(c) Comprehensive interspace models

One of the most important developments in epithelial physiology was the observation that in some low resistance tissues the path through the tight junction and lateral intercellular space is the principal route of transepithelial ion flow during the passage of an electric current [59]. Indeed, it is precisely the paracellular route (tight junction–interspace) that serves as a shunt pathway and is responsible for the low electrical resistance of the epithelium. An intriguing aspect to this observation is that it is generally the low resistance or so-called "leaky" epithelia that transport salt solutions isotonically [10]. Moreover, there is a striking correlation between tissue electrical resistance and tissue water permeability [60]. In sum, active debate has emerged regarding the possibility of water flow across the tight junction in vivo and the role of the junction in achieving isotonic transport.

The tight junction has also been implicated in the regulation of the transepithelial salt flux across proximal tubule. Measurements in *Necturus* proximal tubule have shown that when the animal is in a volume expanded state there can be a 3-fold decline in epithelial resistance due essentially to a change in the paracellular shunt resistance [61]. In the volume expanded state then, although active transport of salt into the lateral interspace is little changed from the control conditions, enhanced backflux of salt across the tight junction into the lumen results in substantially diminished transepithelial salt flux [61–63]. These increases in junctional permeability are likely mediated by increases in peritubular (serosal) and hence, lateral interspace pressure [64,65].

Thus, recent models of the lateral intercellular space have included a permeable apical membrane for the channel, designed to reflect the properties of the tight junction. Hill [66] has used such a model to discuss the possible role of the junction in determining transport tonicity. Sackin and Boulpaep [55] have presented a detailed simulation of Necturus proximal tubule in the control and volume expanded state to illustrate the effects of altered junctional permeability. The proximal tubule model of Huss and Marsh [56] included a compliance relation for the interspace that permitted study of hypothesized relations between channel pressure and junctional permeability.

One may gain a sense of the performance of these more comprehensive interspace models by considering again the model depicted in Fig. 6. This time we make no a priori assumptions about the parameters of the component membranes. The equations for volume flow are

$$J_{vA} = L_A[(P_M - P_E) + RT\sigma_A(C_E - C_M)] \tag{113}$$

$$J_{vL} = L_L[(P_M - P_E) + RT\sigma_L(C_E - C_M)] \tag{114}$$

$$J_{vB} = L_B[(P_E - P_S) + RT\sigma_B(C_S - C_E)] \tag{115}$$

and for solute flux

$$J_{sA} = J_{vA}(1 - \sigma_A)(C_0 + \bar{C}_A) + H_A(C_M - C_E) \tag{116}$$

$$J_{sL} = N + J_{vL}(1 - \sigma_L)(C_0 + \bar{C}_A) + H_L(C_M - C_E) \tag{117}$$

$$J_{sB} = J_{vB}(1 - \sigma_B)(C_0 + \bar{C}_B) + H_B(C_E - C_S) \tag{118}$$

where

$$C_0 + \bar{C}_A = \frac{C_E - C_M}{\ln(C_0 + C_E) - \ln(C_0 + C_M)} \tag{119}$$

$$C_0 + \bar{C}_B = \frac{C_E - C_S}{\ln(C_0 + C_E) - \ln(C_0 + C_S)} \tag{120}$$

are the mean osmolalities of the tight junction and basement membrane. The model is completed by specifying the equations of steady state mass conservation for volume and solute

$$J_v = J_{vB} = J_{vA} + J_{vL} \tag{121}$$

$$J_s = J_{sB} = J_{sA} + J_{sL} \tag{122}$$

As in Section 3a we may simplify the model equations by using the reference osmolality to approximate the mean membrane osmolality:

$$J_{sA} = J_{vA}(1 - \sigma_A)C_0 + H_A(C_M - C_E) \tag{123}$$

$$J_{sL} = N + J_{vL}(1 - \sigma_L)C_0 + H_L(C_M - C_E) \tag{124}$$

$$J_{sB} = J_{vB}(1 - \sigma_B)C_0 + H_B(C_E - C_S) \tag{125}$$

At this point it is useful to set

$$L_M = L_L + L_A \tag{126}$$

$$\sigma_M = \frac{L_L \sigma_L + L_A \sigma_A}{L_L + L_A} \tag{127}$$

$$H_M = H_A + H_L + (\sigma_L - \sigma_A)^2 RTC_0 \frac{L_A L_L}{L_A + L_L} \tag{128}$$

Adding Eqns. 113 and 114, one obtains total transmucosal volume flow

$$J_{vM} = J_{vA} + J_{vL} = L_M[P_M - P_E + RT\sigma_M(C_E - C_M)] \tag{129}$$

and adding Eqns. 123 and 124 yields total transmucosal solute flux

$$J_{sM} = J_{sA} + J_{sL} = J_{vM}(1 - \sigma_M)C_0 + H_M(C_M - C_E) + N \tag{130}$$

In this manner the composite mucosal membrane is represented by the equations of a simple membrane with composite parameters.

We may next analyze this composite mucosal membrane in series with the basement membrane. This means substituting the volume Eqns. 115 and 129 into 121 and the solute Eqns. 125 and 130 into 122. One then solves for the unknowns C_E and P_E in terms of the independent variables C_M, C_S, P_M, P_S and N. Substitution of these expressions in Eqns. 115 and 125 yields the transepithelial fluxes. Again it is useful to define composite parameters

$$L_{MB} = \frac{L_M L_B}{L_M + L_B} \tag{131}$$

$$L_P = \frac{L_{MB}(H_M + H_B)}{H_M + H_B + RTL_{MB}(\sigma_M - \sigma_B)^2 C_0} \tag{132}$$

$$\sigma = \frac{H_B \sigma_M + H_M \sigma_B}{H_B + H_M} \tag{133}$$

$$H = \frac{H_M H_B}{H_M + H_B} \tag{134}$$

With these parameters the transepithelial fluxes take the form

$$J_v = L_P(P_M - P_S) + RT\sigma\hat{C} + RT\sigma(C_S - C_M) \tag{135}$$

and

$$J_s = J_v(1 - \sigma)C_0 + H(C_M - C_S) + \frac{H_B N}{H_M + H_B} \tag{136}$$

where

$$\hat{C} = \frac{(\sigma_M - \sigma_B)N}{H_B \sigma_M + H_M \sigma_B} \tag{137}$$

Thus, this interspace model has been cast in the form of the general phenomenological model of Section 2b (Eqns. 33 and 35), and the relevant parameters of coupled water transport may be calculated as outlined there.

Since there is little reason to suppose the basement membrane reflection coefficient to be any value other than zero [41], the strength of transport for a realistic interspace model will be

$$\hat{C} = \frac{N}{H_B} \quad (\sigma_B = 0) \tag{138}$$

precisely that determined with the elementary compartment model (Eqn. 84). Thus, the maximum adverse salt gradient against which water may be transported should be uninfluenced by junctional properties. With respect to the strength of transport, an important limitation of these interspace models must be indicated. Given the equation for volume flow, Eqn. 135, and the parameters of Table 1 one predicts, for example, that dog intestine should be capable of transporting water against a pressure gradient of nearly 5 atmospheres. It is observed, however, that volume flow across the intestine is reversed by 5 mm Hg [67]. Similarly in rabbit gallbladder Van Os et al. [39] found a mucosal hydrostatic pressure 16 times more effective than an osmotic gradient in driving water flow and a serosal pressure 480 times more effective. Such non-linearities in response to pressure have also been observed in *Necturus* proximal tubule [65]. Viewed in terms of the composite membrane model the reversal of flow with small serosal pressures must mean a value of σ_M close to zero. This in turn is likely due to changes in the junctional complex so that $\sigma_A = 0$ and $L_A \gg L_L$ with water flow reversal through a disrupted junction. Alteration in the junctional reflection coefficient, σ_A, with changing pressures has not yet been incorporated into an epithelial model.

Models of the lateral interspace have been developed that include diffusion within the channel as well as bounding membranes of finite permeability [55,56]. The model of Weinstein and Stephenson [57] extended these efforts by incorporating the three ionic species Na^+, K^+ and Cl^- into a model of the whole epithelium. Within both cell and channel the ions move in response to both chemical gradients and electric fields according to the Nernst–Planck equation

$$J_s = CJ_v - DA\frac{dC}{dX} - uAC\frac{d\psi}{dX} \tag{139}$$

where u is ionic mobility and ψ is electrical potential. Similarly ion flux across membranes is governed by an electrical form of the Kedem–Katchalsky equations [7]. In addition, both cell and channel are permitted to change size in response to variation in intraepithelial pressures.

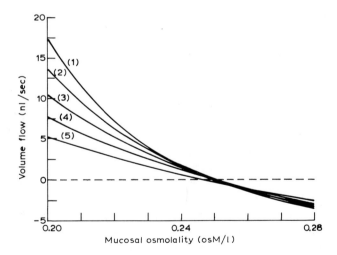

Fig. 8. Volume transport against an adverse osmotic gradient as determined by a comprehensive epithelial simulation. The model includes as variables, the concentrations of Na^+, K^+, and Cl^-, as well as electrical potential; intraepithelial concentration gradients are computed; and the cell and channel are represented as compliant structures. Transepithelial volume flow is plotted against mucosal osmolality (serosa fixed at 0.2 mosM) for five values of mucosal water permeability (spanning two orders of magnitude). As predicted by the compartment model, the strength of transport (intercept with $J_v = 0$) is quite insensitive to mucosal water permeability. (From [15].)

Certainly, the general framework of Sections 2a and 2b should remain applicable to describing the osmotic properties of these comprehensive models. However, the validity of the insight gained from the specific interspace models of Sections 3a and 3b can only be established by examining the numerical output of the large simulations. In particular, one would like to know whether the strong statements of parameter dependence made for the elementary models hold for the full epithelial models. This question has been addressed by Weinstein and Stephenson [15] and Fig. 8 shows output of their electrolyte model in simulating uphill water transport. Model parameters were chosen so as to represent briskly transporting gallbladder epithelium. The serosal bath was fixed at 200 mosM the mucosal osmolality was varied from 200–280 mosM by the addition of NaCl, and transepithelial volume flow was plotted as a function of mucosal osmolality. The five curves correspond to different choices of the water permeabilities of the mucosal structures (apical cell membrane, lateral cell membrane, and tight junction) and range over two orders of magnitude. The nearly common point of intersection of these curves at the point of zero volume flow displays the independence of the strength of transport on mucosal water permeability. This confirms for this comprehensive model the intuition of formulas 84 and 138 derived for the neutral solute interspace models.

4. Conclusion

The phenomena usually included under the heading "coupled water transport" are epithelial transport of water against an adverse osmotic gradient and isotonic transport between nearly equal bathing media. We have presented above a survey of some of the mathematical models that have been used to understand these aspects of solute-solvent interaction in the transporting tissue. A major theme of this chapter has been the use of approximation wherever possible to reveal the intuitive content of a model and to address the issue of model applicability.

We have seen that the basic issues of coupled water transport can be formulated in quite a general way. This general framework has been applied to the several elementary models discussed in this chapter but may equally well be applied to comprehensive computer simulations of epithelial function. For any model, comparison with experimental data necessitates computing the fluxes at equal bathing media, the derivatives of these fluxes with respect to the bath conditions, the bath conditions required for mucosal and serosal transport equilibrium, and the strength of transport. Via the formulation of non-equilibrium thermodynamics the flux derivatives can be related to the experimentally determined epithelial permeabilities. For analytical work, the fluxes and flux derivatives at equal bathing solutions can be used to assess transport isotonicity, and, with less certainty, estimate the strength of transport.

The essential principle responsible for the transport of water against a gradient remains, as Curran spelled it out, the limitation of solute diffusion from the cell surface. This limitation may be due to an external unstirred layer, a basement membrane with finite permeability, or diffusion limitation within the paracellular channel. We have seen that diffusion limitation within the interspace is confounded by high cell-membrane water permeability. By contrast, transport isotonicity depends essentially only on cell membrane water permeability. It seems to be little influenced by diffusion limitation of solute. Finally, solute polarization effects appear to be an inescapable feature of coupled water transport. That is, in the models considered, the measured epithelial water permeability must be less than the membrane water permeability that is relevant to isotonicity. When coupling is tight, in the sense of large volume flows between identical media, solute polarization is substantial. In general, the magnitude of this effect is determined by a ratio of diffusion limitation to water permeability.

The availability of efficient numerical algorithms for solving large systems of equations by computer has made possible the elaboration of comprehensive simulations of epithelial transport. Published models have included several electrolyte species, variable cell sodium transport, and compliant epithelial structures. Nevertheless, the basic features of coupled water transport operating within the large models may be appreciated from a study of the elementary models. One can only suppose that as the detailed physiology of the cell is disclosed, the elaborate models will continue to grow. Features such as the colligative properties of the cell matrix, the regulation of cell transport, and the non-linear properties of cell membranes would all be useful to incorporate in an epithelial simulation. Regardless of model complexity the analysis of simple models will always be useful in gaining intuitive insight and planning experiments. The comprehensive models will always be important in locating the limitations of the analysis.

References

1 Reid, E.W. (1902) J. Physiol. 28, 241–256.
2 Diamond, J.M. (1962) J. Physiol. 161, 442–473.
3 Diamond, J.M. (1962) J. Physiol. 161, 474–502.
4 Diamond, J.M. (1962) J. Physiol. 161, 503–527.
5 Curran, P.F. and Solomon, A.K. (1957) J. Gen. Physiol. 41, 143–168.
6 Windhager, E.E., Whittembury, G., Oken, D.E., Schatzmann, H.J. and Solomon, A.K. (1959) Am. J. Physiol. 197, 313–318.
7 Kedem, O. and Katchalsky, A. (1963) Trans. Faraday Soc. 59, 1918–1930.
8 Kedem, O. and Katchalsky, A. (1963) Trans. Faraday Soc. 59, 1931–1940.
9 Kedem, O. and Katchalsky, A. (1963) Trans. Faraday Soc. 59, 1941–1953.
10 Frömter, E. and Diamond, J. (1972) Nature New Biol. 235, 9–13.
11 Diamond, J.M. (1979) J. Membrane Biol. 51, 195–216.
12 Goldschmidt, S. (1921) Physiol. Rev. 1, 421–453.
13 Diamond, J.M. (1964) J. Gen. Physiol. 48, 15–42.
14 Lee, J.S. (1968) Gastroenterology 54, 366–374.
15 Weinstein, A.M. and Stephenson, J.L. (1981) J. Membrane Biol., in press.
16 Hill, B.S. and Hill, A.E. (1978) Proc. Roy. Soc. Lond. B 200, 151–162.
17 Whitlock, R.T. and Wheeler, H.O. (1964) J. Clin. Invest. 43, 2249–2265.
18 Parsons, D.S. and Wingate, D.L. (1961) Biochim. Biophys. Acta 46, 170–183.

19 Katchalsky, A. and Curran, P.F. (1967) Nonequilibrium Thermodynamics in Biophysics, Harvard Univ. Press, Cambridge, Mass.
20 Essig, A. and Caplan, S.R. (1968) Biophys. J. 8, 1434–1457.
21 Diamond, J.M. (1966) J. Physiol. 183, 58–82.
22 Sauer, F. (1973) in J. Orloff and R.W. Berliner (Eds.), Handbook of Physiology, Renal Physiology, pp. 399–414, Am. Physiological Soc., Washington, DC.
23 Andreoli, T.E. and Schafer, J.A. (1978) Am. J. Physiol. 234, F349–F355.
24 Andreoli, T.E. and Schafer, J.A. (1979) Fed. Proc. 38, 154–160.
25 Andreoli, T.E. and Schafer, J.A. (1979) Am. J. Physiol. 236, F89–F96.
26 Curran, P.F. and Schwartz, G.F. (1960) J. Gen. Physiol. 43, 555–571.
27 Dainty, J. (1963) Adv. Botan. Res. 1, 279–326.
28 Dainty, J. and House, C.R. (1966) J. Physiol. 182, 66–78.
29 Tormey, J.M. and Diamond, J.M. (1967) J. Gen. Physiol. 50, 2031–2060.
30 Curran, P.F. (1960) J. Gen. Physiol. 43, 1137–1148.
31 Curran, P.F. and MacIntosh, J.R. (1962) Nature 193, 347–348.
32 Ogilvie, J.T., MacIntosh, J.R. and Curran, P.F. (1963) Biochim. Biophys. Acta 66, 441–444.
33 Patlak, C.S., Goldstein, D.A. and Hoffman, J.F. (1963) J. Theoret. Biol. 5, 426–442.
34 Durbin, R.F. (1960) J. Gen. Physiol. 44, 315–326.
35 Kaye, G.I., Wheeler, H.O., Whitlock, R.T. and Lane, N. (1966) J. Cell Biol. 30, 237–268.
36 Williams, A.W. (1963) Gut 4, 1–7.
37 Bentzel, C.J., Parsa, B. and Hare, D.K. (1969) Am. J. Physiol. 217, 570–580.
38 Spring, K.R. and Hope, A. (1978) Science 200, 54–58.
39 Van Os, C.H., Wiedner, G. and Wright, E.M. (1979) J. Membrane Biol. 49, 1–20.
40 Henin, S., Cremaschi, D., Schettino, T., Meyer, G., Donin, C.L.L. and Cotelli, F. (1977) J. Membrane Biol. 34, 73–91.
41 Welling, L.W. and Grantham, J.J. (1972) J. Clin. Invest. 51, 1063–1075.
42 Sha'afi, R.I., Rich, G.T., Sidel, V.W., Bossert, W. and Solomon, A.K. (1967) J. Gen. Physiol. 50, 1377–1399.
43 Blom, H. and Helander, H.F. (1977) J. Membrane Biol. 37, 45–61.
44 Diamond, J.M. and Bossert, W.H. (1967) J. Gen. Physiol. 50, 2061–2083.
45 Segel, L.A. (1970) J. Theoret. Biol. 29, 233–250.
46 Lin, C.C. and Segel, L.A. (1974) in Mathematics Applied to Deterministic Problems in the Natural Sciences, pp. 244–276, MacMillan, New York.
47 Weinbaum, S. and Goldgraben, J.R. (1972) J. Fluid Mech. 53, 481–512.
48 King-Hele, J.A. (1979) J. Theoret. Biol. 80, 451–465.
49 Hill, A.E. (1975) Proc. Roy. Soc. Lond. B 190, 99–114.
50 Hill, A.E. (1977) in Gupta, B.L., Moreton, R.B., Oschmen, J.L. and Wall, B.J. (Eds.), Transport of Ions and Water in Animals, pp. 183–214, Academic Press, New York.
51 Diamond, J.M. (1978) in J.F. Hoffman (Ed.), Membrane Transport Processes, pp.257–276, Raven Press, New York.
52 Stirling, C.E. (1972) J. Cell Biol. 53, 704–714.
53 Kyte, J. (1976) J. Cell Biol. 68, 304–318.
54 DiBona, D.R. and Mills, J.W. (1979) Fed. Proc. 38, 134–143.
55 Sackin, H. and Boulpaep, E.L. (1975) J. Gen. Physiol. 66, 671–733.
56 Huss, R.E. and Marsh, D.J. (1975) J. Membrane Biol. 23, 305–347.
57 Weinstein, A.M. and Stephenson, J.L. (1979) Biophys. J. 27, 165–186.
58 Smulders, A.P., Tormey, J.M. and Wright, E.M. (1972) J. Membrane Biol. 7, 164–197.
59 Frömter, E. (1972) J. Membrane Biol. 8, 259–301.
60 Schultz, S.G. (1977) Yale J. Biol. Med. 50, 99–113.
61 Boulpaep, E.L. (1972) Am. J. Physiol. 222, 517–531.
62 Bentzel, C.J. (1974) Am. J. Physiol. 226, 118–126.
63 Bentzel, C.J., Spring, K.R., Hare, D.K. and Paganelli, C.V. (1974) Am. J. Physiol. 226, 127–135.

64 Lewy, J.E. and Windhager, E.E. (1968) Am. J. Physiol. 214, 943–954.
65 Grandchamp, A. and Boulpaep, E.L. (1974) J. Clin. Invest. 54, 69–82.
66 Hill, A.E. (1975) Proc. Roy. Soc. Lond. B 191, 537–547.
67 Wilson, T.H. (1956) J. Appl. Physiol. 9, 137–140.
68 Diamond, J.M. (1964) J. Gen. Physiol. 48, 1–14.
69 Smyth, D.H. and Wright, E.M. (1966) J. Physiol. 182, 591–602.
70 Whittembury, G., Oken, D.E., Windhager, E.E. and Solomon, A.K. (1959) Am. J. Physiol. 197, 1121–1127.
71 Bentzel, C.J., Davies, M., Scott, W.N., Zatman, M. and Solomon, A.K. (1968) J. Gen. Physiol. 51, 517–533.
72 Ullrich, K.J. (1973) in J. Orloff and R.W. Berliner (Eds.), Handbook of Physiology, pp. 377–398, Am. Physiological Soc., Washington DC.
73 Neumann, K.H. and Rector Jr., F.C. (1976) J. Clin. Invest. 58, 1110–1118.

Subject index

Acetamide transport 9, 22, 52, 54
Acetate as pH probe 278
    transport 270
Acetoacetate transport, mitochondrial 245
Acetylcholine receptor 120
Acetyl CoA carboxylase 237, 238
Action potential
    changes in sodium and potassium permeability 77–79
    timing the sodium flux 79
Activation energy 43, 44
Active transport, design principles 155–156
    primary 154–155, 286, 289–291, 299, 307
    secondary 152–154, 259, 263, 267–269, 270, 277, 280, 286, 287, 291–298
Acylcarnitine transport, mitochondrial 246
Adenine nucleotides, compartmentation 244
Adenine nucleotide translocator 238, 242–244
    effect of hormones 249
    isolation 249
    reconstitution 249
    inhibition by fatty acyl-CoA esters 244
Adenylate cyclase 46, 276
Adenylyl imidodiphosphate (AMPPNP) 177, 224
ADP, as phosphate acceptor in Ca-ATPase 196
ADP-ATP exchange, mitochondrial 243
ADP binding, on Ca-ATPase 195
    phospholipid dependence in Na-K ATPase 173
ADP-insensitive intermediate, role in Na-K ATPase 163
ADP-insensitive phosphoprotein, in Ca-ATPase 200–203
ADP-sensitive intermediate, role in Na-K ATPase 163
ADP-sensitive phosphoprotein, in Ca-ATPase 200–203
Adrenaline effect on mitochondrial transport 248
Aggregation-field effect model 100–102
Alamethicin 111, 118
    negative resistance 112
    noise analysis 108
Alcohol transport 25
Alkaline phosphatase 220
Amino acid transport 10, 154, 270, 287, 298, 303, 304, 307

9-Amino-acridine 231, 279
Amphotericin 118
Amphotericin A 39
Amphotericin B 47, 52, 53
Anaesthetics 25
Anilinonaphtosulfonic acid (ANS) 231
Anion-sensitive ATPase 209–222
    anion-dependence 212–215
    assay 209–210
    definition 209
    divalent cation dependence 212
    effect of arsenate 212
    effect of arsenite 212
    effect of bicarbonate 212–215
    effect of borate 212
    effect of oxyanions 212, 220
    effect of selenite 212
    effect of sulfate 212
    effect of sulfite 212
    effect of thiocyanate 213
    in microsomal fractions 216
    in mitochondrial fractions 216
    localization 215–219
    monovalent cation dependence 212
    pH dependence 212
    presence in erythrocytes 220, 221
    presence in tissues 211
    solubilization 215, 218
    substrate dependence 210–212
Anion transport 154, 270
Antibodies, against Na-K ATPase subunits 168
Antidiuretic hormone (ADH) 39, 46, 50, 54, 55, 57
Antiport 267
Arabinose transport 270
Arrhenius plot 229
Arsenate, effect on anion-sensitive ATPase 212
Arsenite, effect on anion-sensitive ATPase 212
Aspartate
    mitochondrial micro-compartmentation 247
    translocator 238, 247, 248
    transport, mitochondrial 237, 247
Aspartic acid, binding site for phosphorylation in Na–K ATPase 162–163
Asymmetry parameter 145–146

ATPase complex
  bacterial 263–265
  reconstitution 264
ATPase inhibitor protein 265
ATP, binding-site on α subunit of Na–K ATPase 168
  binding to Na–K ATPase 161–162
ATP–ADP exchange, in Na–K ATPase 162, 165
ATP–ADP exchange reaction, in Ca–ATPase 195
ATP–ADP phosphate exchange, in Ca–ATPase 199–200
ATP-binding, to Ca–ATPase 193–195
ATP-driven calcium uptake 187–189
  reversal 190
ATP–P_i exchange in Ca–ATPase 198–200
ATP, hydrolysis 259
  synthesis 259
  synthesizing pumps 155
ATP synthetase by calcium efflux 189–190, 196
Atractyloside 236, 238, 243
Aurovertin, effect on anion-sensitive ATPase 217–219

Bacteriorhodopsin 259, 265–267, 281
Band 3, in erythrocyte membrane 49
Benzene 1,2, 3-tricarboxylate 236–238
Benzoic acid, transport 14
  as pH probe 278
Bicarbonate, effect on anion-sensitive ATPase 212
Bicarbonate-ATPase, see Anion-sensitive ATPase
Bile acids transport 307
"Binder" requiring systems 290
Black films 22, 23
Borate, effect on anion-sensitive ATPase 212
Born charging energy 107
8-Bromo-cyclic AMP 55
Bromocresolpurple 236
Brush border membrane 211–213, 219–221
Butanediols, relation hydrogen bonding with density 53
Butanedione 163, 226
n-Butylmalonate 236

Ca-ATPase 155, 222, 223, 226, 281
  ADP as acceptor 196
  ADP binding 195
  ADP-insensitive phosphoprotein 200–203
  ADP-sensitive phosphoprotein 200–203
  ATP–ADP exchange 199–200
  ATP binding 193–195
  ATP–P_i exchange 198–200
  basal activity 187

  Ca binding 191–193
  Ca-independent activity 187
  effect of DTNB 191
  effect of phospholipase A_2 194
  effect of Triton X-100 194
  effect of tryptic digestion 191
  Mg^{2+} as cofactor 196
  Mg–ATP as substrate 196
  Mg^{2+} binding 195–197
  nucleoside diphosphokinase activity 199–200
  phosphate binding 197
  phosphoprotein formation 197–198
  reaction sequence 197–205
Ca^{2+}, intracellular concentration 183
  role in gastric acid secretion 232
  role in muscle physiology 183
Calcium-activated ATPase, see Ca–ATPase
Ca^{2+} binding, deduced from phosphoprotein binding 191, 192
  high and low affinity binding sites 192–193
  on sarcoplasmic reticulum membranes 191
Ca^{2+} efflux, energy source for ATP-synthetase 189–190, 196
  from liposomes 190
  in muscle as compared to sarcoplasmic reticulum membranes 190
$(Ca^{2+} + Mg^{2+})$-ATPase, see Ca–ATPase
Calmodulin 47, 221
Caproic acid transport 14
Carbonic anhydrase 210
Carboxyatractyloside 236, 242
Carboxylic acid transport 270
Carnitine, translocatie 237, 238
  transport 246
Carotenoids 279
Carrier, model 112–113
  model of cotransport 291–292
  terminology 107
Carriers 112–115
  comparison with channels 107
Cation permeabilities, in resting electrically excitable cells 72–75
Cation transport 154
  bacterial 270
  effect of cyanide 160
  effect of 2,4-dinitrophenol 160
  effect of ouabain 160
  in cells 159
  mitochondrial 249–252
  relation to energy metabolism 159
Ca^{2+} transport, ATP-dependence 186
  correlation with ATP hydrolysis 187, 188
  coupling ratio with ATP 188–189

divalent cation specificity 187
effect of oxalate 186
effect of phosphate 186
effect of pyrophosphate 186
measurement 186
mitochondrial 251–252
role of phosphoryl transfer 203–205
substrate specificity 187
Ca^{2+}/nH^+ exchange 252
CCCP 279
C_4-dicarboxylic acid transport 270
Cells, isolated 288
Channel counting 102–104
Channel, gated 294
Channels 115–119
comparison with carriers 107
terminology 107
Chemical potential, definition 62
Chemiosmotic hypothesis 155, 249–250, 258–259, 278
Chloride transport 230
p-Chloromercuribenzene sulfonate (pCMBS), effect on $(K^+ + H^+)$-ATPase 226
pCMBS, effect on non-electrolyte transport 54, 55
p-Chloromercuribenzene sulfonate (pCMBS), effect on water permeability 44, 46, 48
p-Chloromercuribenzoate, effect on anion-sensitive ATPase 215
Chlorophyll 279
Chloropromazine 221
Cholesterol, content in (K^+-H^+)-ATPase 228
effect on membrane fluidity 47
effect on membrane permeability 25
effect on non-electrolyte permeability 47
effect on water permeability 47
Citrate transport 235, 270, 307
Citrulline, synthesis 244, 248
transport 237
Colicins 119
Complex pore, analysis 135
model 135
transport parameters 131
Conductance, dependence on voltage 111
Constant field hypothesis 68–69
Constitutive transport 270
Conventional carrier 286
kinetic analysis 142–143
model with two substrates 149
Conveyer principle 291
Corticosterone transport 15
Cortisol transport 15
Cortisone transport 15

Cotransport, effects on membrane potential 296
by means of a simple carrier 152–154
Counter transport 147–151
by means of a simple carrier 152–154
Coupling degree (q) 286
Coupling principles, in metabolite transport 286, 287
Cupric phenanthroline 168
Current relaxation method 113
Cyanide, effect on cation transport 160
Cyanine dyes 299
α-Cyanocinnamates 236, 237, 238, 246
Cyclic AMP 46, 47, 50
role in gastric acid secretion 232
L-Cysteine, effect on anion-sensitive ATPase 220
Cytochalasin B 1,47
Cytochrome-linked electron transfer 259, 260–263
Cytoplasmic membrane vesicles 280

DCCD 217–219, 263
Dexamethasone transport 15
Dibenzyldimethylammonia 278, 279
Dicarboxylate translocator 235, 238, 239
DIDS 50
effect on anion-sensitive ATPase 221
Diffusion, theoretical treatment 63–66
Diffusional permeability coefficients 38, 39, 42, 43
Diffusion coefficient 43
temperature dependence 44
(5,α)-Dihydrotestosterone 15
Diisopropylfluorophosphate, effect on anion-sensitive ATPase 215
Dimer-monomer transition in Na–K ATPase, role of phospholipids 174
Dimethyl-3, 3′-dithiobis-propionimidate 168
5,5-Dimethyl-2,4-oxazolidinedione 242, 278
Dimethylsuberimidate 168
1, 3-Dimethylurea transport 52,54
2,4-Dinitrophenol 160, 279
2,4-Dinitrophenylphosphate 167
Diphenylhexatriene 229
Dipicrylamine 227
5,5′-Dithiobis-(2-nitrobenzoic acid) (DTNB) 191, 226
D_2O 215

Einstein equation 66
Efflux of metabolic end products 259
Electrochemical potential, definition 62–63
difference, effect on metabolite transport 289
Electrogenic pump 289, 301

Electron transfer 277
Energization of metabolite transport 289–298
    of the membrane 307
Energy source for coupled transport 285–286
Energy transducing processes, interaction between 277
    in bacteria 260–267
Epithelial cells, in tissue culture 288
Epithelial transport 301–305
Equilibrium exchange procedure, description 125–126
    for studying competition 152
Erythrocyte membrane, presence of band 3 49
    presence of glycophorin 49
    presence of PSA-1 49
$(17,\beta)$-Estradiol transport 15
Ethanol transport 54, 270
Ethylacetate transport 14
Ethylene glycol transport 8, 54
Ethylurea transport 54
Everted sac technique 288
Exchange diffusion 146–147
Excitability-inducing material (E.I.M.) 109, 119

Facilitated diffusion, competition 151–152
    complex pore model 135
    conventional carrier 142–143
    counter transport 147–151
    definition 123
    effect of unstirred layer 127–128
    equilibrium exchange procedure 125–126
    exchange diffusion 146–147
    infinite cis procedure 127
    infinite trans procedure 126
    kinetic analysis 123–128
    methods 123–128
    molecular interpretation of transport parameters 143–146
    problems with initial rate measurements 127
    relation to primary active transport 155
    secondary active transport 152–154
    simple carrier model 136–142
    simple pore model 129–134
    zero trans procedure 124–125
F_1-ATPase 216, 218, 263–265
Fatty acid composition, effect on membrane permeability 26
Fatty acid transport 10, 22
Fick's law 2–4, 66
Filipin 220
Filtration (permeability) coefficient 32, 39
Flagellar movement 259, 277
Fluctuation analysis 107–109

Fluorescence-quenching techniques 306
Fluoride as inhibitor of (K^+-H^+)-ATPase 227
Flux equations 63–66
Flux ratio 70
Formamide transport 9, 22, 52, 54, 56, 57
Friction coefficients 40
β-(2-Furyl)acryloyl phosphate 167
Fructose transport 270
Fumarate transport 270

Galactose transport 270
Gastric acid secretion 232
    fusion model 232
Gastric mucosal vesicles 222, 223, 229–232
    anion permeability 231–232
    Cl^- transport in 230
    Rb^+ transport in 230
Gated channel 294
Gating mechanisms, aggregation-field effect model 100–102
    channel counting 103–104
    single channel conductance 102–104
    two-state transition model 96–100
Glisoxepide 236
Glucagon, on mitochondrial transport 248, 249
Gluconate transport 270
Glucose-6-phosphate transport 270
Glucose transport 58, 270
Glucuronate transport 270
Glutamate-aspartate translocator 246–248
Glutamate translocator 240
Glutamate transport, mitochondrial 235, 237
γ-Glutamyl cycle 305
γ-Glutamyl peptides 305
Glutamyl transferase system 287, 289–290, 305
Glycerol-3-phosphate transport 270
Glycerol transport 2, 8, 14, 270
Glycol transport 8
Glycophorin 49, 50
Glycoproteins in Na–K ATPase 168
Goldman equation, for calculation of ion permeabilities 74–75
    for measurement of ion selectivity ratios 81
    theoretical derivation 70–71
Gradient hypothesis 298, 299
Gramicidin 111, 115–118
    analogues 116
    dimer model 116
    NMR studies 117
Group translocation of solute in unmodified form 267, 272–277

H^+-ATPase 281

H^+/ATP ratio 251
Hemocyanin 119
Hepatocytes 243
$HgCl_2$, effect on anion-sensitive ATPase 215
Hille's selectivity filter 94–95
$(H^+ + K^+)$-ATPase, see $(K^+ + H^+)$-ATPase
Hodgkin–Huxley channels 86–90
H^+/O ratio 250
Hormones, effect on Ca^{2+} 252
    effect on mitochondrial transport 248–249
H^+-oxidoreductases 281
H^+-transhydrogenases 281
H^+ transport, effect of nigericin 230
    effect of TCS 230–231
    effect of valinomycin 229, 230
Hydraulic conductivity 36, 43, 48
Hydrogen bonding, relating with density 53
Hydrostatic pressure 35
β-Hydroxybutyrate transport, mitochondrial 245

Infinite *cis* procedure, description 127
Infinite *trans* procedure, description 126
    for studying competition 152
Influx kinetics of cotransport 293–294
Inhibitory protein, effect on anion-sensitive ATPase 221
Intercellular space, solute-solvent coupling 331–347
Interspace models 343–348
Intraepithelial solute-solvent coupling 314
Intramembrane diffusion coefficient 4, 12
Intrinsic dissociation constant 144–145
Iodide as potential probe 278
Ionic channel, concept 86–94
    molecular transitions in activation of 90–94
Ionic permeability measurements 72–85
Ionophores 279, 297, 306
Ion permeability, activation of channels 90–94
    gating mechanisms 95–104
    measurement by tracer fluxes 72–75, 84–85
    mechanisms 86–104
    monovalent cation selectivity ratios 79–85
    theoretical aspects 62–71
    changes in excitable cells 75–79
Ion transport 10
Isocitrate transport, mitochondrial 235
Isolated cells 288
Isolated plasma membranes 288
Isotonic convection approximation 339
Isotonic transport 312

K^+-activated ATPase, see $(K^+ + H^+)$-ATPase

K^+ channel, blocking by tetraethylammonium 88
    difference with sodium channel 88
    permeability ratios 84
Kedem–Katchalsky equations 347
$(K^+ + H^+)$-ATPase 222–232
    activator 226
    Arrhenius plot 229
    binding of adenylylimidodiphosphate (AMP-PNP) 224
    bound Mg^{2+} 225
    carbohydrate composition 223, 227
    catalytic subunit 223
    cholesterol content 228
    divalent cation dependence 224
    effect of antibodies 223
    effect of butanedione 226
    effect of dipicrylamine 227
    effect of DTNB 226
    effect of EEDQ 227
    effect of fluoride 227
    effect of hydroxylamine 226
    effect of ionophores 222, 223, 229
    effect of 2-methoxy-2,4,-diphenyl-3-dihydrofuranone 227
    effect of N-ethylmaleimide 226
    effect of nigericin 223
    effect of N,N'-dicyclohexylcarbodiimide 227
    effect of ouabain 227
    effect of p-chloromercuribenzene sulfonate (pCMBS) 226
    effect of phospholipases 228–229
    effect of thiocyanate 227
    effect of valinomycin 223
    effect of vanadate 227
    effect of Zn^{2+} 227
    electroneutrality 230–231
    monovalent cation dependence 224
    pH optimum 224
    phospholipid composition 228
    phospholipid dependence 228
    phosphorylation by ATP 223–225
    presence in gastric mucosa 222
    purification 222–223
    substrate dependence 224
    subunit composition 223, 227
    tryptic digestion 223, 227
K^+-channels, from sarcoplasmic reticulum 120
Ketogenic conditions 246
$K^+ - H^+$ exchange 222, 252
Kinks hypothesis 50, 51
$K^+ - K^+$ exchange transport 177–178
K^+-stimulated phosphatase 222, 223

lipid dependence 173-174
role in Na-K ATPase 167
role in transport 178-179
K^+ transport, mitochondrial 252

Lactate dehydrogenase 237
Lactate, efflux 272
    transport 270, 307
Lactose transport 270
Light-dependent cyclic electron transfer 281
Light driven H^+-pump 281
Lipid bilayer 9, 12, 22, 39, 40, 43, 44, 46, 47, 51, 53, 56, 107-120, 288
Lipid fluidity, role in Na-K ATPase 173
Lipid removal, effect on phosphorylated intermediates in Na-K ATPase 163
Lipid solubility, effects on transport 2
Lipid soluble ions as probe for membrane potential 289
Lipid viscosity 229
Liposomes 22, 249, 281, 288, 305-307
    calcium efflux 190
Lysine, as inhibitor of the ornithine translocator 236

Mg^{2+}, as cofactor for Ca-ATPase 196
Mg^{2+} binding to Ca-ATPase 195-197
Mg-ATP as substrate for Ca-ATPase 196
Malate-aspartate shuttle 246
Malate transport 235, 270
Mass selectivity index 12
Mediated transport, concepts 123-157
Membrane conductance 71
Membrane potential, effect on cotransport 294
Membrane, 5-region model 18-20
Membrane vesicles 280, 305-307
Mercury compounds, as inhibitors of water transport 44
Merocyanins 279
Mersalyl 236, 237
Metabolite distribution 238-241
Metabolites, coupled transport 285-307
Methane sulfonyl chloride 215
Methanol transport 13, 18, 54
2-Methoxy-2,4-diphenyl-3-dihydrofuranone 227
Methylamine as pH probe 278
Methylurea transport 52, 54, 56, 57
Michaelis-Menten equation 125
Middle compartment scheme 331-337
Mitchell's chemiosmotic hypothesis 155, 249-250, 258-259, 278
Mitochondrial inhibitor protein, effect on anion-sensitive ATPase 217-218

Mitochondrial ion transport 235-252
Mitochondrial metabolite transport 235-248
Mitochondrial transport, inhibitors 237-238
Mobile carrier model 269
Molecular size, effects on transport 2
Monacetin transport 8
Monocarboxylic acid transport 270
Monozomycin 118
Muscle physiology, role of calcium 183-184

Na^+ binding to Na-K ATPase 161
Na^+-Ca^{2+} exchange 183
Na^+ channel, blocking by tetrodotoxin 88
    conductance 104
    density in nerves 103
    difference with potassium channel 88
    maximum rate of transport 102-104
    negative resistance 112
    permeability ratios 81
    selectivity filter 85, 94-95
Na^+ dependence of metabolite transport 298-307
Na^+-dependent Ca^{2+} flux 252
Na^+-H^+ exchange system 252
Na^+ inflow in axons under voltage clamp 75-77
Na-K ATPase 50, 155, 185, 222, 223, 226, 227, 281, 296, 303
$(Na^+ + K^+)$-ATPase, see Na-K ATPase
Na-K ATPase, ADP-ATP exchange 162
    ADP-insensitive intermediate 163
    ADP-sensitive intermediate 163
    ATP binding 161-162
    ATP effect on E_2-E_1 transition 165
    amino acids involved in ATP binding centre 162-163
    antibodies against both subunits 168
    conformational changes 163-164, 165-166, 170-171, 173
    cross linkage 168
    discovery 160
    divalent cation specificity 166
    effect of butanedione 163
    effect of 7-Cl-4-NO_2-benzo-2-oxa-1,3-diazole (NBD-Cl) 163
    effect of thimerosal 170
    effect of tryptic digestion 169-171
    enzyme properties 160
    function of α-subunit 168-169
    function of β-subunit 169
    localization in epithelia 340
    minimal lipid content 173
    molecular weight by radiation inactivation analysis 169

monovalent cation specificity 166
Na$^+$/ATP ratio 161
Na$^+$ binding 161
nucleotide specificity 166
phosphoenzyme hydrolysis 164
phosphoenzyme transition 163–164
phospholipid dependence 171–174
phosphorylation 162–163
rate limiting steps 166
reaction mechanism 161–167
role of phospholipids in phophoenzyme transition 173
structural aspects 168–171
subunit composition 168–169
subunit ratio 168
transport mechanism 174–179
Na$^+$–K$^+$ transport, mechanism 174–176
reversed transport 174–176
Na$^+$–Na$^+$ exchange transport 176
Na$^+$-transport, mitochondrial 252
NBD-Cl, effect on Na–K ATPase 163
Negatively charged phospholipids, role in Na–K ATPase 171–173
Nernst equation 159, 289, 299
Nernst-Planck equation 67–68, 347
N-Ethylmaleimide 226
as inhibitor of phosphate translocator 236
as inhibitor of the glutamine translocator 236, 247
effect on phosphorylated intermediates in Na–K ATPase 163
use in metabolic studies 237
Nigericin 230, 279
effect on (K$^+$ + H$^+$)-ATPase 223
p-Nitrophenylphosphate, as substrate for K$^+$-stimulated phosphatase activity 167
Nitroxides, spin-labeled derivatives of 24
N-Methyldeptropine 278
N,N'-dicyclohexylcarbodiimide 227
N,N'-ethoxycarbonyl-2-ethoxy-1,2-dihydroquinoline (EEDQ) 227
Noise analysis 108–109
Noise, high-frequency 109
low frequency 108
Nonactin 111, 114
Non-electrolyte permeation, temperature dependence 53
Non-electrolytes, permeability of small polar 51–58
Non-equilibrium thermodynamics 314–315
Nonpolarized cells, transport in 298–301
Nuclear Magnetic Resonance (N.M.R.) 34

Nucleoside diphosphokinase, activity in Ca-ATPase 199–200
Nucleoside transport 10
Nystatin 39, 52, 53, 118

Oestriol transport 15
Oligomycin, effect on anion-sensitive ATPase 217–219
effect on Na$^+$-transport 177
effect on phosphorylated intermediates in Na-K ATPase 163
Onsager reciprocal relation 35
Optical methods, as probe for membrane potential 289
Organic cations, permeability of 80–81
Ornithine transport, mitochondrial 237
Osmotic coefficient 39
Osmotic coupling coefficient 316, 320, 327, 330
Osmotic volume changes, use in water transport 34
Osmotic water flow 45, 46
activation energy 44
relation to water diffusion 38, 39
Osmotic water permeability coefficient, see Hydraulic conductivity
Ouabain 227, 297, 303
binding site on α-subunit of Na–K ATPase 168
effect on ATP binding to Na–K ATPase 162
effect on cation transport 160
effect on phosphoenzyme hydrolysis in Na–K ATPase 164
Overshoot phenomenon 305
Oxalate, effect on calcium accumulation 186
transport 270
α-Oxoglutarate translocator 237, 239
α-Oxoglutarate transport, mitochondrial 235

Partition coefficient 2, 6, 13
p-Chloromercuribenzene sulfonate (pCMBS), effect on (K$^+$ + H$^+$)-ATPase 226
PCMBS, see p-Chloromercuribenzene sulfonate (pCMBS)
Pentanediols, relation hydrogen bonding with density 53
Peptidoglycan 257
Perfusion studies 288
Permeability coefficient 37, 46
definition 69
effect of ADH 47
for water 47
Permeability, effect of cholesterol content 25

effect of fatty acid composition 26
effect of temperature 25
Permeability ratios, for monovalent cations in K^+ channel 84
for monovalent cations in Na^+ channel 81
Permselectivity, mechanism for 94–95
Perturbation analysis 36, 37
l-Phenylalanine, effect on anion-sensitive ATPase 220
Phenyldicarbo-undecaborane 278
Phloretin, effect on non-electrolyte transport 54, 57, 58
Phlorizin 295, 304
Phosphate, effect on calcium accumulation 186
Phosphate binding, to Ca–ATPase 197
Phosphate translocator 239
Phosphate transport, mitochondrial 235
Phosphatidylinositol, role in Na–K ATPase 171–173
Phosphatidylserine, role in Na–K ATPase 171–173
Phosphoenolpyruvate carboxykinase 237
Phosphoenolpyruvate phosphotransferase system, see PTS
Phosphoenolpyruvate transport, mitochondrial 237
Phospholipase A_2, effect on Ca–ATPase 194
effect on $(K^+ + H^+)$-ATPase 228
Phospholipase C, effect on $(K^+ + H^+)$-ATPase 228–229
effect on Na–K ATPase 172
Phospholipid dependence, of Na–K ATPase 173–174
of $(K^+ + H^+)$-ATPase 228–229
Phosphoprotein formation, in Ca–ATPase 198–199
Phosphoryl transfer, role in Ca^{2+} movement 203–205
Phosphorylation, of Na–K ATPase by ATP 162
Photophosphorylation 281
Phthalonate 236, 237
Pitressin 47
Plasma membranes, isolated 288
Poiseuille's law 32, 40
Polar head group specificity of phospholipids, role in Na–K ATPase 171–173
Polyol transport 10
P/O ratio 251
Pore concept 40–43
Potential, effects on cotransport 295–296
probes 278
Primary active transport 154, 155, 186
of metabolites 289–291, 299, 307

Progesterone transport 15, 18
Propanediols, relation hydrogen bonding with density 53
1, 3-Propanediol transport 54
Proton-motive force 259, 272, 307
maximal value of 262
Proton-motive Q cycle 261, 262
Protonophores 279
Proton pump, bacteriorhodopsin 266
Proton translocation mechanism 261
Proton transport, mitochondrial 249–251
PSA-1 in erythrocyte membrane 49
PS-decarboxylase, effect on Na–K ATPase 172
Pseudocompetition 297–298
PTS 267, 272–277, 287, 290, 291
PTS components, cellular localization 274
PTS, glucose transport 273
lactose transport 274
mannitol transport 273, 274
purification 274
specificity for phosphoenolpyruvate 276
Pump-leak concept 159
Purple membranes 267
Pyrophosphate, effect on calcium accumulation 186
Pyruvate carboxylation 244, 248
Pyruvate metabolism 245–246
Pyruvate translocator 240, 248
Pyruvate transport 235, 245–246, 270

Q-cycle 251, 261–262
Quercetin, effect on anion-sensitive ATPase 217
effect on phosphorylated intermediates in Na–K ATPase 163
Q_{10} values of transport 25

Rapid flow technique 33
Rb^+ transport 230
Reconstitution of ATPase complexes 264
Rectification of water flow 48
Redox loop model 261
Reflection coefficient 40–43, 52, 321
Resistance parameters 144
Respiratory control 263
Retinal 265
Reversal potential 111
for measurement of selectivity ratios 79–84
Rhamnose transport 270
Rojas aggregation-field effect model 100–102
Ruthenium red, effect on anion-sensitive ATPase 221
effect on mitochondrial calcium uptake 252

Sarcoplasmic reticulum membranes, isolation 185
  low passive calcium permeability 189–190
  organization in muscle 184–185
  permeability for Mg^{2+} 197
  presence of ATP-driven Ca^{2+} transport system 184
  role in muscle physiology 185
  tryptophan fluorescence 191
Saxitoxin (STX) 94–95
Schiff base 265–267
Secondary (active) transport 152–154, 259, 263, 267–269, 270, 277, 280, 286, 287, 291–298
Selectivity filter 85, 94–95
Selectivity sequence 111
Selenite, effect on anion-sensitive ATPase 212
Simple carrier, analysis 135
  analysis by equilibrium exchange procedure 138
  analysis by infinite *cis* procedure 139–140
  analysis by infinite *trans* procedure 138–139
  analysis by zero *trans* procedure 138
  characterizing and testing 141
  comparison with simple pore 140
  kinetic analysis 136–142
  model 136
  model with two substrates 149
Simple diffusion, criteria 1
Simple pore, analysis by equilibrium exchange procedure 132
  analysis by infinite *cis* procedure 133
  analysis by infinite *trans* procedure 132–133
  analysis by zero *trans* procedure 132
  comparison with simple carrier 140
  interpretation of transport parameters 134
  kinetic analysis 129–133
  model 130
  transport parameters 131
Single cells, transport in 298–301
Single channel recordings 109
Single file mechanism 74
Sodium-potassium activated ATPase, see Na–K ATPase
Solute-solvent coupling 314
  in the lateral intercellular space 331–347
Solute transport, bacterial 267–269
Solvent drag 40–43
Source and sink principle 286, 291
Standing gradient interspace models 337–343
Steady-state assumption 130
Steroid transport 13–18, 21
Stokes–Einstein relation 66
Stokes law 66
Substrate level phosphorylation 272, 277

Succinate transport 270
Sugar transport 10, 154, 270, 298, 303, 304, 307
Sulfate, effect on anion-sensitive ATPase 212
Sulfite, effect on anion-sensitive ATPase 212
Sulfhydryl groups, absence in glycophorin 50
Sulfhydryl-reactive reagents, effect on Ca–ATPase 191
  effect on non-electrolyte transport 54, 55, 58
  effect on water movement 44, 50, 51
Sulfhydryl reagents, effect on anion-sensitive ATPase 215
Sutherland–Einstein equation 66
Symport 267

Temperature, effect on water permeability 43
  effects on transport 25
Testosterone transport 15
Tetrachlorsalicylanilide (TCS) 230, 231
Tetraethylammonium (TEA), blocking of potassium channel 88
Tetraphenylphosphonium 278, 279, 299
Tetrodotoxin (TTX) 49, 81
  binding protein 100
  blocking of sodium channel 88
  for density of sodium conductance units 103
  role of guanidinium group 94
Thermodynamic formulation, of epithelial transport 319
Thermodynamics, non-equilibrium 314
Thimerosal, effect on Na–K ATPase 170, 173
Thiocyanate 227
  as potential-sensitive probe 231, 278
  effect on anion-sensitive ATPase 213
  effect on H^+ transport 221
Thiourea transport 56, 57
Thyroid status 249
Tissue culture, of epithelial cells 288
Transepithelial water flow, effect of osmotic gradient 311–312
  primacy of solute flux 312
Transhydrogenase reaction 259
Translocators 235–248
  isolation 249
  kinetic properties 238
Transmembrane gradients, methods for determination 278–279
Transport, active primary 154–155, 286, 289–291, 299, 307
  channel model 269
Transport-deficient mutants 279
Transport, model systems for 279–282
Transport parameters in facilitated diffusion; molecular interpretation 143–146

Transport, phenomenological models 315–331
  secondary 152–154, 259, 263, 267–269, 270, 277, 280, 286, 287, 291–298
Transport studies, materials for 288
Triacetin transport 13
Tricarboxylate translocator 235, 238, 239
Tricarboxylic acid cycle intermediates, transport 154
Trimethylaminoacylcarnitine 236
Triphenylmethylphosphonium 278
Triton X-100, effect on anion-sensitive ATPase 215, 220
  effect on Ca–ATPase 194
Tryptic digestion, effect on $(K^+ + H^+)$-ATPase 223
  effect on Ca–ATPase 191
  effect on phosphorylated intermediates in Na-K ATPase 163
  of Na–K ATPase 169–171
Tryptophan fluorescence, of sarcoplasmic reticulum membranes 191
Two-state transition model 96–100

Uncoupled Na^+ transport 177
Uncouplers 279
Uniport 267
Unit conductance steps 109
Unstirred layer, effect on facilitated diffusion 127–128
  effect on water transport 37, 38, 323
Urea transport 9, 22, 51, 52, 54
  carrier mediation of 55–57
Ussing chamber 288

Valinomycin 49, 111, 114, 229, 230, 279
  analogues 115
  effect on $(K^+ + H^+)$-ATPase 223
  noise analysis 108
Vanadate 227
Vasopressin, see Antidiuretic hormone (ADH)
Vesicular transport 229–232
Vitamine A aldehyde 265
Voltage-dependent anion conductance (VDAC) 119, 120
Voltage-dependent channels 118–119

Water diffusion, activation energy 44
  relation to osmotic flow 38–39
Water filled channels 43
Water filled pores 40–43
Water permeability 29–51
  effect of cholesterol 47
  effect of pH 48
  effect of temperature 43
Water transport 10, 54
  coupling to ion transport 311–349
  inhibitors 50
  methods 30–38
  net flow under pressure 31
  osmotic volume changes 34
  radioactive methods 30–31, 33
  use of NMR 33

Zero *trans* procedure, description 124–125
  for studying competition 152
Zinc, effect on $(K^+ + H^+)$-ATPase 227